DISCOVERING INTELLIGENT DESIGN

DISCOVERING INTELLIGENT DESIGN

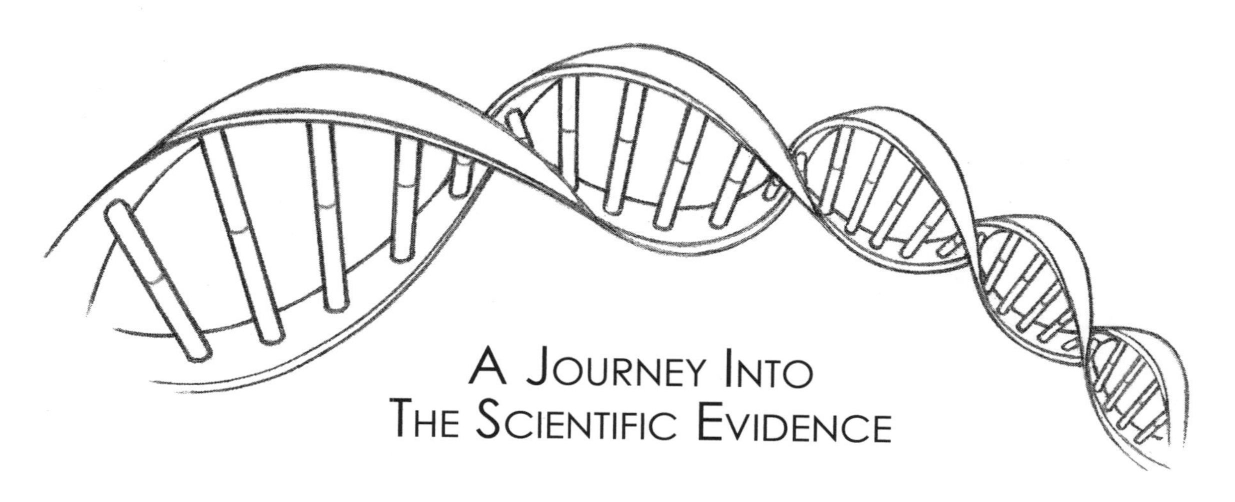

A Journey Into The Scientific Evidence

DISCOVERY INSTITUTE PRESS, SEATTLE, WA

Description

The *Discovering Intelligent Design* textbook is part of a comprehensive curriculum that presents both the biological and cosmological evidence in support of the scientific theory of intelligent design. Developed for middle-school-age students to adults, the curriculum also includes a workbook with learning activities and a DVD with video clips keyed to the content of the textbook. Produced by Discovery Institute in conjunction with Illustra Media, the curriculum is intended for use by homeschools and private schools. The curriculum is divided into six modules that explore topics such as the origin and development of the universe, the origin of biological complexity, the fossil record's evidence (or lack thereof) for universal common descent, and the broader cultural debate over intelligent design. More information can be obtained by visiting the curriculum's website, **http://discoveringid.org.**

Copyright Notice

Library Cataloging Data

Discovering Intelligent Design: A Journey into the Scientific Evidence by Gary Kemper, Hallie Kemper, and Casey Luskin

286 pages, 8.5 x 11x 0.6 inches or 280 x 216 x 15 mm, 1.5 lbs or 680 g.

Library of Congress Control Number: 2013934288

BISAC: SCI015000 SCIENCE/Cosmology

BISAC: SCI027000 SCIENCE/Life Sciences/Evolution

BISAC: SCI034000 SCIENCE/History

BISAC: SCI075000 SCIENCE/Philosophy & Social Aspects

ISBN-13: 978-1-936599-08-0

ISBN-10: 1936599082

Publisher Information

Discovery Institute Press, 208 Columbia Street, Seattle, WA 98104

http://www.discoveryinstitutepress.com

Published in the United States of America on acid-free paper.

First Edition, First Printing, May 2013.

TABLE OF CONTENTS

SECTION I: DEFINING MOMENT

SECTION II: COSMIC DESIGN

SECTION III: COMPLEXITY OF LIFE

SECTION IV: COMMON DESCENT

SECTION V: SUMMARIZING THE EVIDENCE

SECTION VI: ACADEMIC FREEDOM

APPENDICES AND ENDNOTES

ACKNOWLEDGMENTS

This book would not have been possible without the support of Lad Allen and Jerry Harned of Illustra Media and John West of the Discovery Institute.

We also thank our reviewers and editors: Douglas Axe, Michael Behe, Charles Garner, Michael Egnor, Ann Gauger, Bruce Gordon, Jens Jorgenson, David Klinghoffer, Andrew McDiarmid, Paul Nelson, Edward Peltzer, Jay Richards, Geoffrey Simmons, Richard Stevens, and Jonathan Wells. For layout and design, we thank Brian Gage and Mike Perry, and for proof-reading we thank Cameron Wybrow.

We are grateful to Valerie Gower for her excellent illustrations, and our thanks also go to Danielle and Nicholas, who made this book necessary.

INTRODUCTION

Over the past century, tremendous discoveries and insights have been made in many scientific fields—information that can help us better understand our bodies and the world around us. We live in an exciting era, with unparalleled opportunities to learn about these new discoveries.

But do scientists always agree? While all scientists deal with the same facts, they often have different interpretations. Not everyone is aware of these controversies, and most people only learn about the evidence after it has been interpreted to fit prevailing theories. For example, scientists actively debate fundamental questions about origins, such as:

- How did the universe begin and develop?
- Is there anything special about our planet?
- How did the first life originate?
- Where did the information in DNA come from?
- How did complex biological systems like the eye develop?
- What does the fossil record say about the origin of species?

For answers to questions like these, much of the scientific evidence points away from unguided causes and toward purposeful design. In our culture, however, scientists and educators who mention the possibility of design are often accused of having religious motives and are labeled "anti-science."

This book will explore the evidence and follow it wherever it leads. The authors think the facts point compellingly toward a scientific theory of intelligent design (ID). The book is divided into six sections that explain this evidence:

- Section I provides an overview of viewpoints and terms common to this controversy.
- Section II examines theories about the origin and development of the cosmos.
- Section III investigates the origin of life, including the origin of information in DNA, and also critiques Darwinian evolution while presenting the case for biological design.
- Section IV explores the evidence for common descent, and many problems with the theory that all living organisms are related.
- Section V summarizes the evidence for design.
- Section VI reviews the tactics used by critics to silence debate, responds to some common objections, and gives ideas about how the reader can get further involved.

The intent behind *Discovering Intelligent Design* (*DID*) is to provide an overview rather than an in-depth analysis of the subjects, simplifying technical jargon and complex concepts without sacrificing accuracy. The basic goal has been to provide a scientifically current and technically correct presentation of the science supporting ID at a level that a student or layperson can comprehend.

Who can use *DID*? In a word (or two), most anyone. While best classified as a textbook, *DID* reads like a book and is intended for a wide range of ages—from teen to adult—in settings such as:

- Private schools.
- General family and home setting.
- Homeschools.
- Church environments.
- Community groups.
- Extracurricular school organizations (such as IDEA Clubs[1]).
- Personal use.

When used as a textbook, *DID* is not intended to replace standard subject science texts, but instead can supplement them by presenting information not available in many standard textbooks. Readers are encouraged not to just take our word for this material, but to use it as a starting point for further research.

WHY *DISCOVERING INTELLIGENT DESIGN*?

DID was written to fill a specific niche. There are other excellent textbooks about intelligent design, such as *The Design of Life: Discovering Signs of Intelligence in Biological Systems*, which provide a compelling and thorough presentation of the scientific evidence. However, such textbooks are written at an advanced level and do not comprehensively cover both the biological and cosmological evidence for design.

Another superb textbook is *Explore Evolution: The Arguments For and Against Neo-Darwinism*—however this supplemental textbook focuses only on biology and evolution, and does not cover cosmology or intelligent design. Finally, there are numerous other high-quality books out there on intelligent design—but they either are not written as a comprehensive study tool, or include religious components.

While we highly recommend the two books mentioned above (as well as many other resources), there is a reason why *DID* stands out. At the time of its publication, *DID* is the only strictly scientific textbook that comprehensively introduces both the cosmological and biological evidence for intelligent design at a layperson's level.

DID has been written as a stand-alone text; however, it can also be used in conjunction with a companion DVD to create an integrated multimedia learning experience. Throughout the book, prompts encourage readers to watch DVD segments, which have been drawn from four highly regarded documentaries: *Unlocking the Mystery of Life*, *The Privileged Planet*, *Darwin's Dilemma*, and *Icons of Evolution*. For more information regarding the DVD, see the next section, Creating a Multimedia Experience.

At the beginning of each chapter, readers will find a thought-provoking question to help introduce the material. Each chapter then concludes with multiple discussion questions to allow reflection and further investigation. Instructors or small discussion group leaders will find these questions valuable in encouraging critical thinking.

In addition, a separate workbook has been created to enhance the book's educational value for younger readers / students. The workbook provides additional questions for each chapter. It can be purchased at **www.discoveringid.org**.

A NOTE ON USE IN PUBLIC SCHOOLS

This supplemental textbook is not intended for use in public schools. ID is a scientific theory and is not religiously based, but we live in a highly-charged political climate that is often hostile to ID. While ID should be perfectly legal to discuss in public schools, there are strong reasons not to push ID into the public school curriculum.

In particular, the priority of the ID movement is to see the theory progress and mature as a science. However, when the subject is forced into public schools, it tends to generate controversy, changing the topic from a scientific investigation into an emotional, politicized debate. This can result in persecution of ID proponents in the academy, ultimately preventing ID from gaining a fair hearing within the scientific community. The policy of the Discovery Institute, the leading organization supporting ID, states:

> As a matter of public policy, Discovery Institute opposes any effort to require the teaching of intelligent design by school districts or state boards of education. Attempts to mandate teaching about intelligent design only politicize the theory and will hinder fair and open discussion of the merits of the theory among scholars and within the scientific community. Furthermore, most teachers at the present time do not know enough about intelligent design to teach about it accurately and objectively.
>
> Instead of mandating intelligent design, Discovery Institute seeks to increase the coverage of evolution in textbooks. It believes that evolution should be fully and completely presented to students, and they should learn more about evolutionary theory, including its unresolved issues. In other words, evolution should be taught as a scientific theory that is open to critical scrutiny, not as a sacred dogma that can't be questioned.[2]

The uniqueness and utility of *Discovering Intelligent Design* are seen in its inspiration: *DID*'s authors are highly involved in the ID movement and have received many requests from educators, church leaders, discussion-group leaders, and home-schoolers for an ID curriculum that is simple, strictly scientific, and integrated with multimedia activities. The authors regretted that no such ID curriculum existed, so they decided to create it.

We hope that you find *DID* both educational and inspirational as you discover the scientific evidence supporting intelligent design.

Gary Kemper
Hallie Kemper
Casey Luskin

CREATING A MULTIMEDIA EXPERIENCE

Through arrangements with Illustra Media and Discovery Institute, a *Discovering Intelligent Design* DVD has been created for use with the *DID* textbook. Notations throughout this book encourage readers to watch segments from the DVD, and those segments will add significantly to your learning experience. If you have not already purchased the DVD, you can order it at **www.discoveringid.org**.

Material for the DVD was selected and compiled with permission from the following four documentaries:

Using state-of-the-art computer animation, *Unlocking the Mystery of Life* (Illustra Media) takes you inside the living cell to explore systems and machines that bear the unmistakable hallmarks of design. For details see: **www.unlockingthemysteryoflife.com**.

Current astronomical evidence suggests that a rare and finely tuned array of factors makes Earth suitable for life. *The Privileged Planet* (Illustra Media) includes stunning computer animation, interviews with leading scientists, and spectacular images of Earth and the cosmos. For details see: **www.privilegedplanet.com**.

Filmed on four continents, *Darwin's Dilemma* (Illustra Media) explores one of the most spectacular events in the history of life: the Cambrian explosion. In a short window of geological time, an abundance of new animals—and new body designs—were fully formed without evidence of any evolutionary ancestors. For details see: **www.darwinsdilemma.org**.

Icons of Evolution (Coldwater Media) explores the scientific evidence regarding several icons of Darwinian evolution. It investigates whether students are learning the whole truth about Darwin's theory. For details see: **www.discovery.org/a/2125**.

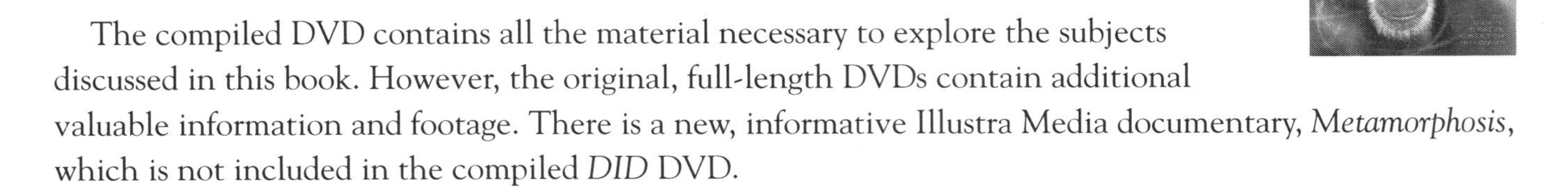

The compiled DVD contains all the material necessary to explore the subjects discussed in this book. However, the original, full-length DVDs contain additional valuable information and footage. There is a new, informative Illustra Media documentary, *Metamorphosis*, which is not included in the compiled *DID* DVD.

Illustra DVDs can be purchased through Randolf Productions: (800) 266-7741 or **www.go2rpi.com**. *Icons of Evolution* can be purchased at **www.discovery.org/a/2125**.

SECTION I

DEFINING MOMENT

CHAPTER 1

DESIGN DECODED

Question:

What have you already heard about intelligent design?

DESIGN DECODED

"We must follow the argument wherever it leads."
— Socrates[1]

Since the dawn of human history, people have looked up at the night sky and wondered how the cosmos came to be.

Turning our eyes toward earthly matters, we've pondered questions like:

- How did the first life arise?
- How did intricate abilities like sight and hearing develop?
- Where did our own species come from?
- Does the complexity of life and the universe point to unguided causes or intelligent design?

Discovering Intelligent Design (DID) examines modern scientific evidence to answer such fundamental questions.

INTELLIGENT DESIGN

The theory of **intelligent design** (ID) studies indications of design in nature.

> **Intelligent Design:**
> ***A scientific theory that holds that many features of the universe and life are best explained by an intelligent cause.***

ID scientists seek to distinguish between intelligently-caused objects and events, and those which originated via unguided processes.

Consider an object that everyone knows is designed—a car. Car parts were originally in the form of their raw materials, such as metal ore and petroleum. Obviously, unguided causes alone will never organize these materials into a car. Something else is needed.

Changing the metal ore into an engine block requires machines that were built by intelligence. Likewise, converting petroleum into fuel, plastic parts, or synthetic tires requires intelligent intervention.

Shaping these materials into parts and then assembling them into a car requires an immense amount of intelligent guidance, a process that generates information. In fact, the presence of information is the primary feature distinguishing objects produced by unguided processes from those produced by intelligent design.

Experience shows that many elements of our everyday lives—such as language, codes, and machines—depend upon information produced by **intelligent agents**.

> **Intelligent Agent:**
> ***A being with the ability to plan ahead and think with a goal in mind.***

Blind and unguided natural processes cannot think ahead to see what is necessary to build a car. Intelligent agents are different. They can use foresight to design a blueprint, collect the proper materials, construct the necessary tools and machinery, and then process and assemble everything in the right order.

ID theorists start by studying the type of information that is created by intelligent agents. Next, they examine natural objects to see if those objects contain such information. When such information is found, it leads to the inference that an intelligent cause was at work in the origin of that object.

We use ID reasoning in our everyday lives. Imagine you are driving along a road and come to a place where the asphalt is covered by a random splatter of paint. You would probably ignore the paint and keep driving onward.

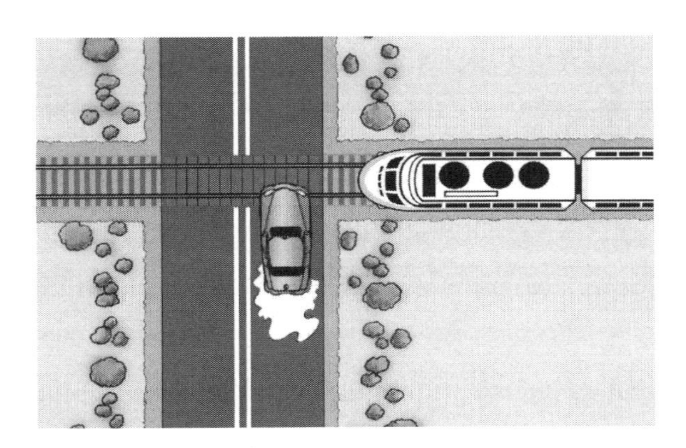

But what if the paint is arranged in the form of a warning? In this case, you would probably make a design inference that could save your life. You would recognize that an intelligent agent was trying to communicate an important message.

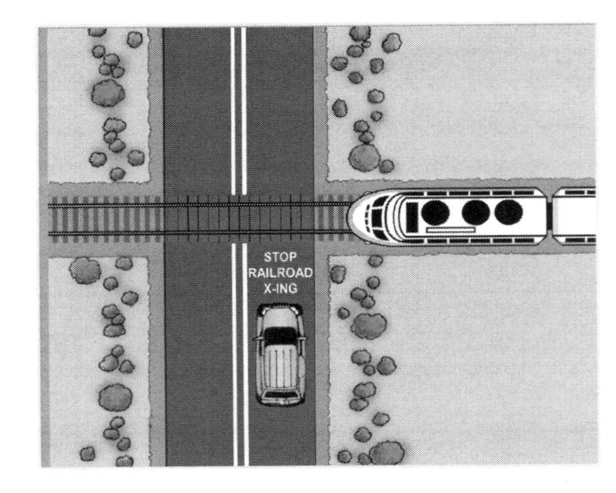

Only an intelligent agent can use foresight to accomplish an end-goal—such as building a car or using written words to convey a message. Recognizing this unique ability of intelligent agents allows scientists in many fields to detect design.

- Archaeologists discriminate between rock formations that have been shaped by natural geological forces and those shaped by intelligence—such as walls, roads, or aqueducts.

- Forensic scientists distinguish between naturally caused deaths, and intelligently caused deaths, such as *murder*. These are important questions that our legal system must answer with great accuracy.

- Environmental scientists seek to determine whether environmental problems result from natural causes or human action. Finding dead fish in a river, scientists might determine that a nearby chemical plant is responsible, rather than a disease.

Following such logic, design theorists ask a simple question: If we can detect design in other scientific fields, why should it be controversial when we detect it in biology or cosmology?

THE SCIENTIFIC BASIS OF ID

Imagine you are a tourist visiting the mountains of North America. You first come across this mountain:

There are features of this mountain that differentiate it from any other mountain on Earth. In fact, if all possible combinations of rocks, peaks, ridges, gullies, cracks, and crags are considered, this exact shape is extremely unlikely.

Nonetheless, there is nothing special about this mountain that would lead to the conclusion it was carved by human hands. Its shape can easily be explained by the natural geological processes of uplift and erosion. Unlikelihood alone is not enough to infer design.

Then, you come across this mountain:

What's different about this mountain? It too has a very unlikely shape—but that shape matches a well-known pattern: the heads of four U.S. presidents.

Observing it, you realize:

- Information was used in shaping it. The artist knew how to shape rock, and he knew the facial patterns of those four presidents.

- The sculpted images also present information to observers—four recognizable faces.

You conclude that this mountain was shaped by an intelligent agent. Why?

As noted, ID begins with observations about what happens when intelligent agents act. Design theorists have observed that intelligent agents are uniquely able to produce high levels of **specified complexity**.[2]

> **Specified Complexity:**
>
> ***A property exhibited by event, object, or sequence of information that is unlikely and matches an independent pattern.***

In the context of science and information theory, something is *complex* if it is unlikely. It is *specified* if it matches an independent pattern. The information inherent in such an object is called **complex and specified information**, or "CSI." From experience, we recognize that finding complexity and specification together is a hallmark indicator of design.

Figure 1-1 on the next page shows the difference between complexity and specification, and how they work together to indicate design.

Let's consider a few real-world examples.

If a structure is readily created by natural laws, then it will not be both complex and specified. For example, a crystal has a repeating pattern of atoms

Complex but Not Specified	Specified but Not Complex	Both Complex and Specified
...ahplyrlerrerhtuavitol daehnma erliaolrgkn larnwcoari kfiv g reebtsi luhisaatt bddlesncm ukle hrdoel oeznigbs... **Explanation:** The exact ordering of the letters of this text is unlikely, making it complex. However, it was typed as random gibberish, meaning it matches no independent pattern, and is not specified. This text was not designed.	*...12121212121212122121212121212 121212121212121212121212121 212121212121212121212121212 121212121212121212121212121 21212121212121212...* **Explanation:** This text matches a simple, repeating pattern, making it specified. However, its precise ordering is easy to predict, meaning it is not complex. Design cannot be inferred.	*...all men are created equal, that they are endowed by their Creator with certain unalienable Rights...* **Explanation:** The precise ordering of these letters is not just unlikely, but it also matches a known pattern—the English language. Intelligent design is evident in the ordering of these letters.

Figure 1-1: Complexity and Specification.

which makes it specified, but since these patterns are ordered by simple natural laws, they aren't complex. Conversely, a random mixture of organic polymers might be very complex, but would lack specificity since it does not conform to any pattern.

Living organisms, on the other hand, are both complex and specified. They are complex because their structures are highly unlikely, and are specified since their structures are specially suited to perform biological processes.

How does such specified complexity arise? We will explore potential answers to that question throughout this book, but a clear possibility was mentioned previously: intelligent agents. They can think with a goal in mind—resulting in solutions that are specified and complex. ID theorist Stephen Meyer explains that language, machines, and computer codes are prime examples of designed objects with large quantities of CSI:

> Our experience-based knowledge of information-flow confirms that systems with large amounts of specified complexity (especially codes and languages) invariably originate from an intelligent source—from a mind or personal agent.[3]

As this book will show, specified complexity—an indicator of design—is found throughout nature, from the constants and laws of the universe down to tiny molecular machines inside the **cell**.

Not everyone, however, accepts that ID provides a valid scientific explanation.

PLAYING THE RELIGION CARD

Despite the scientific utility of ID, its critics have worked overtime to convince the public and the courts that ID is a stealth form of "creationism." But is that claim true?

There are many definitions of **creationism**, but in our culture, the common understanding is as follows:

> **Creationism:**
> ***The belief that the universe and life were created by God as described in the book of Genesis in the Bible.***

Some creationists accept the mainstream scientific view that the Earth and the universe are billions of years old. In contrast, **"young earth" creationists** believe that the Earth and universe are on the order of six to ten thousand years old. What all creationists have in common is that they start with religious texts like the Bible and end with religious conclusions.

Intelligent design is different from creationism because it begins with our observations of nature rather than the Bible, and it limits its scientific claims to what can be learned from the scientific method. As a science, ID refers only to an intelligent cause and does not attempt to establish whether or not the source of intelligence is God.

It's not that asking about the nature of the designer is unimportant. In fact, many people feel that such questions are very important. However, answers to those questions reach beyond the scope of science, and into philosophy and religion.

Many ID theorists have interests in broader questions such as making a case for design from philosophy or theology. All ID theorists agree, however, that an argument for design can be made using the scientific evidence alone. This book explores only the scientific case for intelligent design.

NATURAL AND SUPERNATURAL

Beyond the issue of "creationism," a related question is whether ID requires supernatural intervention. One needs to read no further than Chapter 3 to see why this may be a reasonable question. However, in the debate over ID, those who raise questions about the supernatural are often attempting to shut down the discussion by refusing to address the evidence.

In this book, we use the term "natural" to describe objects or events which originated by unguided mechanisms or processes. However, in some instances natural laws themselves may show evidence of design. As we will discuss in Chapters 4 and 6, ID proponents often infer design from the fine-tuning of the laws that govern the universe and make it friendly for life.

The bottom line is simply this: All that intelligent design scientifically detects is the prior action of intelligence. It does not venture further, beyond the boundaries of science. There are several reasons for this position.

- Perhaps most importantly, ID scientists want to keep the discussion focused on the evidence.
- When we challenge naturalistic explanations, we don't necessarily advocate the supernatural. Although mountains are natural objects, the shape of Mount Rushmore did not arise by natural causes. Rather, the intelligent agents who designed and shaped it were humans.
- ID theorists respect the limits of science, and they don't believe in overstating what can be known from science alone. Science doesn't answer all questions. For example, the DNA

encoding complex molecular machines may indicate that they arose by intelligent design. But it does not tell us whether the intelligence was God, Buddha, Yoda, or some other type of intelligent agent.

This book explores many reasons why ID's critics might wish to avoid debating the evidence. But there is an interesting, related question: *Why are so many critics concerned about potential philosophical and even theological implications of ID?*

MATERIALISM

In our culture there is a great deal of resistance to the evidence for intelligent design.[4] Much of this resistance comes from those who accept an idea known as **materialism**.

> **Materialism:**
> ***The philosophical belief that the material world is the only reality that exists.***

According to this philosophy, the universe and all life are products of long chains of blind, unguided, and chance events. No intelligent design, no designer. Under materialism, everything is ultimately purposeless.

There are two terms related to materialism that should be defined because they occasionally appear in the debate over ID.

> **Philosophical Naturalism:**
> ***Essentially the same meaning as materialism.***

> **Methodological Naturalism:**
> ***The belief that, whether or not the supernatural exists, we must pretend that it doesn't when practicing science.***

To clarify the different viewpoints, the chart in Figure 1-2 shows how their proponents would answer two simple questions.

With regard to their scientific claims about origins, methodological naturalism, philosophical naturalism, and materialism are essentially identical. *DID* uses these terms interchangeably, generally using the term "materialism." Since this book is about scientific investigation, it doesn't examine creationism.

Since so much of the resistance to ID comes from materialists, we will examine that philosophy

Viewpoint	Is there scientific evidence of intelligent design in nature?	What about God?
Creationism	Yes	God created all things.
Materialism / Philosophical Naturalism	No	There is no God.
Methodological Naturalism	No	God is irrelevant.
ID Theory	Yes	ID scientifically infers only an intelligent cause. To determine whether that cause is God would require additional investigation beyond science.

Figure 1-2: Different Viewpoints about Intelligent Design.

THE SEVEN TENETS OF MATERIALISM

1. Either the universe is infinitely old, or it appeared by chance, without cause.
2. The physical laws and constants of the universe were produced by purposeless, chance processes.
3. Life originated from **inorganic** material through blind, chance-based processes.
4. The information in life arose by unguided, blind processes.
5. Complex cellular machines and new genetic features developed over time through purposeless, blind processes.
6. All species evolved by unguided natural selection acting upon random mutations.
7. All living organisms are related through universal common ancestry.

Figure 1-3: The Seven Tenets of Materialism.

in detail in the coming chapters. For materialism to be valid, all of the seven **tenets** in Figure 1-3 must be true.

Materialists and naturalists believe that they own science. After many decades of growing influence in academia, they have gained control of public education, the courts, and the mainstream media.

Materialists impose philosophical restrictions upon science which prohibit any reference to intelligent causes. As one evolutionary biologist argued in the world's leading scientific journal, *Nature*:

> Even if all the data point to an intelligent designer, such an hypothesis is excluded from science because it is not naturalistic.[5]

ID seeks to remove these restrictions because they block scientific advances and hinder the search for truth. A discussion of how ID advances scientific knowledge can be seen in Appendix A.

CONCLUSION

ID challenges materialism by proposing that:

- Intelligent causes can be studied by science.
- Scientific investigations demonstrate that many aspects of nature are best explained by intelligent design.

We are at an important time in scientific history. In spite of the widespread influence of materialism, many scientists, teachers, and policymakers are raising serious questions about its validity and influence on research and education.

How well do the tenets of materialism stand up to scrutiny? Is intelligent design a superior explanation? As we explore the concept of design, we will attempt to answer those questions. Before examining the evidence, however, the next chapter will briefly explore some history leading to the current controversy.

DISCUSSION QUESTIONS

1. If someone wanted to build a house, would it be better to start building immediately, or first create a blueprint? Why? Discuss your answer.

2. Identify a design inference that you made today. Explain how you made that inference.

3. Look around the room. Describe five things you see that have high CSI, and five things that don't.

4. How is intelligent design different from creationism? Does it have any similarities to creationism? If so, does that logically mean that ID is the same as creationism?

5. Materialists would have you believe that the theory of intelligent design is unscientific. Do you agree? Why or why not? What might change your mind one way or the other?

6. Based on this chapter, do you think it is reasonable to have doubts about the philosophy of materialism? Explain your answer.

7. Look at the seven "Tenets of Materialism," and discuss a few of them. Which one do you think is most likely to be valid? Which one might be most difficult to validate? Why?

CHAPTER 2

SURVIVAL OF THE MATERIALIST

Question:

What have you heard about Darwin's theory of evolution?

SURVIVAL OF THE MATERIALIST

"A fair result can be obtained only by fully stating and balancing the facts and arguments on both sides of each question."

— Charles Darwin[1]

Materialists have been around for millennia, but for the last 150 years, they have relied heavily on the theories of British natural historian Charles Darwin to validate their philosophy. To examine materialism, we must examine Darwin's ideas.

As a pigeon breeder, Darwin was familiar with the concept of **artificial selection**.

> **Artificial Selection:**
> ***The selection of certain animals by a breeder in order to increase desirable traits in a population or eliminate undesirable ones.***

Darwin hypothesized that, given enough time, nature could alter populations of organisms in a similar way through a blind and unguided process he called **natural selection**.

> **Natural Selection:**
> ***An unguided natural process whereby organisms that are better suited to survive and reproduce tend to leave more offspring, passing on their traits to the next generation.***

Closely related to natural selection is the phrase "**survival of the fittest**." This simply means that organisms that survive and reproduce are "fit" and therefore naturally selected to pass on their traits to the next generation.

Figure 2-1: Charles Darwin.

When Darwin published his book *Origin of Species* in 1859, he proposed a general theory of evolution that came to be known as **Darwinism**.

> **Darwinism (original theory):**
> ***The theory that all life shares common ancestry and evolved through descent with modification, driven by an unguided process of natural selection acting upon random variation.***

Darwin didn't understand how traits arise or are inherited. Like many others in his day, he accepted Jean-Baptiste Lamarck's theory of the **inheritance of acquired characteristics**. Under this now-refuted theory, organisms were thought to pass on traits they acquired during their lifetime.

For example:

- If a lizard's tail got bitten off by a predator, its offspring might lack tails.
- If a person lost a limb, the child of that person would have an increased chance of being born missing a limb (Figure 2-2).

Figure 2-2: Lamarckian Evolution.

In the twentieth century, the discovery of genes revealed problems with Lamarck's theory. A trait must be present in the parental genes to be passed along to offspring. Changes acquired during the lifetime of a parent, such as loss of a limb, do not affect those genes.[2]

Darwin's followers then updated his theory to include discoveries about genes. This modified theory is called **neo-Darwinism**.

> **Neo-Darwinism:**
>
> ***The theory that all life shares common ancestry, and evolved through descent with modification, driven by unguided natural selection acting upon random genetic mutations in DNA.***

At this point, a reasonable person might ask, "Genetic mutations occur, and natural selection seems possible; species can change over time to adapt to their environment—so what's wrong with Darwinism?"

For the remainder of this book, when we use the term "Darwinism," we will be referring to the modern, neo-Darwinian definition.

The problem is that biological organisms contain many complex features that cannot be built through numerous, successive small-scale changes—a requirement of Darwinian evolution. To help separate opinion from fact, two more terms must be defined.

> **Microevolution:**
>
> ***Small-scale changes in a population of organisms.***

> **Macroevolution:**
>
> ***Large-scale changes in populations of organisms, including the evolution of fundamentally new biological features. Typically this term also means that all life forms descended from a single common ancestor through unguided natural processes.***

No one doubts that species change over time. **Microevolution** has been observed in nature and is an undisputed fact. However, there is considerable disagreement over the validity of **macroevolution**.

To clarify the difference between the two, we must consider the way scientists classify life forms.

The most basic division of life is called the domain. The three domains are Archaea, Eukarya, and Bacteria. The plant and animal kingdoms are in the domain Eukarya.

Each domain is in turn divided into kingdoms, such as plants (Plantae) and animals (Animalia).

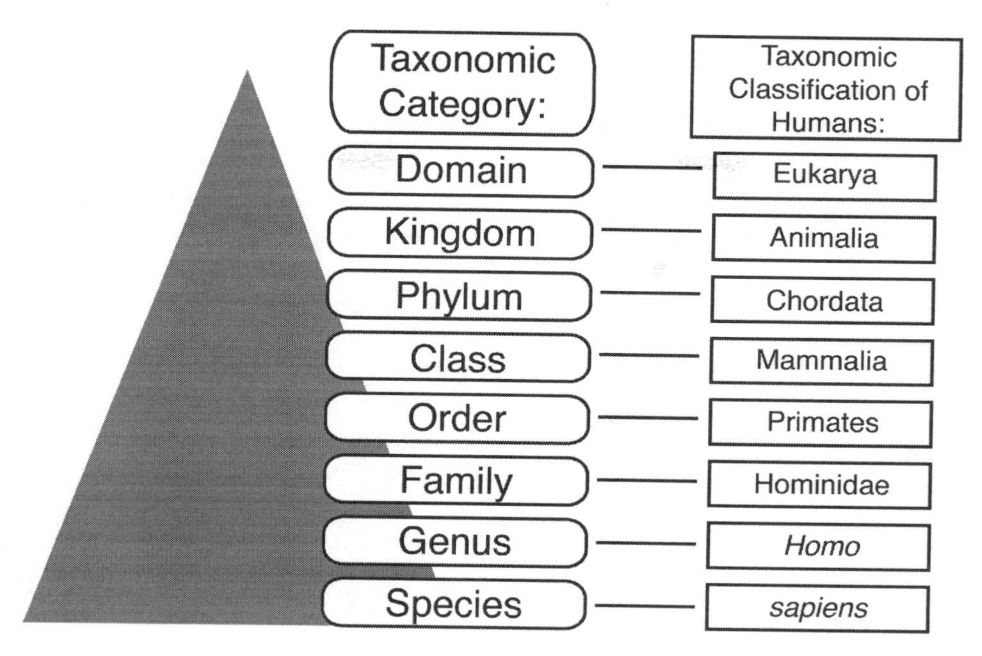

Figure 2-3: Taxonomic Classification.

To illustrate how kingdoms are divided into taxonomic categories, Figure 2-3 shows the classification of our own species.

> **Taxonomic Classification:**
>
> ***The grouping of organisms into categories based on features that scientists consider similar.***

As one looks higher up the taxonomic classification list, fewer groups are found. For example, near the top of the list there are only a few dozen animal phyla. Each level down the list separates animals into a larger number of groups. At the bottom, millions of different animal species have been identified (many of which are now extinct).

Microevolution relates only to small-scale changes at the bottom of the list.[3] Macroevolution asserts that changes occur between categories spanning the entire list, particularly changes between the higher taxonomic categories.

Such an assertion was made by Darwin. He observed evidence of minor variations within species and, based on the limited information available, expanded it to theorize that if populations could change at the species level (microevolution), then, given enough time, they could also change into fundamentally different forms of life at the highest levels (macroevolution).

Following Darwin's approach, a common tactic of **evolutionists** is to take evidence for microevolution and then claim it demonstrates macroevolution. But are such extrapolations warranted?

A **species** is a population of organisms that can interbreed in nature. For example, wolves and most domestic dogs can naturally interbreed, so they are classified in the same species, *Canis lupus*.

> **Evolutionist:**
> ***A person who believes in macroevolution, typically a supporter of neo-Darwinism.***

Below is the first DVD checkpoint in this book. You'll be watching a segment in the DVD that was created for this book, taken (with permission) from *Unlocking the Mystery of Life*. Such segments are suggested throughout the book and, although they are not essential, viewing them will significantly enhance your learning experience.

If you do not yet have the *Discovering Intelligent Design* DVD and are interested in obtaining a copy, see "Creating a Multimedia Experience" preceding Chapter 1. This first DVD segment is a brief introduction to the origin of the ID movement and to Darwin's theory.

DVD SEGMENT 1
"Introduction" (10:39)

From: ***Unlocking the Mystery of Life***

ALFRED RUSSEL WALLACE

Darwin wasn't the only scientist who discovered natural selection. Another British naturalist, Alfred Russel Wallace, co-discovered and co-published with Darwin on the same idea. While Wallace agreed with many of Darwin's ideas, he believed natural selection could not explain many aspects of biology. In fact, he held that many features of nature were best explained by a form of guided evolution similar to intelligent design.

Figure 2-4: Alfred Russel Wallace

Historian Michael Flannery explains why Wallace poses a problem for ID critics:

> Wallace [was] non-Christian, non-creationist, a thorough evolutionist, who not only independently discovered natural selection but came to a view of nature imbued with intelligent design. So an unamended Wallace would really force the Darwinists to admit three important things. First, it's absolutely possible to believe in evolution without subscribing to Darwinism.... Second, ID is not nor was it from the beginning creationist. And third, ID was derived from scientific observations more than a hundred years ago, drawn from Alfred Russel Wallace.[4]

For more on Wallace's life and views, visit **www.alfredwallace.org** and watch the video *Darwin's Heretic* at **www.darwinsheretic.com**.

THINKING TOOLS

Critical thinking is a vital part of exploring the disagreements between materialism and design. Here are a few tips for avoiding logical errors as you investigate this debate:

Tip 1: Question Assumptions

Though materialists rarely (if ever) acknowledge it, nearly all of their claims are based upon assumptions. They typically assume that their theories are correct, and then interpret all evidence based on their assumptions. Very quickly, assumptions become "facts," and future data must be interpreted in a way that is consistent with those "facts."

Treating assumptions as facts will often lead to **circular reasoning**.

> **Circular Reasoning:**
> ***An invalid type of reasoning where the rules or starting assumptions permit only the desired results.***

An example of this type of reasoning is as follows:

- Materialists presume that the only reality is the material world.
- Materialists then claim that anything beyond material causes is unscientific.
- Therefore, any evidence that seemingly points toward intelligent causes is being misinterpreted, because ID, by their definition, is labeled "unscientific."
- They then conclude that there is no evidence of design, and all rational explanations lie within the material realm.

We're back where we started—not due to the evidence, but due to circular reasoning (see Figure 2-5).

The challenge for students who are seeking the truth is to separate assumptions from evidence. Critical thinking will help you see materialist philosophy for what it is: a set of highly questionable assumptions, not a compelling (or even reasonable) conclusion.

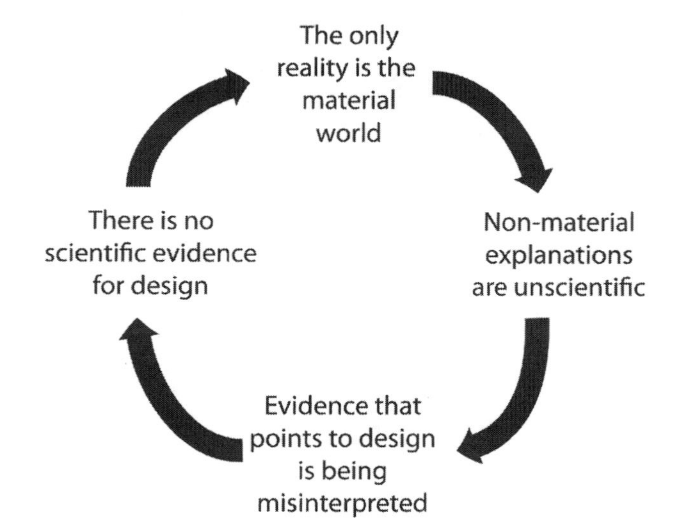

Figure 2-5: The Circular Logic of Materialism.

Tip 2: Demand Evidence

Critical thinkers need to determine whether a conclusion is justified. They should demand: "Show me the evidence!" Molecular biologist Douglas Axe explains how materialists use meager evidence to justify the results of speculation:

> Instead of asking what *needs* to be explained naturalistically, [materialists] concentrate on what *can* be explained. Specifically, you look for some small piece of the real problem for which you can propose even a sketchy naturalistic solution. Then, once you have this mini-solution, you present it as a small but significant step toward the ultimate goal of a full credible story.[5]

The tactic Axe describes can be seen in many mainstream media articles, which use modest evidence to promise that a full materialistic explanation is just around the corner. By assuming that any gap in our knowledge will be filled by material causes, they can be said to use materialism-of-the-gaps reasoning.

Tip 3: Define Your Terms

When terms are not carefully defined, miscommunication and false leaps of logic can result. For instance, when you see the word "**evolution**," you should ask, "Which definition is being used?" Typically, there are three common meanings.

> **Evolution #1:**
> ***Microevolution (defined earlier): Small-scale changes in a population of organisms.***

Macroevolution (also defined earlier) can be divided into two parts.

> **Evolution #2 Universal Common Descent:**
>
> ***The view that all organisms are related and are descended from a single common ancestor.***

> **Evolution #3 Natural Selection:**
>
> ***The view that an unguided process of natural selection acting upon random mutation has been the primary mechanism driving the evolution of life.***

Sometimes evolutionists purposefully confuse these definitions, hoping you won't notice that they overstated their case. It's not uncommon for an evolutionist to take evidence for microevolution (evolution #1), and claim it supports common descent (evolution #2) or development solely through unguided mechanisms (evolution #3).

Tip #4: Seek the Best Explanation

There are few, if any, things we can know with complete certainty. In fact, science never claims to provide absolute proof. Most scientific questions require a choice between competing possible answers. In those situations, one should choose the explanation that best fits the evidence. But what if two or more explanations can explain the data—then what?

Seek more evidence.

Eventually, given enough evidence, one should be able to narrow the possibilities down to the single answer that best explains all of the known evidence. This is called making an inference to the best explanation.

Throughout this book, we will provide scientific evidence that we believe points strongly toward intelligent design. We will also usually present counter-claims from ID critics. It will be up to you, the reader, to evaluate the alternatives and seek the best explanation.

And of course, like all good scientists, we draw our scientific conclusions only tentatively, subject to future discoveries.

INTELLIGENT DESIGN VS. EVOLUTION

Is intelligent design compatible with evolution? The answer depends on which definition is being used.

Intelligent design does not conflict with microevolution (evolution #1).

ID does not even necessarily conflict with common descent (evolution #2)—so long as the mechanism of change is not considered wholly blind and unguided. However, many ID proponents, including the authors of this book, are scientifically skeptical of common descent. This controversial theory is explored more fully in Section IV.

The third definition of evolution is where Darwinism and ID directly conflict. Neo-Darwinism claims that an unguided process of random mutation and natural selection was the driving force in the evolution of life (evolution #3).

CONCLUSION

ID acknowledges that mutation and selection occur, but claims they aren't the entire story. ID contends that many complex biological features could not have been built by random mutations and natural selection. In many instances, intelligent design provides better explanations than Darwinian processes.

Now that we have defined the conflict between ID and materialism, we're ready to begin exploring the evidence. This chapter has discussed biological evolution, but the subject of intelligent design also includes other significant areas, such as physics and cosmology.

To fully explore the evidence, we must go back to the beginning. Not just the beginning of life or our planet, but the beginning of the universe.

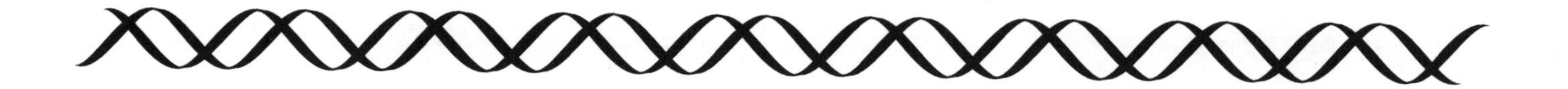

DISCUSSION QUESTIONS

1. Darwin used artificial selection to justify the idea of natural selection. How is artificial selection different from natural selection? Does the former idea truly support the latter? Explain your answer.

2. Give an example of a microevolutionary change and a macroevolutionary change. Can the microevolutionary change realistically be extrapolated to explain macroevolution? Explain your answer.

3. Have you ever played a game where the rules were set up so you could never win? Was the game fair? How is this similar to the rules established by materialists for investigating origins?

4. Describe a claim you heard in a news story recently that was not fully supported by the evidence. Why were you skeptical of that claim? What evidence should the story have presented to make the claim more believable?

5. Give an example of a situation where you had to choose between two alternative explanations. What evidence helped you make that decision?

6. What are some of the benefits of using the thinking tools described in this chapter?

7. Discuss the three possible types of "evolution." Which do you find more plausible, and which less plausible? Explain.

8. Scientific theories aren't written in stone. They can be supported, refuted, or altered by new scientific evidence. Why might this be important to keep in mind as you consider Darwin's theory of evolution?

SECTION II
COSMIC DESIGN

CHAPTER 3

THINK BIG… NO, BIGGER

Question:

What would you say to someone who claims that the universe had no beginning and has always existed?

THINK BIG...NO, BIGGER

"[A]lmost everyone now believes that the universe, and time itself, had a beginning at the big bang."
***— Theoretical physicists Stephen Hawking and Roger Penrose*[1]**

The scale of the universe is almost too large to comprehend.[1] While there is disagreement on the numbers, a common scientific estimate is that the observable universe contains at least 100 billion galaxies[2] and has a radius of over 46 billion light-years.[3]

> **Light-year:**
> ***The distance light travels in one year (5.88 trillion miles).***

Astronomers now estimate that our galaxy, the Milky Way, may contain as many as 400 billion stars.[4]

> **Astronomy:**
> ***The study of celestial bodies and phenomena.***

> **Cosmology:**
> ***The study of the universe as a whole—its structure, origin, and development.***

Our local galactic group consists of two major galaxies—the Milky Way and Andromeda—and a number of smaller ones. When astronomers and **cosmologists** look at another galaxy, they are actually looking into the past. How can that be?

The answer has two parts: (1) The distances between Earth and other galaxies are enormous. Our nearest major galactic neighbor, Andromeda, is approximately 2 million light years away. (2) Even though light travels incredibly fast, by the

Timelines, the Big Bang and Materialism

This book endorses the Big Bang theory and adopts the standard cosmological timeline. Some people oppose these scientific claims, mistakenly thinking they support materialism.

In actuality, the Big Bang model shows that the universe had a beginning, which presents tremendous problems for the philosophy of materialism. Some who argue in favor of a "Young Earth" viewpoint may dismiss ID because:

- ID is based solely on an examination of the scientific evidence, and
- ID finds evidence for design from scientific data that implies the universe is billions of years old.

In the authors' opinion, such rejection of ID is mistaken because it may result in dismissing significant evidence for design. Some of the strongest evidence for design comes from modern physics and cosmology.

time light from another galaxy reaches us, we are not seeing that galaxy as it is, but the way it was when the light we observe today left it.

CONFLICTING WORLDVIEWS

Arguments for a beginning of the universe go back thousands of years. In fact, most cultures throughout history have believed the universe was created by an intelligent agent. For example, in ancient Greece, Plato (Figure 3-1) and his followers argued that the universe was the product of a powerful supreme being. Others, however, claimed the universe is eternal in age and constant in size, and its origin required no explanation. This model later came to be called the **static universe**.

Figure 3-1: Plato.

In the sixth through tenth centuries, early Christian and Muslim philosophers attempted to settle the issue with the ***kalam* argument**.[5] This argument for a first cause has three parts:

- Anything that begins to exist has a cause.
- The universe began to exist.
- Therefore the universe has a cause.[6]

Refusing to accept that the universe had a cause, materialists fought against the *kalam* argument by claiming that the universe never began to exist. Many scientists tended to agree with the materialist philosophy by adopting the belief in an infinitely old universe.

Figure 3-2: Albert Einstein.

Scientific evidence was not able to address this controversy one way or another until 1915, when Albert Einstein (Figure 3-2) proposed his **general theory of relativity.**

> **General Relativity:**
>
> ***A description of gravity as it relates to space and time.***

When Einstein developed his theory, he accepted the view common among his scientific peers that the universe was infinitely old. But his equations conflicted with the idea of an eternal universe by predicting an expanding universe, thus implying a beginning of space and time.

To offset that expansion he inserted a numerical constant (the **cosmological constant**) into his equations. This allowed Einstein to preserve belief in an eternal, static universe.

This still done a lot to make a theory work. It has been done several times on the Big Bang to keep it viable.

When Einstein later became convinced that the universe was not eternal, he referred to his introduction of the cosmological constant as "the greatest blunder of my life."[7] However, scientists now believe that Einstein's error was not in the use of such a constant. His error was giving it a value that supported an eternal universe, and letting philosophical biases influence his science.

We will have more to say about the cosmological constant in the next chapter, but what convinced Einstein and others that the universe is expanding, and has not existed eternally?

EVIDENCE FOR A BEGINNING

In 1927, Belgian astronomer Georges Lemaître theorized that the universe began with a single explosion from a densely compacted state.[8] That explosion eventually became known as the **Big Bang**.

> **Big Bang:**
>
> ***A model of the universe's origin that holds it is finite in size and age. According to this theory, the universe—including all space and time—originated with a single powerful expansion event, and is still expanding.***

Two years after Lemaître's theory, astronomer Edwin Hubble published a study supporting it. Hubble's study indicated that all galaxies are receding from one-another and that the universe is, in fact, expanding. How did Hubble make this discovery?

The next time an ambulance drives past with its siren blaring, pay attention to the pitch of the sound. As the ambulance approaches, the pitch is high, but then as it screams past, the pitch suddenly drops. This is called the **Doppler effect**.

Figure 3-3: In 1931 Albert Einstein Visited Edwin Hubble, Who Verified the Redshift.

The Doppler effect states that sound waves are heard with a higher frequency when the source of the sound is moving toward you, but with a lower frequency when it is moving away from you. Although light waves behave differently than sound waves, a similar effect takes place—also called the Doppler effect.

Light waves coming from an approaching object will have their frequency shifted up toward the blue end of the spectrum of visible light. Correspondingly, light waves coming from a receding object are stretched to a lower frequency, and thus shifted down toward the red end—a phenomenon known as the **redshift**.

Hubble's research confirmed that galaxies are receding from one another by discovering a disproportionately high level of red light coming from virtually every galaxy. If every observable galaxy is moving away from every other, the universe is expanding.[9]

There are other possible explanations for the observed redshift.

PHILOSOPHICAL OBJECTIONS

Faced with evidence supporting an expanding universe, scientists accepted the expansion but many still resisted the concept of a beginning. As physicist Hubert Yockey reports:

> In spite of other successes of the general theory of relativity, the Big Bang, and in particular the idea that the universe had a beginning, was fought bitterly every step of the way.[10]

Yockey might have had in mind astronomers such as Sir Arthur Eddington, who in 1931 stated:

> Philosophically, the notion of a beginning of the present order of Nature is repugnant… I should like to find a genuine loophole.[11]

For a time, the **Steady State theory** seemed to provide such a loophole. This philosophically motivated theory accepted the evidence that the universe is expanding, but proposed that there was no actual beginning because the universe has been expanding eternally. This seemingly impossible claim was justified by theorizing that matter was constantly being created to fill the new space.

THE END OF THE DEBATE OVER THE BEGINNING

In 1948, physicist George Gamow provided a way to settle the controversy between the Big Bang and Steady State theories. He and other cosmologists theorized that if the universe began with a Big Bang, there would be radiation left over from this explosive event.

This radiation was discovered in the 1960s,[12] but the debate continued because the measurements were made using earthbound instruments with limited accuracy.

Finally, in the early 1990s, precise measurements from NASA's Cosmic Background Explorer (COBE) satellite[13] indicated that the universe was filled with radiation having the exact properties predicted by the Big Bang theory.

The COBE measurements confirmed that all matter in the early universe exploded from a densely compacted state. Scientists now had conclusive evidence that the universe had a beginning. As astrophysicist Neil F. Comins explained it:

> Detection of the cosmic microwave background is a principal reason why the Big Bang is accepted by astronomers as the correct cosmological theory.[14]

Theoretical physicist Stephen Hawking called the COBE research "the discovery of the century, if not of all time,"[15] and physicist George Smoot, who won a Nobel Prize for this work, elaborated on the implications of the data:

> What we have found is evidence for the birth of the universe.… [I]t is like looking at God.[16]

CONCLUSION

It is now widely accepted that the universe had a beginning. According to the *kalam* argument, this implies that it has a cause. What does this mean for the philosophy of materialism?

- The cosmological evidence suggests that a first cause, outside the universe, caused the beginning of the universe. As philosopher and former materialist Antony Flew observed, "the universe is something that begs an explanation," and therefore "If the universe had a beginning, it became entirely sensible, almost inevitable, to ask what produced this beginning."[17]

- The Big Bang forces materialists to confront and explain the evidence for cosmic fine-tuning. With an infinitely old cosmos, it seemed logical to many that, sooner or later, the ideal conditions for life would arise by chance.

(These conditions will be discussed further in Chapters 4 and 6.) Once time constraints were imposed by the Big Bang model, however, the origin of life had to be explained. Subsequent inquiries led cosmologists to discover that the laws and constants of the universe are finely-tuned to an extraordinary degree.

- A beginning also imposes time constraints on theories of unguided biological evolution, such as Darwinian evolution.

A beginning to the universe, and the related implication of a first cause, points towards cosmic design, and directly challenges the first tenet of materialism.

MATERIALISM TENET #1:
Either the universe is infinitely old, or it appeared by chance, without cause.

Needless to say, materialists disagree with this outcome. Forced to accept a beginning to the universe, they try to explain it without appealing to a first cause. In Chapter 5, we will explore these arguments. Meanwhile, the next chapter will investigate the development of the universe, and ask whether its physical laws and constants could have arisen by chance, without intelligent design.

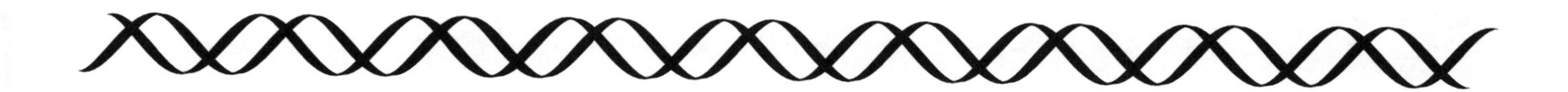

DISCUSSION QUESTIONS

1. Do you find the evidence for the Big Bang persuasive? Why or why not?

2. What impact does the Big Bang model have on the theory of Darwinian evolution?

3. The *kalam* argument poses an important question for materialism: If the universe had a cause, what could it have been? Discuss.

4. The first tenet of materialism postulates that "either the universe is infinitely old, or it appeared by chance, without cause." What reasons have you learned in this chapter to doubt Tenet #1?

5. Some materialists try to refute the *kalam* argument by disputing the statement that "everything that begins has a cause." Can you imagine anything that began without a cause?

CHAPTER 4

DON'T TOUCH THAT DIAL

Question:

Could life exist if the physical laws and constants of the universe were different?

DON'T TOUCH THAT DIAL

"Intelligent design, as one sees it from a scientific point of view, seems to be quite real. This is a very special universe: it's remarkable that it came out just this way. If the laws of physics weren't just the way they are, we couldn't be here at all." — **Nobel Prize-winning physicist Charles Townes**[1]

The fact that the universe had a beginning and a cause is not the only evidence for design of the cosmos. As suggested in the quotation above, the universe appears to meet a special set of requirements necessary for the existence of life. In fact, the requirements are so specific that many scientists describe the universe as **finely tuned.**

This chapter will examine a few of those finely tuned parameters, and will consider the validity of the second tenet of materialism.

MATERIALISM TENET #2:
The physical laws and constants of the universe were produced by purposeless, chance processes.

To provide a background for the discussion of fine-tuning, let's first consider what scientists have discovered about the development of the universe.

IT'S ABOUT TIME

Life could not have existed at the Big Bang. Writing about the infant universe, physicist Steven Weinberg explains:

> At about one-hundredth of a second… the temperature of the universe was about a hundred thousand million (10^{11}) degrees Centigrade. This is much hotter than in the center of even the hottest star…[2]

Following the Big Bang, the universe was filled primarily with hydrogen and helium.[3] Heavier elements necessary for life—such as iron, carbon, or oxygen—did not exist at the Big Bang or even when the first galaxies formed. Instead, these necessary elements have been manufactured in stars.[4]

Since the early days of the universe, there has been an ongoing cycle of stars burning out and dying, and of new stars being born. When a star dies, the heavy elements it has created are dispersed into the universe. Later stars may then incorporate that material, generating even heavier elements.

Approximately three generations of stars—taking at least 9 billion years—were necessary to generate the elements necessary for life. Though our universe is about 13.7 billion years old, the elements necessary for life have only been available for about the last 4.5 billion years, when our **solar system** formed (Figure 4-1).

However, just because the necessary elements existed doesn't mean that life was possible. For example, life on Earth could not have thrived until

Figure 4-1: History of Time.

Event	*Billions of Years Ago*
Big Bang	**13.7**
Formation of our solar system	**4.5**
Earth becomes habitable	**3.8**

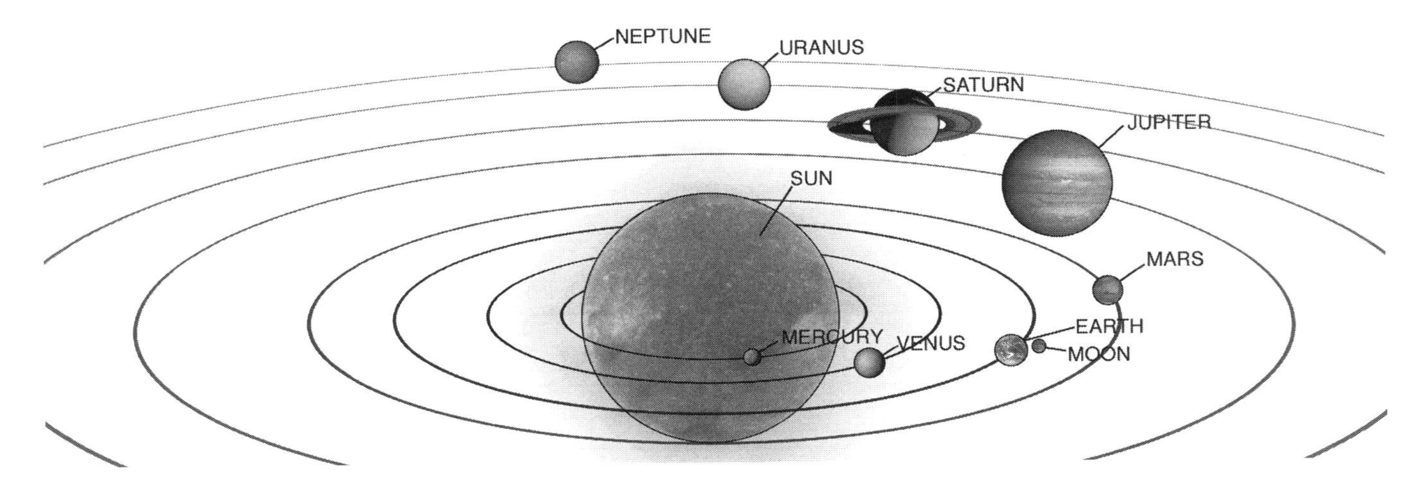

Figure 4-2: Our Solar System (not to scale).

the planets in our solar system (see Figure 4-2) had stable orbits, the Earth's crust solidified and cooled, and major meteorite impacts subsided. Scientists now think that the Earth could have begun sustaining life at about 3.8 billion years ago.[5]

Now that we have a basic understanding of how cosmologists think the universe developed, let's consider a few of the essential parameters.

FINE-TUNING

Imagine that in your basement is a machine for creating universes.[6] On Saturday mornings, you go downstairs, turn it on, and design your own personal cosmos (Figure 4-3).

On the front of the machine are dials for controlling the fundamental laws and constants. How much fine-tuning of the dials would be necessary to construct a universe that could house advanced forms of life? Physicist Paul Davies explains:

> If we could play God, and select values for these natural quantities at whim by twiddling a set of knobs, we would find that almost all knob settings would render the universe uninhabitable. Some knobs would have to be fine-tuned to enormous precision if life is to flourish in the universe.[7]

What exactly are the knobs that must be finely tuned to allow for the existence of life? There are dozens of such parameters, including the speed of light, and the masses of protons, neutrons, and electrons.

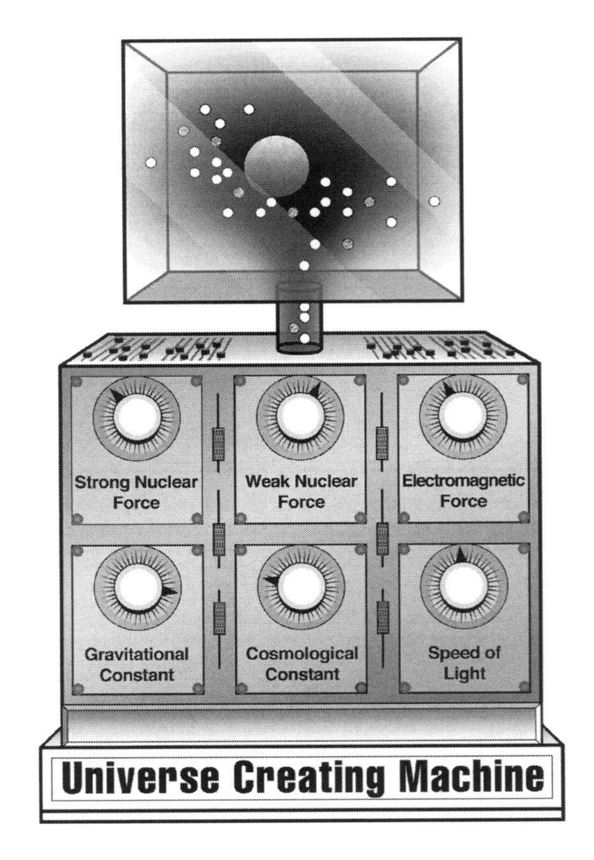

Figure 4-3: A Universe Creating Machine.

If these **universal constants** had only slightly different values, life as we know it would be impossible.

DVD SEGMENT 2
"Fine-Tuning"
(3:07)

From:
The Privileged Planet

2

The DVD segment you have just seen identifies a number of factors whose fine-tuning is necessary for a universe capable of supporting complex life. To elaborate on some of those factors, let's consider the four fundamental forces in nature.[8]

- If the **strong nuclear force** were slightly more powerful, then there would be no hydrogen, an essential element of life. If it was slightly weaker, then hydrogen would be the only element in existence.

- If the **weak nuclear force** were slightly different, then either there would not be enough helium to generate heavy elements in stars, or stars would burn out too quickly and supernova explosions could not scatter heavy elements across the universe.

- If the **electromagnetic force** were slightly stronger or weaker, atomic bonds, and thus complex molecules, could not form.

- If the value of the **gravitational constant** were slightly larger, one consequence would be that stars would become too hot and burn out too quickly. If it were smaller, stars would never burn at all and heavy elements would not be produced.

The finely tuned laws and constants of the universe are an example of specified complexity in nature. They are complex in that their values and settings are highly unlikely. They are specified in that they match the specific requirements needed for life.

As mentioned earlier, there are dozens of parameters that must be finely tuned if advanced life is to exist. Two of them, the cosmological constant and the initial arrangement of mass and energy, stand out because the degree of fine-tuning is extraordinary.

COSMOLOGICAL CONSTANT

Since the Big Bang 13.7 billion years ago, the universe has been expanding, with two competing forces acting on that expansion. Gravity, a force that causes matter to attract, slows down the expansion. An effect called **dark energy** speeds it up. Astronomers are currently studying this effect to determine its exact nature and cause, but an approximate definition is:

> **Dark Energy:**
> ***The effect in the universe that acts to push galaxies away from one another, accelerating the expansion of the universe.***

Astronomers believe that for billions of years after the Big Bang, the universe was expanding, but gravity caused the expansion rate to slow. As the distance between the galaxies increased, the effect of gravity became weaker and the dark energy effect took over. As a result, for about the last 6 billion years, the expansion rate has been accelerating.[9]

This rate of expansion of the universe must be finely tuned if life is to exist. If the universe expands too quickly, matter becomes thinned out and would never clump to form galaxies, stars, or planets. If the universe expands too slowly, then the universe would collapse back in on itself.

The **cosmological constant**, discussed briefly in Chapter 3, is a measure of the effect of dark energy on the expansion of the universe. The probability of the constant having the value necessary for a life-friendly universe has been estimated at $1 / 10^{120}$. Written out, this number is:

$$\frac{1}{1{,}000}$$

Philosopher of physics Robin Collins uses an analogy to demonstrate the likelihood of the cosmological constant having the required value. He compares it to the probability of throwing a dart at a dartboard the size of our entire galaxy, and hitting a bull's-eye less than the size of a quarter.[10]

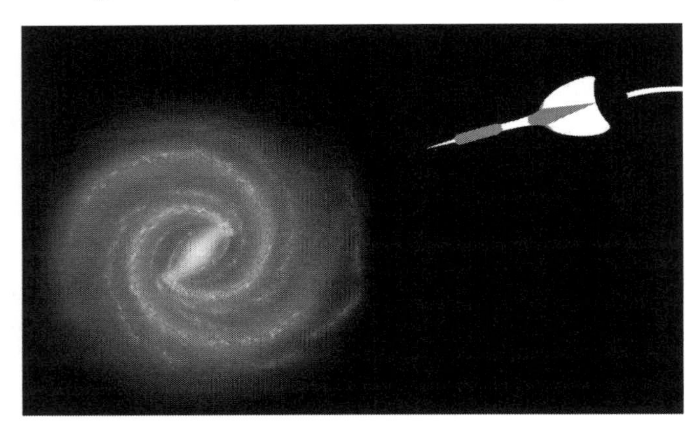

Figure 4-4: What's the Chance of Hitting the Target?

ORDER AND DISORDER

A striking example of fine-tuning relates to the fact that the universe exhibits far more order than it should, given the way it began. Highly respected mathematical physicist Sir Roger Penrose determined the required fine-tuning by calculating the initial **entropy** of the universe.

> **Entropy:**
> ***A measure of the amount of disorder in a system.***

The fine-tuning calculated by Penrose almost defies numerical comprehension. In order for matter in the universe to condense into galaxies or stars, the initial entropy must be accurate to within:[11]

$$\frac{1}{10^{10^{123}}}$$

That's not a mere 1 in 10 followed by 123 zeroes—it's 1 in 10 raised to the power of 10^{123}, or 1 in 10 followed by 10^{123} zeroes! This number should be put in perspective by Figure 4-4.

According to Penrose, "[o]ne could not possibly even write the number down in full," even when making use of all the fundamental particles in the universe.[12] Clearly, the probability of a random occurrence of the initial entropy is such an unimaginably small number that it strains belief to claim that the universe arose by chance. For more information about the calculations of the cosmological constant and the initial entropy, see Appendix B.

NECESSARY BUT NOT SUFFICIENT

The finely tuned factors discussed in this chapter, as well as many others, were *necessary* to make our planet habitable. But there is a distinction between *necessary* and *sufficient* conditions. For example, an engine and many other parts are necessary for a car to function, but they are not sufficient. One must assemble the parts and provide fuel before the car will move.

In a similar way, these finely tuned parameters of the universe are insufficient to explain how life arose. We'll have much more to say about this

$$\frac{1}{10^{10^{1}}} = \frac{1}{10^{10}} = \frac{1}{10{,}000{,}000{,}000}$$

$$\frac{1}{10^{10^{2}}} = \frac{1}{10^{100}} = \frac{1}{\text{(number below)}}$$

10,000,000,000,000,000,000,000,000,000,
000,000,000,000,000,000,000,000,000,000,
000,000,000,000,000,000,000,000,000,000,
000,000,000,000,000,000

$$\frac{1}{10^{10^{3}}} = \frac{1}{10^{1{,}000}} = \frac{1}{10\ \textit{(Plus 999 more zeroes)}}$$

$$\frac{1}{10^{10^{123}}} = \frac{1}{\text{(number below)}} = \frac{1}{10\ \textit{(Plus } 10^{123} \textit{ more zeroes)}}$$

10,000,000,000,000,000,000,000,000,000,000,000,000,
000,000,000,000,000,000,000,000,000,000,000,000,000,
000,000,000,000,000,000,000,000,000,000,000,000,000,
10 000,000,000,000,000,000

Figure 4-4: Exponents and Small Numbers.

in upcoming chapters, but we'll give a hint: it *involves information.*

CONCLUSION

As we have seen in this chapter, the physical laws and constants of the universe are extremely unlikely, and they are specified to precisely match a very narrow range of parameters and settings that are necessary for life to exist. Proponents of intelligent design feel that this fine-tuning is an example of complex (unlikely) and specified (life-friendly) information—CSI.

In our experience, high CSI only comes from the action of an intelligent agent. In fact, some believe that, if the cosmological constant were the only finely tuned factor in the universe, that degree of accuracy alone would indicate design.

Many scientists think that this highly unlikely and life-friendly configuration of our universe offers powerful evidence for design of the cosmos. But there are still holdouts. Materialists often claim our universe just got lucky, and the next chapter investigates their attempts to explain away the evidence for fine-tuning.

DISCUSSION QUESTIONS

1. The second tenet of materialism holds that the "physical laws and constants of the universe ultimately occurred by purposeless, chance processes." After reading about the fine-tuning discovered by scientists, does this change your thoughts on Tenet #2?

2. Which of the finely tuned elements presented in this chapter offers the most convincing argument against an unguided origin? Explain why.

3. Scientists posit that life could not have begun at the time of the Big Bang. Why not? What had to happen first?

4. Give an example of something in your daily life that is necessary, but not sufficient, to produce some desired outcome. The finely tuned parameters of the universe are necessary, but not sufficient, to explain the existence of life. Why do you think this is the case?

5. How is the fine-tuning of the physical laws and constants of the universe an example of specified complexity? Could there be an explanation other than design?

CHAPTER 5

THE EMPIRE STRIKES BACK

Question:

Regarding the origin and fine-tuning of the universe, are there any reasonable alternatives to intelligent design?

THE EMPIRE STRIKES BACK

"If you can't convince them, confuse them."
—President Harry S. Truman[1]

In Chapter 3, we investigated evidence showing that the universe had a beginning, which, according to the *kalam* argument, indicates it had a cause. In Chapter 4, we discussed how evidence shows that the universe is finely tuned to allow for the existence of advanced forms of life. This chapter will assess the attempts of materialists to explain this evidence without intelligent design.

THE FIRST CAUSE

A number of theories have been proposed to explain a materialistic origin of the universe.

Self-creation

Some materialists have claimed that the universe created itself. As Stephen Hawking argues, "Because there is a law like gravity, the universe can and will create itself from nothing."[2]

But for anything to create itself, it would have to exist before it was created. Most people would agree this is logically absurd. Oxford mathematician John Lennox observes that Hawking confuses physical laws—which merely describe how the universe works—with ultimate explanations:

> The laws of physics can explain how the jet engine works, but someone had to build the thing, put in the fuel and start it up. The jet could not have been created without the laws of physics... Similarly, the laws of physics could never have actually built the universe. Some agency must have been involved.[3]

What options are left for materialists? Since they are unwilling to accept intelligent design as a first cause, materialists hold that ultimately the universe came into being by chance for no reason at all.

Betting on Chance

Oxford University scientist and author Peter Atkins parodies the book of Genesis with a summary of the materialistic view:

> In the beginning there was nothing. Absolute void, not merely empty space. There was no space; nor was there time, for this was before time. The universe was without form and void. By chance there was a fluctuation...[4]

Atkins goes on to argue that this random, theoretical, primordial fluctuation spawned a chain of events that caused everything else—the chance universe.

While Atkins is correct that before the universe there was nothing, not even space or time, his argument does not account for the very beginning of everything. He says, "By chance there was a fluctuation." But if absolutely nothing existed, some questions arise.

- What was it that fluctuated?
- Why was there an environment that allowed for such "fluctuations"?
- What *caused* that non-existent something to fluctuate?

For many years, materialists have been attempting to answer such questions without much success. Is "chance" an appropriate final explanation in science?

When a person says that something happened "by chance," there may seem to be an implication that chance actually *caused* the event. But "chance" is not the true cause.

For example, we often think of a coin toss before a football game as an example of "chance." When a referee flips the coin, there are a number of factors that will cause it to come down heads or tails, such as the weighting of the coin, the placement of the coin in his hand, the amount of applied force, wind, and gravity.

Because many of these factors are difficult to predict or control beforehand, we attribute the outcome to "chance." But "chance" is not really the cause at all. That term is an expression of probability and is used simply to predict and describe events. It is not a causal agent.

Yet Atkins attributes the origin of the universe to chance. In this context, chance is not an explanation. It is the absence of an explanation.

Everybody Say: "Quantum"

Though chance is not an explanation for the origin of the universe, some materialists try to give it a scientific facade by appealing to quantum theory.

Most people know little about quantum theory and would probably like to keep it that way. However, a basic knowledge of the concept and terminology can be helpful in understanding one of the materialists' most popular theories regarding the cause of the universe.

Quantum theory describes matter and energy in terms of subatomic units called "quanta." According to the theory, quantum events take place in a quantum vacuum where waves and particles seem to pop in and out of existence randomly without any apparent physical cause.

The *Kalam* Argument Refuted?

Because there is no physical explanation for the origin of quanta, some physicists claim that this refutes the first part of the *kalam* argument: "Anything that begins to exist has a cause."

Of course, there being no *physical* explanation for something does not mean that there is *no* explanation. After all, quantum theory predicts that quantum events will occur within a range of well-defined statistical probabilities—a highly ordered set of circumstances, which is unlikely if there is no underlying cause. This suggests that there is a cause for quanta, but the cause is *not physical.*[5]

Quantum theory does not invalidate the *kalam* argument. Rather, quantum theory suggests that some events may originate from causes outside the physical universe—countering the philosophy of materialism.

Cosmic Design without *Kalam*?

Although the *kalam* argument still stands, what if it did not? Could materialists then avoid the inference to cosmic design?

Whether or not quanta have a cause, materialists would still have to explain how the vast, finely tuned universe suddenly came to exist—bringing with it the Big Bang origin of time and space, as well as the universal laws and constants. The enormity and complexity of that event points toward design, not unguided mechanisms.

Just Another Quantum Particle?

Some materialists postulate that the universe arose from nothing, as a random event, in the same way that quantum particles do. There are at least three problems with this claim.

First, quantum particles arise in a quantum vacuum, but a quantum vacuum is not the complete absence of everything; quantum theory does not address the origin of the particle-producing quantum vacuum. As philosopher and cosmologist William Lane Craig writes with James D. Sinclair, a quantum vacuum "is not nothing but is a sea of fluctuating energy endowed with a rich structure and subject to physical laws."[6] Thus Craig asks, "what is the origin of the whole quantum vacuum itself?" adding, "You've got to account for how this very active ocean of fluctuating energy came into being."[7]

Second, philosopher of physics Bruce Gordon explains that quantum theory is merely a description of observed phenomena, not an explanation for it:

> A common misunderstanding… is that quantum theory… actually explains the phenomena it describes. It does not. Quantum theory offers mathematical descriptions of measurable phenomena with great facility and accuracy, but it provides absolutely no understanding of why any particular quantum outcome is observed.[8]

Third, if the universe arose in the same way quanta do, materialists should realize that quanta may be best explained by a non-physical cause. In the same way, the beginning of our universe some 13 billion years ago also requires an explanation outside of the universe.

Much like Peter Atkins's inadequate appeal to a random fluctuation in the void, quantum theory does not explain why a fluctuation-producing void exists in the first place nor why quantum particles suddenly appear. As we will see, this is a common theme among materialistic theories: materialists are forced to argue that natural phenomena appear abruptly, without any clear explanation.

After years of effort, materialists have not explained the origin of the universe. The materialist arguments we have discussed here do not refute the evidence for a beginning of the universe, nor remove the necessity of a first cause. A beginning and a first cause refute the first tenet of materialism.

MATERIALISM TENET #1:
Either the universe is infinitely old, or it appeared by chance, without cause.

The mere existence of the universe is not the only thing that materialists struggle to explain. They must also answer the question, *Why is it is so finely tuned?*

FINE-TUNING: WINNING THE COSMIC LOTTERY?

Materialists recognize that the odds of our universe having life-supporting parameters are extremely low. They must find a way to overcome staggeringly unlikely odds without appealing to intelligent design. An illustration from lotteries helps explain their strategy.

Imagine that the odds of winning the lottery are 1 in 10 million—vastly better odds than finding a life-friendly universe. One way to improve those odds is simply to buy more tickets.

Similarly, materialists have sought ways to improve the odds of our finely tuned universe by increasing the total number of universes. As the reasoning goes, if there are enough universes, chances are that eventually one will happen to get the right conditions for life.

This thinking has led to some highly imaginative and outlandish hypotheses.

Oscillating Universe

Confronted with the special properties of our cosmos, some astronomers have promoted the **oscillating universe** theory. Under this cosmological model, the universe perpetually expands, then collapses, and then expands again, thereby allowing vast opportunities for a universe like ours to eventually appear.

Some scientists believe that during such a hypothetical collapse, the universe would return to the extremely condensed state it was in immediately prior to the Big Bang. This state is called a **singularity**.

> **Singularity:**
> ***A point where matter is so condensed by gravity that it has infinite density and almost no volume.***

There are numerous problems with the oscillating universe theory. To name a few:

1. **Passing through singularity:** William Lane Craig writes that "there is no known physical mechanism for producing a cyclic 'bounce'" making it "impossible for a universe to pass through a singularity."[9]

2. **Overcoming acceleration:** Astronomers now widely believe that the rate of expansion of the universe is accelerating. The loss of gravity caused by this expansion would make it impossible for the universe to contract back on itself. Matter in the universe is simply too far apart to ever pull itself back together.[10] There can be no contraction, and hence no oscillation, of the universe.

3. **Decreasing energy:** The loss of energy at each bounce means a limited number of oscillations. According to the **second law of thermodynamics**, the amount of usable energy in the universe will decrease over time and thus every successive bounce of the universe would have less energy to rebound.[11] Much as a bouncing basketball will eventually stop bouncing, the number of rebounds would be limited.

Clearly, the oscillating universe theory is inadequate to explain the origin of our finely tuned universe.

Multiple Universes (a.k.a. the Multiverse)

The way to increase the odds of winning the cosmic lottery without resorting to oscillation is to simply proclaim that there are more universes outside our own. This approach, called **multiverse theory**,[12] is currently popular with materialists. It postulates that there are potentially a near-infinite number of universes, each with different values for its physical laws and constants—and we simply occupy the one that permits complex life.

Theories as bizarre as the multiverse provoke observations and questions, such as:

- What is the mechanism that keeps cranking out all of these hypothetical universes, and how did it originate? There is no explanation for the cause of the multiverse.

- The theory relies on the assumption that the universes would be different from each other. If there are multiple universes, why wouldn't many or all of them have the same characteristics?

- Since we cannot observe anything outside our universe, these theories are 100% philosophical speculation, not science.[13]

That last point leads to a deeper reason to reject the multiverse hypothesis. Scientific analyses include a requirement that is violated by the multiverse concept: they should be **falsifiable** (see Chapter 20).

> **Falsifiable:**
>
> ***A common requirement of the scientific method that there must be a means to test whether a theory is false.***

The inability of science to observe anything outside our universe does not seem likely to change anytime soon, or ever. As cosmologist George F. R. Ellis explained in *Scientific American*, any potential parallel universes would "lie outside our horizon and remain beyond our capacity to see, now or ever, no matter how technology evolves," and thus "we have no hope of testing it observationally."[14] Therefore, multiverse theories cannot be part of proper scientific explanations.

Another danger of "multiverse thinking" is that it would effectively destroy the ability of scientists to study nature. A short hypothetical example shows why.

Imagine that a team of researchers discovers that 100% of an entire town of 10,000 people got cancer within one year—a "cancer cluster." For the sake of argument, say they determine that the odds of this occurring just by chance are 1 in $10^{10,000}$. Normally, scientists would reason that such low odds establish that chance cannot be the explanation, and that there must be some physical agent causing cancer in the town.

Under multiverse thinking, however, one might as well say, "Imagine there are $10^{10,000}$ universes, and our universe just happened to be the one where this unlikely cancer cluster arose—purely by chance!" Should scientists seek a scientific explanation for the cancer cluster, or should they just invent $10^{10,000}$ universes where this kind of event becomes probable?

The multiverse advocate might reply, "Well, you can't say there *aren't* $10^{10,000}$ universes out there, right?" Right—but that's the point. There's no way to test it, and science should not seriously consider untestable theories. Multiverse thinking makes it impossible to rule out chance, which essentially eliminates the basis for drawing scientific conclusions.

Multiverse reasoning leads to one final important question.

Ockham's razor is a logical principle, often used by scientists, that holds that the simplest explanation tends to be the correct one. What is the simplest explanation: (1) that the fine-tuning of the universe is the result of a near-infinite number of unobservable universes spawned by an unknown mechanism of unexplained origin, or (2) that the special, life-friendly conditions of our cosmos are the result of intelligent design?

For a discussion of some additional theories from materialists attempting to explain away cosmic fine-tuning, see Appendix C.

CONCLUSION

The precarious condition of the philosophy of materialism is indicated by its imaginative speculation and logical errors regarding the origin and fine-tuning of the universe. Problems include:

- Pushing back the question of origins by appealing to a quantum vacuum which itself remains unexplained.

- Resorting to far-fetched ideas such as self-creating universe, oscillating universe, and multiverse, for which there is no evidence.

The fundamental problem for materialism is that it will never be able to account for the first cause of the universe—it will always push the question of origins back to some chance event that remains unexplained.

Meanwhile, this and the previous two chapters have presented scientific evidence refuting the second tenet of materialism.

MATERIALISM TENET #2:
The physical laws and constants of the universe were produced by purposeless, chance processes.

The evidence of fine-tuning is found not just in the universe at large, but also in our local neighborhood. In the next chapter, we will examine the location and characteristics of our planet to uncover further indications of design.

DISCUSSION QUESTIONS

1. Which of the materialist scenarios claiming a chance origin of the universe do you find to be the most plausible? Explain your answer. Now discuss the arguments against it.

2. Some of the hypotheses for the chance origin of the universe depend on the idea that some entity exists—such as quantum particles or gravitational attractions—that produced our universe. But how did those entities come to exist? Would their origins be truly chance events?

3. If other universes were somehow discovered outside ours, would it disprove ID, or simply push the design back to a deeper level? Is it possible for materialists to escape the problem of fine-tuning?

4. Is it possible to falsify the multiverse hypothesis? How does it negate our ability to do science?

5. What is Ockham's razor, and how does it relate to theories about the origin of the universe? Explain your answer.

6. Can materialism explain why there is something rather than nothing?

CHAPTER 6

HOME, SWEET HOME

Question:

Is our planet just an insignificant ball of rock floating aimlessly in the galaxy, or is there something special about Earth's location and composition?

HOME, SWEET HOME

"Be it ever so humble, there's no place like home."
—Playwright and actor John Howard Payne[1]

In 1966, astronomers Carl Sagan and Iosef Shklovskii estimated that there are at least a billion planets in our galaxy that could sustain life, and up to one million with advanced civilizations.[2] Inspired by this belief, Sagan later wrote, "We live on an insignificant planet of a humdrum star lost between two spiral arms in the outskirts of a galaxy which is a member of a sparse cluster of galaxies, tucked away in some forgotten corner of a universe."[3]

Sagan and others believed that life on Earth evolved easily via unguided natural processes. Combined with their belief in our insignificant location in the universe, this led to the assumption that life is plentiful throughout the cosmos.

Until his death in 1996, Sagan was a leading advocate of the **Search for Extraterrestrial Intelligence (SETI)**,[4] a science research program that has now spent tens of millions of dollars in public and private funds in an unsuccessful attempt to find extraterrestrial life.[5]

I.D. MEETS E.T.

The theory of intelligent design does not claim that life must exist only on Earth. However, many hypotheses about extraterrestrial life are based upon dubious—and often hidden—materialistic assumptions that life evolves easily in the universe.

The belief that we occupy no special place in our galaxy (let alone our universe) is called the **Copernican Principle**, based on Nicolaus Copernicus's 16th century discovery that Earth is not the center of the solar system.

Figure 6-1: Planet Earth.

But how well does this principle stand up against modern evidence? Are there any indications that our planet is extraordinary, and astronomers like Sagan are wrong?

LOCATION, LOCATION, LOCATION

Advocates of the Copernican Principle claim that because the Earth is not in the center of the galaxy, our planet does not have a special position. However, nothing could be further from the truth.

Our Milky Way galaxy is flat and disk-shaped with spiral arms. At its center is a giant black hole that rips apart any star system that gets too close. Additionally, the area around the galactic core is densely packed with stars and filled with intense radiation that would destroy Earth's atmosphere and any life. The center of the galaxy is clearly not a desirable location.

On the other hand, a position too far from the center would also be inhospitable to life because

Time is assumed to be the same on earth and the Cosmos

the outskirts of the galaxy lack sufficient heavy elements necessary for complex life.

The optimal location for life within our galaxy is a narrow band in the middle that escapes the large zones of deadly radiation at the core, yet contains the necessary elements. This region, called the **galactic habitable zone**, is precisely where our solar system exists (see Figure 6-2).

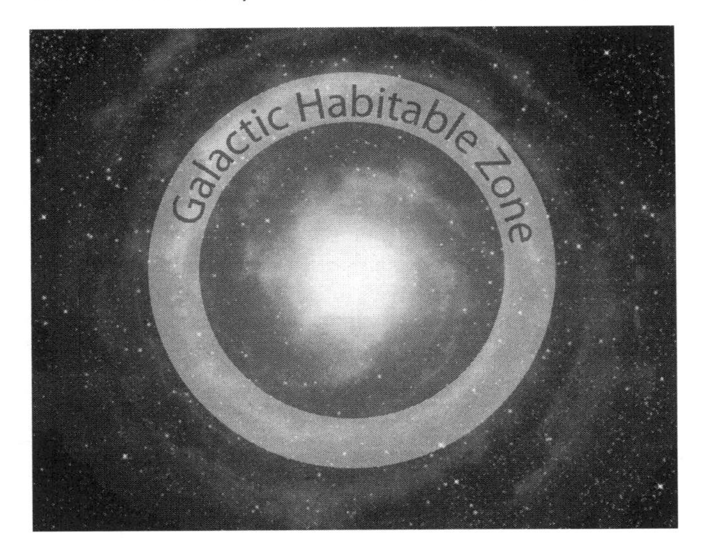

Figure 6-2: Galactic Habitable Zone

In addition to our distance from the center, our position between the galactic arms is also important. Were our solar system located inside the arms, extreme radiation from supernovae and "star nurseries" would again become a problem for life.

Contrary to Dr. Sagan's belief that we are "lost between two spiral arms," we are placed exactly where a life-friendly solar system would want to be.

This special location isn't just optimal for life; it also provides an ideal position to view and learn about the universe. Spiral arms are full of dust and light that, much like city lights and clouds, would obscure astronomical observation. Between the spiral arms, our planet has a clear view of not just the galaxy but much of the universe.

Ironically, for all of Sagan's complaints about our position in the galaxy, were it not for our privileged location, none of us—including Sagan—would have existed, much less been able to study the stars.

IT'S A BEAUTIFUL DAY IN THE NEIGHBORHOOD

Just as there is a galactic habitable zone, there's also a special region within our solar system that permits complex life.

As far as we know, life requires liquid water, which can only exist in a narrow band of temperatures. A planet's surface temperatures are determined by the type of star it orbits and the planet's distance from that star.

Our sun is the type of star (a middle-aged single star with the right mass and brightness) that is ideally suited for complex life.

In addition, our planet's distance from the sun is ideal. If Earth were too close to the sun, any water would only exist as a gas; too far away and water would freeze. Liquid water is essential for life because it:

- Acts as a "universal solvent," providing an ideal medium for the chemistry of life.
- Is denser as a liquid than as a solid, meaning ice floats, a property that prevents Earth's oceans from freezing solid.
- Boils at a high temperature, allowing it to remain liquid over a wide range of temperatures and permitting bodies of water to easily form.
- Has a high surface tension, fostering many biological processes.

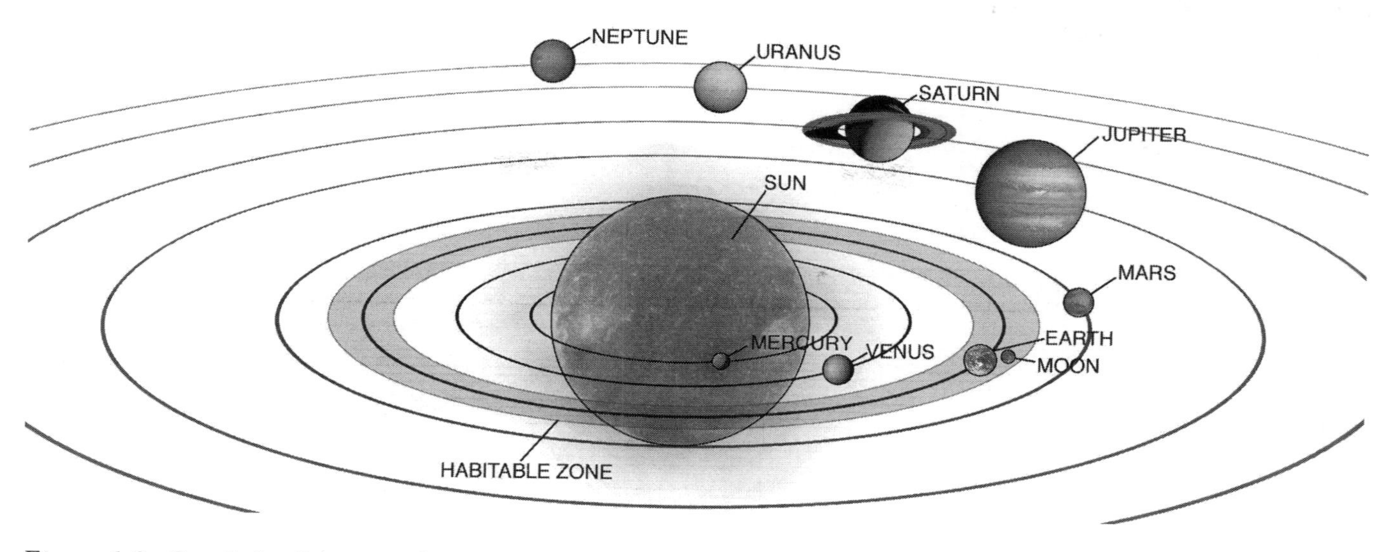

Figure 6-3: Our Solar System and its Circumstellar Habitable Zone.

- Retains heat well, curtailing extreme temperature swings in Earth's climate.[6]

Earth's orbit is well-positioned and stable enough to constantly reside in the **circumstellar habitable zone** (see Figure 6-3), the special band that permits temperatures that are *just right* for liquid water.

When Sagan and Shklovskii suggested that life would be abundant throughout our galaxy, they primarily considered two parameters that define the habitable zone—the type of star and the planet's distance from that star.[7] But there are many other necessary factors they failed to consider. Over 20 essential parameters are required for life to be possible on a planet, including:

Terrestrial planet: Complex life requires heavy elements, such as iron and copper. But some planets, such as gas giants like Jupiter, do not have large concentrations of these elements. The presence of heavy elements on a rocky, terrestrial planet like Earth satisfies many requirements for life.

Plate tectonics: The plates on the Earth's surface are in constant motion. This motion recycles carbon, mixes essential elements, and regulates Earth's interior temperature.

Liquid iron core: The movement of the liquid iron creates a magnetic field that protects life from harmful radiation and prevents the atmosphere from being stripped away.

Atmosphere composition: The ideal combination of 78% nitrogen, 21% oxygen, and a small amount of carbon dioxide and other gases in Earth's atmosphere guarantees a temperate climate, protects us from the sun's radiation, and permits liquid water and complex life.

Large moon: Our moon is about one-quarter the size of our planet. Its gravitational pull stabilizes the tilting of Earth's axis at 23.5 degrees, guaranteeing relatively mild seasonal changes.

Some scientists have estimated that there are many additional factors necessary for a life-friendly planet, including the origin of the moon and the sizes of other planets in the solar system.

Many problems with all moon formation theories.

FORMATION OF THE MOON

Based on the evidence, including Apollo lunar rock samples, a theory has emerged that a planet roughly the size of Mars impacted early Earth and was mostly absorbed into the Earth's core. This impact would have ejected debris into space that eventually clumped together to form our moon.[8]

If this theory is correct, many aspects of this collision event had to be just right—including the composition, mass, and density of the colliding planet, as well as the exact location of the impact—to result in our large moon with its special ability to stabilize the tilt of the Earth's rotation axis.

It is also hypothesized that this impact melted Earth's core and added iron, resulting in the molten iron that now generates our protective magnetic field.

GIANT PLANETS

Jupiter, Saturn, Uranus, and Neptune are all significantly larger than Earth. Two of these planets, Jupiter and Saturn, are so massive that they are often referred to as giants. For comparison, the diameter of Jupiter is more than 11 times that of Earth, and Jupiter has more than twice the mass of all the other planets in our solar system combined.

Due to their strong gravitational pull, the giant planets protect Earth by either attracting asteroids to themselves or deflecting them out of the solar system. Because of their enormous size, they can absorb the impact of very large asteroids without sustaining damage.

Scientists have estimated that if Jupiter weren't protecting us, then more than 1,000 times as many large asteroids would have hit our planet.[9] Additionally, without the stable orbits of Jupiter and Saturn, the orbit of our planet would be so chaotic that complex life could not exist.[10]

DVD SEGMENT 4
"Habitability"
(11:42)

From:
The Privileged Planet

A PRIVILEGED PLANET

As mentioned earlier, Earth is not just well-situated to allow for life; it also sits in an ideal location to study the cosmos. Similarly, our atmosphere allows both for life and for observation of the universe. Some ID theorists feel this adds another strong argument for design.

As astronomer Guillermo Gonzalez and philosopher Jay Richards point out:

> The fact that our atmosphere is clear; that our moon is just the right size and distance from Earth, and that its gravity stabilizes Earth's rotation; that our position in our galaxy is just so; that our sun is its precise mass and composition—all of these facts and many more not only are necessary for Earth's habitability but also have been surprisingly crucial to the discovery and measurement of the universe by scientists.[11]

They conclude that this evidence overturns the Copernican Principle and shows that "our place in the cosmos is designed for discovery."[12]

CONCLUSION

It's apparent that estimates of abundant life throughout our galaxy were wrong, and that we truly live on a privileged planet. The Big Bang, the fine-tuning of the universal constants, the ideal laws of physics, and the extremely favorable conditions on Earth all combine to point strongly to an intelligent cause.

Materialists have long used a fallacious argument for extraterrestrial life. Stephen Hawking provides a good example of this logic:

> The life we have on earth must have spontaneously generated itself. It must therefore be possible for life to be generated spontaneously elsewhere in the universe.[13]

The problem with the argument, as we will discuss in the next chapter, is that it has not been established that life arose on Earth by unguided natural processes. Even Hawking admits that materialists "don't understand how life formed."[14] The only logical basis for stating that it "must" have formed naturally would be the deliberate exclusion of the possibility of intelligent design.

If scientists haven't yet explained how life could have spontaneously generated on Earth, how do they know that it arose naturally elsewhere in the universe? The gaping flaw in the argument demonstrates that even a mind as great as Hawking's is not immune to the philosophical bias of materialism.

This section of our book has shown that the laws of the universe are finely tuned to fulfill many conditions that are required for life. The next section will show that while these laws are necessary for life, they are not sufficient to explain the origin of life. Something else is required—what is it?

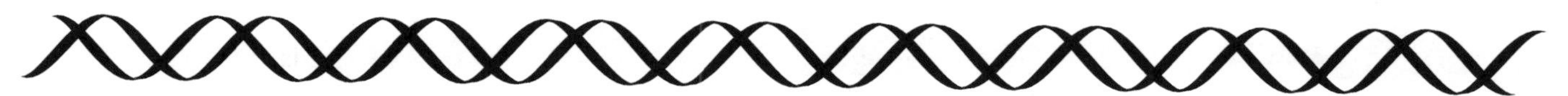

DISCUSSION QUESTIONS

1. Citing the Copernican Principle, some scientists would have us believe that our placement in the universe is insignificant. Discuss what you think about our location in the galaxy, and why you hold that belief.

2. Like the universe, Earth has important features that are finely tuned for life. Choose one of the parameters, and explain what would happen if it didn't exist, or had a different quality.

3. What would happen to the Earth if our moon were smaller, closer or farther away, or had a different orbit?

4. The type of star we orbit, and our location in both the circumstellar and galactic habitable zones, are ideal for life. Does this knowledge give more support to the theory of materialism or intelligent design? Elaborate on your answer.

5. Give an example of a parameter that is necessary for life, but also allows scientific observations of the universe.

6. Would the presence of extraterrestrial life refute intelligent design? Why or why not?

SECTION III

COMPLEXITY OF LIFE

CHAPTER 7

THINK SMALL… NO, SMALLER

Question:

Could life arise by unguided chemical processes?

THINK SMALL... NO, SMALLER

"[I]t has to be true that we really don't have a clue how life originated on Earth by natural means."

—Evolutionary biologist Massimo Pigliucci [1]

The previous section explored the evidence for intelligent design from the life-friendly fine-tuning of the cosmos down to the special properties of Earth. Scientific advances of the past century now show that this fine-tuning also extends to life itself.

Nevertheless, materialists believe that the first living cell developed as follows:

MATERIALISM TENET #3:

Life originated from inorganic material through blind, chance-based processes.

But could it really have been that simple? After being repeatedly asked why the Yankees were losing, baseball legend Yogi Berra made a statement that some materialists may be whispering privately among themselves: "I wish I had an answer to that, because I'm tired of answering that question."[2]

Materialists hypothesize that random chemical reactions—cheered on by electricity, heat, other forms of energy, and vast eons of time—spontaneously formed a self-replicating molecule that then evolved through unguided processes into life as we know it. Origin-of-life theorist George Wald captured the spirit of this perspective in a paper written in 1955:

> Given so much time, the "impossible" becomes possible, the possible probable, and the probable virtually certain, One only has to wait: time itself performs the miracles.[3]

Decades later, fossil evidence of early life led theorists to say things like "we are left with very little time between the development of suitable conditions for life on the earth's surface and the origin of life"[4] and "we are now thinking, in geochemical terms, of instant life..."[5]

Nevertheless, materialists believe that the first living cell developed through blind processes.

But is there any evidence that even the simplest cell could arise via purely unguided, natural processes, in any reasonable amount of time? To help answer this question, let's embark on a journey into the complexity of the cell.

THE CELL

All life is fundamentally based upon a common requirement—information—which is so significant that the next chapter will be devoted entirely to it. But before information is considered, we first need an understanding of the components that make up living cells.

Cells are miniature factories, composed of sub-microscopic molecular machines that are built of **proteins**.

> **Proteins:**
>
> ***Worker molecules that carry out many key biological functions.***

Proteins are composed of long chains of smaller organic compounds known as **amino acids** (Figure 7-1). Using the analogy of written language, amino

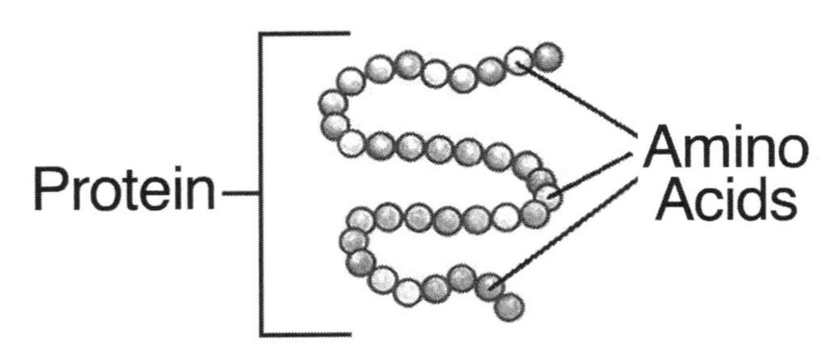

Figure 7-1: Proteins are composed of amino acids.

acids are like individual letters of the alphabet. Creating a cell would be like writing an entire book, as shown in Figure 7-2.

In fact, the number of amino acids used to make proteins is close to the number of letters in our alphabet. As one science news article stated, "With few exceptions, all known forms of life use the same common 20 amino acids—and only those 20—to keep alive organisms as diverse as humans, earthworms, tiny daisies, and giant sequoias."[6]

To continue the analogy, proteins would be like words or sentences. However, since a protein with 100 amino acids would be considered small, even the shortest "words" in the cell are very long compared to words in human language. And just as languages have thousands of words, there are thousands of different types of proteins in many living cells.[7]

Just as sentences combine to form paragraphs, proteins combine to form multi-protein molecular machines (see Figure 7-2). Just as paragraphs are arranged together to form chapters, molecular machines coordinate to form major systems in the cell.

MOLECULAR MACHINES

Long before the advent of modern technology, biologists compared the workings of life to machinery.[8] In recent decades, this comparison has become even stronger.

In 1998, former president of the U.S. National Academy of Sciences Bruce Alberts wrote an article celebrating molecular machines:

> [T]he entire cell can be viewed as a factory that contains an elaborate network of interlocking assembly lines, each of which is composed of a set of large protein machines.... Why do we call the large protein assemblies that underlie cell function protein *machines*? Precisely because, like machines invented by humans to deal efficiently with the macroscopic world, these protein assemblies contain highly coordinated moving parts.[9]

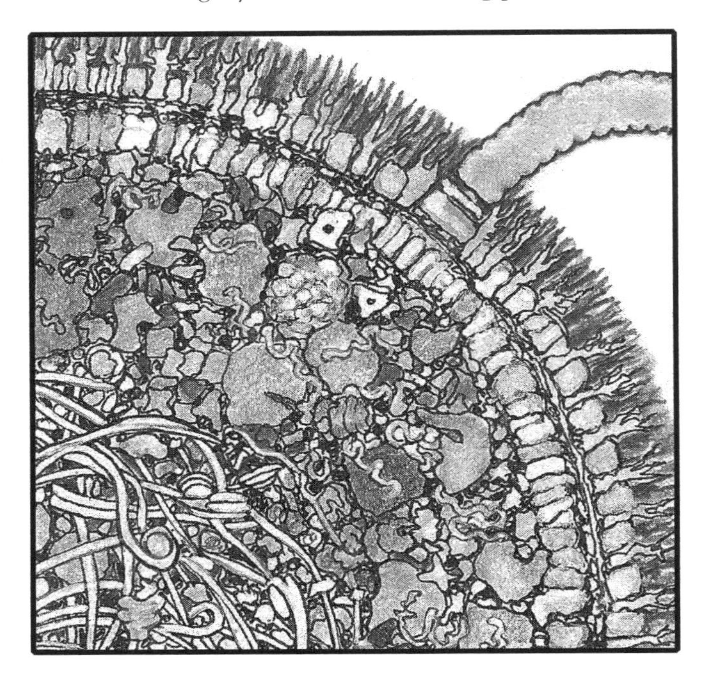

Figure 7-3: The Cell as a Miniature Factory.

Figure 7-2: Comparing Life to Language.

Amino Acids	>	Proteins	>	Molecular Machines	>	Major Cellular Systems	>	Living Cell
Letters	>	Words & Sentences	>	Paragraphs	>	Chapters	>	Book

There are hundreds, if not thousands, of molecular machines in cells. The following two machines are found in all living cells.

Ribosome: The ribosome is a multi-part machine responsible for translating the genetic instructions during the assembly of proteins.[10] According to Craig Venter, a widely respected biologist, the ribosome is "an incredibly beautiful complex entity" which requires a minimum of 53 proteins.[11] Bacterial cells may contain up to 100,000 ribosomes, and human cells may contain millions.[12] Biologist Ada Yonath, who won the Nobel Prize for her work on ribosomes, observes that they are "ingeniously designed for their functions."[13]

ATP Synthase: ATP (adenosine triphosphate) is the primary energy-carrying molecule in all cells. In many organisms, it is generated by a protein-based molecular machine called ATP synthase. This machine is composed of two spinning rotary motors connected by an axle. As it rotates, bumps on the axle push open other protein subunits, providing the mechanical energy needed to generate ATP. In the words of cell biologist David Goodsell, "ATP synthase is one of the wonders of the molecular world."[14]

These are only two examples of molecular machines, and multiple machines are required for each of the major cellular systems.

MAJOR CELLULAR SYSTEMS

Materialists have argued that the first cells may have been primitive compared to modern ones, but there are certain functions that even the most basic cells must exhibit, including:

- Protein production (using many molecular machines, including the ribosome).
- Transportation of parts within the cell.
- Energy production (which in many cases relies upon the ATP synthase machine).
- Disposal of waste.
- Replication.
- Protection from elements outside the cell.

Returning to the book analogy, if all the "chapters" listed above (and more) are combined in one system, you have a very complicated book—a living cell.

Considering the complexity of even the most basic cell, how did so many people come to believe that it arose by unguided, chance processes? A bit of history may help answer that question.

GET A LIFE

Chapter 2 discussed Charles Darwin's mistaken belief in the inheritance of acquired characteristics. But that was not the only erroneous thinking in the mid-nineteenth century.

In Darwin's day, many scientists still believed in **spontaneous generation**.

> **Spontaneous Generation:**
> ***The sudden development of organisms from non-living matter. In essence, life can spontaneously come from non-life.***

To be more specific, it was thought that flies would spontaneously develop from within rotting meat, mice would suddenly come to life out of dirty rags (see the box on the next page), and that liquid could spontaneously create bacteria.

In the 1600s, Italian scientist Francesco Redi challenged spontaneous generation by showing that maggots would not grow in a sealed jar containing rotting meat. But his results were

Recipe for Mice

Nineteenth century Flemish scientist Jan Baptista van Helmont offered this recipe for creating mice:

"Place a dirty shirt or some rags in an open pot or barrel containing a few grains of wheat or some wheat bran, and in 21 days, mice will appear."

not widely accepted. It wasn't until 1864, five years after Darwin published *Origin of Species*, that French scientist Louis Pasteur finally disproved spontaneous generation to the scientific community. One reason for such outlandish beliefs is that the scientific tools in Darwin's time were primitive by modern standards. Lacking powerful microscopes, Darwin and his contemporaries could not understand the complexity of life. They thought that the cell was a simple blob of **protoplasm** surrounded by a cell membrane, perhaps looking something like Figure 7-4:

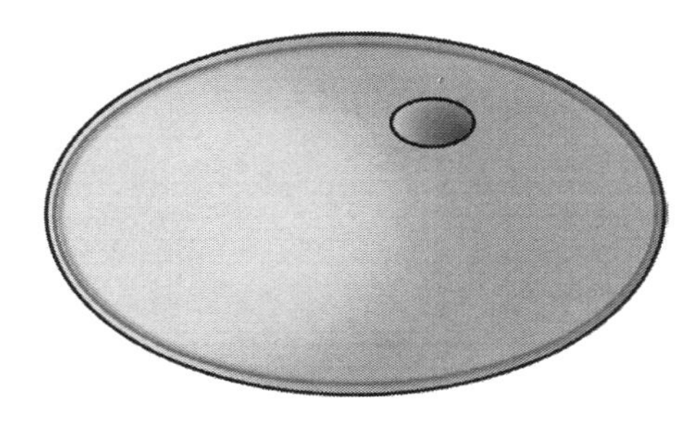

Figure 7-4: The Cell as seen by Darwin.

Though Darwin did not address life's beginning in *Origin of Species*, his private notes and letters show that he took for granted a purely natural origin of life.[15] In an 1871 letter to a colleague, Darwin speculated that life might have started "in some warm little pond with all sorts of ammonia and phosphoric salts, —light, heat, electricity etc. present."[16]

How has thinking about the origin of life progressed since Darwin's day?

CHEMICAL EVOLUTION

In the 1920s, two scientists seeking to support the idea of a purely natural origin of life—Aleksandr Oparin and J. B. S. Haldane—independently proposed similar theories. They theorized that an energy source such as light, heat, or lightning might have interacted with the Earth's early atmosphere—thought to be rich in ammonia—to produce simple molecules.

These molecules then condensed into what Haldane called the **prebiotic soup**,[17] a primordial sea from which life sprang.

> **Prebiotic (a.k.a. Primordial) Soup:**
> ***A hypothetical sea of simple molecules on the early Earth where life formed by chance.***

Sound familiar? These ideas were essentially glorified versions of Darwin's "warm little pond" hypothesis. Oparin and Haldane gave more elaborate details, but used the same simple logic: mix the right stuff, add some energy, and voila—life comes from non-life!

Oparin built upon Darwin's ideas by speculating that after this **primordial soup** arose, some molecules would excel at copying themselves. Competition for the resources needed for replication would cause a form of Darwinian **chemical evolution**, eventually leading these primitive, self-replicating molecules to evolve into the first living cells.[18]

> **Chemical Evolution:**
>
> ***The theory that chemicals in nature assembled through blind, unguided, chance chemical reactions to create life.***

In the scientific community, chemical evolution is now the dominant theory for the origin of life.

DVD SEGMENT 5
"Origin of Life"
(7:44)

From:
Unlocking the Mystery of Life

In the DVD segment you just viewed, biochemist Dean Kenyon described how he had begun doubting his own theories about the chemical origin of life. In the next chapter, we will see where those doubts led him. Meanwhile, we need to examine other attempts to validate ideas about chemical evolution.

MILLER-UREY EXPERIMENTS

As the twentieth century progressed, experiments were attempted to show that life could have arisen from non-life via unguided processes. Perhaps the most famous experiments were the **Miller-Urey experiments**, performed in 1953 by a graduate student, Stanley Miller, and his professor, Harold Urey.

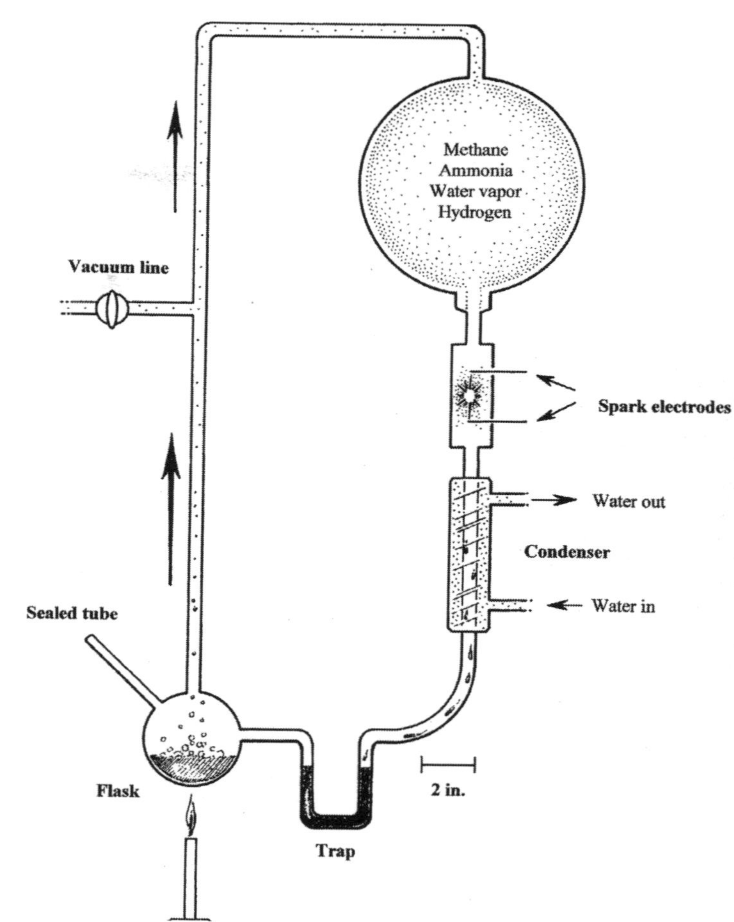

Figure 7-5: The Miller-Urey Apparatus.

Miller and Urey constructed an apparatus to show that the building blocks of life could have developed if lightning sparked gases thought to be present in Earth's early atmosphere. (See Figure 7-5.)

When sparks of electricity (simulating lightning) struck the mixture of gases in the closed container, some amino acids were formed.

This work was celebrated as a huge success for chemical evolutionary theory, and widely publicized as proof of a "prebiotic soup."

Much less publicized, but now acknowledged by many in the scientific community, were significant problems with the experiment.

EARTH'S EARLY ATMOSPHERE

Miller and Urey understood that any newly created organic molecules would be destroyed by **oxidation** if **free oxygen** were present.

> **Free Oxygen:**
>
> ***Oxygen that is not combined with other elements (unlike the combination of oxygen with hydrogen in H_2O).***

Their experiment used methane, ammonia, and a significant amount of hydrogen, but no free oxygen. However, geological evidence now suggests that the Earth's early atmosphere was probably very different from the gases they used. It was most likely composed of carbon dioxide, nitrogen, and only a small amount of hydrogen.

Not only was Earth's early atmosphere probably lacking in methane,[20] but scientists also believe that there was some free oxygen present—enough to have caused immediate oxidation, destroying any organic compounds in the "soup."[21] As an article in the journal *Science* acknowledged, "the early atmosphere looked nothing like the Miller-Urey situation."[22] Consistent with this, there is no geological evidence in the rock record showing that a primordial soup ever existed.[23]

In later writings, Miller revealed the reason for his choice of atmospheric components. Some were selected *not* because there were hard scientific reasons to believe they were present, but rather because they led to the desired amino acid products.[24] In other words, it appears that he stacked the deck in favor of his experiments.

When scientists performed these experiments using conditions likely to have been present on the early Earth, no amino acids were produced. Yet, in spite of contrary evidence, the popular media and textbooks still cite Miller-Urey as a valid experiment.[25]

But let's assume, for the sake of argument, that Miller and Urey's work *had* used the correct gases and explained the origin of the primordial soup. Would this explain the origin of life? Not in the least.

In fact, there are no really good ways to get amino acids to link in nature; only a few rather implausible scenarios have been suggested. Using our book analogy, the "words," even nonsense words, would have difficulty forming at all, much less forming sentences that had meaning or function.

For example, the last place you'd want to link amino acids into chains would be a vast water-based solution like the primordial soup. As the National Academy of Sciences acknowledges, "Two amino acids do not spontaneously join in water. Rather, the opposite reaction is thermodynamically favored."[26] In other words, water breaks protein chains back down into amino acids.

Additionally, amino acids can exist in two shapes that are mirror images, commonly referred to as "right"- and "left"-handed forms. Virtually all life uses only left-handed amino acids to make proteins. But natural processes, like those in the Miller-Urey experiments, have no way of separating right-handed amino acids from left-handed ones. Instead, they produce more-or-less evenly mixed reaction products—useless for living organisms trying to synthesize proteins.

Without the help of complex molecular machinery—not available in the primordial soup—amino acids would form chemical dead-ends, not functional proteins.

MAKING COMPOUNDS, NOT LIFE

The amino acids produced in the Miller-Urey experiment are insignificant compared to the complexity of the proteins and molecular machines in the cell. Again returning to our book analogy, the most that Miller and Urey's experiment can explain is the origin of mere letters. But how did the letters become organized into words, sentences, paragraphs, and chapters joined in a meaningful order?

CHEMICAL EVOLUTION HITS A BARRIER

As an illustration of how mere amino acids alone could never generate complex cellular systems, consider the cell's protective barrier—the **cell membrane** (also called the **plasma membrane**).

Without this extremely important system, cell contents would be vulnerable to harmful molecules and chemical reactions in the outside environment, such as oxidation. The membrane also keeps the cell's components together to allow for cellular processes to take place.

But the protective membrane is no mere wall—it's a gatekeeper capable of allowing water and nutrients in, and letting waste products out. Specialized machines embedded in the membrane discriminate between helpful and harmful substances through a variety of biochemical pathways and molecular pumps.

How could such a vital yet intricate structure arise by chance in the **prebiotic** world? Some researchers have suggested that amino acids formed hollow "microspheres," creating primitive membrane-like structures. But there are a number of problems with this **hypothesis**.

For one, the production of microspheres requires a carefully choreographed set of chemical reactions in the laboratory that would not take place in nature. Their chemistry must be intelligently designed.

Secondly, the microspheres are chemically different from true cell membranes. Indeed, the most glaring problem is that microspheres would have no ability to discriminate among nutrients, waste products, and toxic chemicals. They could not perform the most basic protective function of cell membranes.[27]

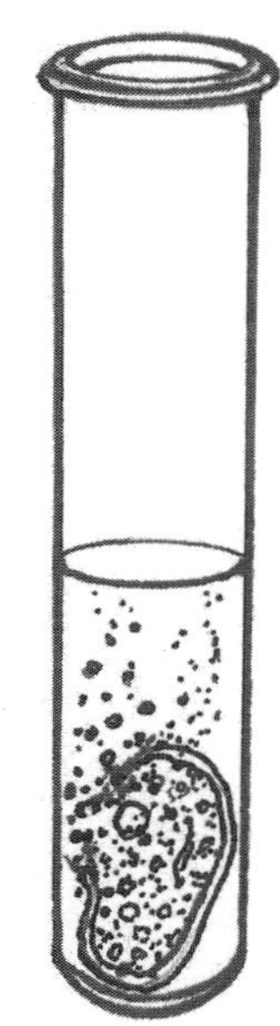

HUMPTY DUMPTY BITES THE DUST

In *The Design of Life*, mathematician William Dembski and biologist Jonathan Wells write about a real experiment testing the idea of the unguided, chance assembly of a cell.

First, a living cell is placed into a test tube filled with the appropriate nutrients. Then the cell is poked with a sterile needle so that its contents spill out into the solution. The test tube now contains all the materials needed for life—not just the amino acids, but the fully assembled proteins.

Nevertheless, even with all the right materials present, the cell cannot reassemble itself. As Dr. Wells puts it, origin-of-life researchers "have been spectacularly unsuccessful in putting Humpty-Dumpty back together again."[28] Even Stanley Miller himself admitted that "making compounds and making life are two different things."[29] But what about more recent media reports that synthetic life has been created in the laboratory?

IT'S ALIVE! IT'S ALIVE!!

In 2010, scientists at the J. Craig Venter Institute presented to the media what they called "the first synthetic cell."[30] The group of 25 scientists worked for 15 years at a cost of $40 million to create a version of the bacterium *Mycoplasma mycoides* with sections of man-made **DNA**.

> **DNA (deoxyribonucleic acid):**
>
> ***A molecule, found in every living organism, which carries the information necessary to produce proteins in that organism.***

But did they really *create* a living cell? First, they read the DNA of a living bacterium and stored this information in a computer. Next, they constructed a new strand of DNA that contained the same information they had *copied* into the computer. In effect, they plagiarized the information in their "synthetic" cell from the DNA of a real living cell.

Genetic information in DNA wasn't the only thing they borrowed. The next step in their experiment was to take the DNA strand they copied and insert it into an *existing bacterial cell* from a related bacteria species, *Mycoplasma capricolim*, to host the new genome. This host cell already contained the necessary proteins, cell membrane, and related materials—without which there would not have been a living cell.

Admitting that they had not actually created original life, the scientists wrote, "We transformed existing life into new life. We also did not design and build a new chromosome from nothing… we synthesized a modified version of the naturally occurring *Mycoplasma mycoides* genome."[31]

Media reports hyping the experimental results as the "first synthetic cell" have misled the public. Suffice it to say, life with all of its required complexities has *not* been created in the laboratory.

CONCLUSION

Cells are like miniature factories full of molecular machines that perform a multitude of necessary functions. As biochemist Michael Denton observed, even the simplest bacterial cells are "far more complicated than any machine built by man and absolutely without parallel in the non-living world."[32]

The complexity of even the most basic living cell poses great difficulty for materialistic explanations and points strongly towards design. And we have not yet discussed a vitally important requirement for all of the structure and functions within every cell—information.

To this day, origin-of-life thinkers have yet to explain how even a simple, self-replicating molecule might originate under natural conditions.[33] As Harvard chemist George M. Whitesides stated when accepting the highest award of the American Chemical Society:

> Most chemists believe, as do I, that life emerged spontaneously from mixtures of molecules in the prebiotic Earth. How? I have no idea.… We need a really good new idea.[34]

A really good new idea already exists. But until biologists learn to appreciate the complexity of biology—and the fact that life is built fundamentally upon information—they will not be able to understand its origin.

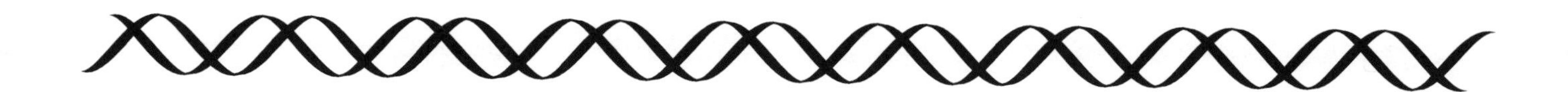

DISCUSSION QUESTIONS

1. How did biologists in the time of Darwin view the cell? How has our understanding of cellular complexity changed since his time?

2. Darwin wrote about a "warm little pond" where the first life arose. What are the pros and cons of this hypothesis? Has origin of life thinking progressed significantly since Darwin's day?

3. Many biology textbooks cite the Miller-Urey experiment to claim that the building blocks of life could have been created on the early Earth through blind chemical processes. Is that a valid claim? Discuss the strengths and weaknesses of this experiment.

4. What important functions are provided by the cell membrane? Could a primitive cell survive with just an inert microsphere?

5. Are cellular components like machines? How is the cell like a factory? Elaborate with examples.

6. If many molecular machines must exist simultaneously in order for cells to function, would it be feasible for them to all develop without forethought and design?

7. Do you think scientists will one day create life in the laboratory? If that occurs, would it disprove intelligent design?

8. The third tenet of materialism holds that "life originated from inorganic material through blind, chance-based processes." After learning about the complexity of a single living cell and the systems within it, does Tenet #3 seem more or less realistic? Why?

CHAPTER 8

INFORMATION, PLEASE

Question:

How does life use information?

INFORMATION, PLEASE

"Human DNA is like a computer program but far, far more advanced than any software we've ever created."

—Microsoft founder Bill Gates[1]

One of the greatest scientific discoveries of the past century is that life is built upon information. It's all around us. As you read a book, your brain processes information stored in the shapes of ink on the page. When you talk to a friend, you communicate information using sound-based language, transmitted through vibrations in air molecules. Computers work because they can receive information, process it, and then give useful output.

No doubt everyday life would be difficult without information. But could there even *be* life without it? Materialist scientist Carl Sagan observed that the "information content of a simple cell" is "around 10^{12} bits, comparable to about a hundred million pages of the Encyclopedia Britannica."[2]

Information forms the chemical blueprint for all living organisms, governing the assembly, structure, and function at essentially all levels of cells (see Figure 8-1). Without this blueprint, life could not exist. As origin-of-life theorist Bernd-Olaf Küppers explains, the origin-of-life question is "basically equivalent to the problem of the origin of biological information."[3]

In Chapter 7 we discussed how the various cellular components work together (comparable to creating a book) starting with the assembly of amino acids into proteins. But how are proteins assembled—more specifically, how is information used in the assembly?

DNA

All living organisms carry biological information in DNA. Located in nearly every cell,[4] DNA is the hereditary molecule that carries **genes**.

> **Gene:**
>
> ***A basic unit of heredity, typically understood as a section of DNA that contains assembly instructions for a particular protein.***

Generally speaking, every protein has its origin in a gene, which orders the sequential arrangement of the protein's amino acids.

A living organism contains anywhere from thousands to billions of bits of digital information in its DNA. As an information storage molecule, DNA functions much as a computer's hard drive, as suggested by the comparison in Figure 8-2.

Figure 8-1: Levels of Information in Living Cells.

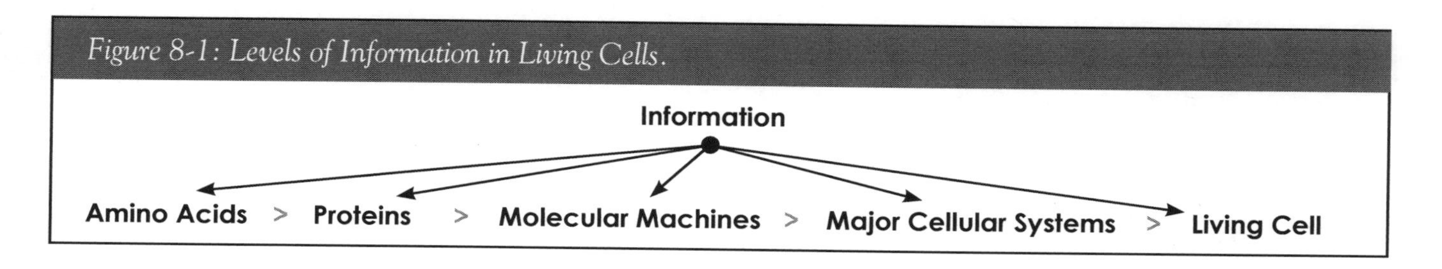

Computer Component	Cell Component	Function
Hard Drive	**DNA Molecule**	**Serves as a physical structure to store information**
Software	**Genetic information (instructions for building proteins)**	**Programs commands**
Hardware	**Molecular machines**	**Executes commands**

Figure 8-2: Computer and Cell Similarities.

Computers use software that contains information (programming instructions), and hardware that executes the programmed commands. In a similar sense, cells use DNA to store genetic information with instructions for building proteins, and molecular machines to carry out those instructions and perform cellular functions.

On a hard drive, data is stored digitally in the form of 0s and 1s. The precise order of the 0s and 1s determines the information content. In a similar fashion, the information in DNA is stored by the ordering of four molecules: adenine (A), cytosine (C), thymine (T), and guanine (G). These molecules are called **nucleotide bases,** that are combined with a sugar molecule (ribose) and a phosphate group to form a **nucleotide**.[5] The ordering of nucleotides makes DNA a digital code.[6] The biochemical language of DNA uses strings of three nucleotide bases (called **codons**) as commands (see Figures 8-3 and 8-4).

Figure 8-3: Nucleotides Bases and Codons.

A codon is a series of three nucleotide bases that encode a particular amino acid, or signify some other command.

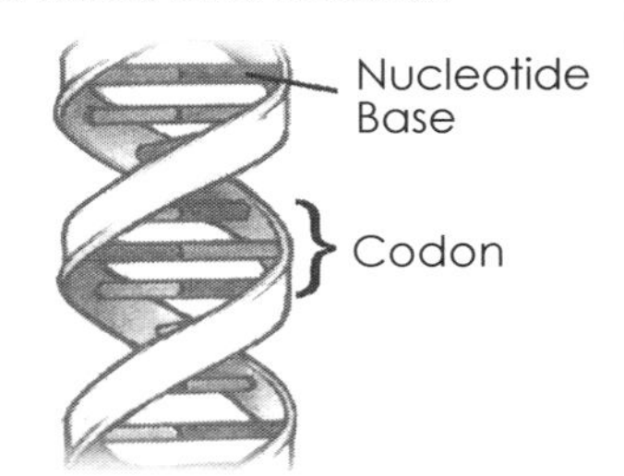

Four codons are shown in the following DNA sequence:

ATG CAC CGA AGC

The first codon, ATG, instructs the cell to begin constructing a protein. The next three codons signify three specific amino acids. An actual DNA sequence might continue for hundreds or even thousands of codons which encode the rest of the amino acids in the protein.

Now imagine that the first nucleotide base of the third codon mutates from a C to a T, as seen in the two DNA sequences below:

ATG CAC CGA AGC

↓

ATG CAC TGA AGC

The first sequence produces a functional section of a protein. However, by changing a C to T a TGA "stop" command is created and protein assembly is prematurely terminated.

In this way, DNA uses codons to indicate start and stop commands, as well as codons that signify

Figure 8-4: DNA and Its Nucleotide Language.

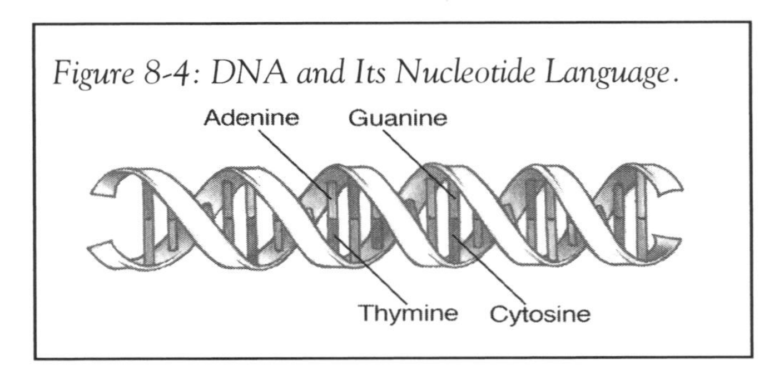

each of the 20 amino acids. The unlikely ordering of the nucleotide bases to match the pattern of a biochemical language represents specified complexity, a hallmark of design.

In the stable, double-helix shape of DNA, A only pairs with T, and C only pairs with G. These combinations are called base pairs. The sequential ordering of **base pairs** in DNA instructs cellular machinery to link amino acids in the correct order to produce functional proteins.

Through a system of programmed commands, molecular machines (such as the ribosome discussed in Chapter 7) in the cell interpret the instructions in DNA and execute them. Even the arch-materialist biologist Richard Dawkins acknowledges that "[t]he machine code of the genes is uncannily computer-like."[7]

It is widely recognized by scientists that no physical or chemical laws dictate the ordering of the base pairs into usable protein-building codes. In unguided natural chemical reactions, each nucleotide has an equal probability of occurrence.[8]

So what is ordering the complex and specified information in the genes? We know from experience of only one cause that generates sequence-specific lines of code—intelligence.

The following DVD segment continues the story of how the need to explain genetic information in the first cells led biochemist Dean Kenyon to become convinced of intelligent design:

DVD SEGMENT 6
"Information" (7:27)

6

From:
Unlocking the Mystery of Life

TIGHT GENES

The free-living organism with the smallest known **genome** is the bacterium *Mycoplasma genitalium*.[9]

> **Genome:**
> ***The full complement of genetic information carried by an organism in its DNA.***

Yet, even this "simple" living organism has over 450 genes, which carry the instructions to make proteins for such activities as DNA replication, transport within the cell, and metabolism.

If the smallest known free-living cell requires so many genes and includes many complex molecules, could the information encoding those molecules spontaneously arise via blind chance? In addition, what if some of those components are dependent on each other for their function or even their existence?

WHICH CAME FIRST: THE CHICKEN OR THE EGG?

Even the simplest living cells today use a system where DNA carries biological information, and proteins carry out cellular functions. But the DNA and proteins cannot exist apart from one another.

DNA needs special proteins (called enzymes) in order to replicate and produce new copies of proteins in the cell—but these enzymes themselves are encoded by DNA. Likewise, DNA needs protection provided by the cell membrane, but again, vital components of the membrane are encoded by the DNA. Many origin-of-life theorists have recognized that this DNA / protein basis of cells presents a chicken-or-egg type of problem: Which came first: DNA, the cell membrane, or proteins (Figure 8-5)?

The answer to the riddle is that none of these came "first," for all are required in DNA-based life.

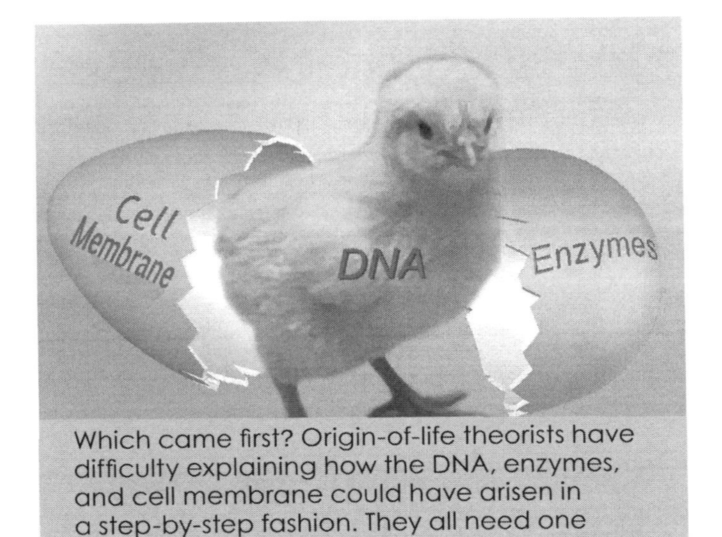

Which came first? Origin-of-life theorists have difficulty explaining how the DNA, enzymes, and cell membrane could have arisen in a step-by-step fashion. They all need one another to exist.

These fundamental components form a complex system in which all components must have been present from the start for a cell to survive. This seems very unlikely, as two origin-of-life theorists admit:

> It is virtually impossible to imagine how a cell's machines, which are mostly protein-based catalysts called enzymes, could have formed spontaneously as life first arose.[10]

But if ID is ruled out, as it always is by materialists, what is the alternative? To put it another way, given such glaring realities, how can the philosophy of materialism possibly survive?

Faced with this unsettling dilemma, materialists did what they have been doing for decades—they proposed a philosophically motivated theory supported by little or no evidence. In cosmology, those efforts brought us the oscillating universe and multiverse theories. In the context of biological information and its origin, they brought us the **RNA world** hypothesis.

IT'S AN RNA WORLD AFTER ALL

In the absence of a mechanism for the origin of life based upon DNA and proteins, materialists have devised a theory that another molecule, **RNA**, could have formed a precursor to DNA-based life.[11]

> **RNA (ribonucleic acid):**
>
> ***A molecule that carries genetic information much as does DNA, but replaces the nucleotide base thymine with uracil. Unlike DNA, it usually exists only as a single strand, and serves as a mobile information transport molecule within cells.***

In living cells, RNA serves as an information-transportation molecule. The transportation involves complicated processes called **transcription** and **translation** to convert the information in the genome into physical parts of the cell—proteins.

> **Transcription:**
>
> ***The first main step in making proteins, cellular machinery copies the information in a gene-coding section of DNA to a strand of messenger RNA (mRNA).***

> **Translation:**
>
> ***After transcription, the mRNA molecule is sent to the ribosome, the molecular machine that assembles the amino acid chain that forms a protein. To do this, the ribosome reads, follows, and "translates" the instructions carried on the mRNA strand to make a protein.***

During translation, another molecule called **transfer RNA (tRNA)** ferries needed amino acids to the ribosome so that the protein chain can be assembled.

If the preceding terms are unfamiliar, help is on the way. The following DVD segment provides a visual depiction of the processes.

DVD SEGMENT 7
"The Cell Factory"
(3:52)

From:
Unlocking the Mystery of Life

Since RNA can both carry information and perform a few cellular functions, materialists reasoned that perhaps it solved the chicken-and-egg riddle. This hypothesis is known as the **"RNA world."**

Proponents of the RNA world hypothesis are at the forefront of modern origin-of-life thinking. They generally claim the RNA world began when a self-replicating RNA molecule appeared by pure chance. While speculation abounds, this is the most popular theory about the origin of life today. In the introduction to a recent *Scientific American* article, the editors boldly stated:

> [S]tudies have supported the hypothesis that primitive cells containing molecules similar to RNA could assemble spontaneously, reproduce and evolve, giving rise to all life.[12]

Makes it sound easy, doesn't it? However, there are major problems with the RNA world hypothesis, any of which could derail the theory.

1. RNA can't form without intelligent design.

RNA has not been shown to assemble in a laboratory without the help of a skilled chemist intelligently guiding the process. Robert Shapiro, professor emeritus of chemistry at New York University, has critiqued the efforts of those who tried to make RNA in the lab:

> The flaw is in the logic—that this experimental control by researchers in a modern laboratory could have been available on the early Earth.[13]

2. RNA can't fulfill the roles of proteins.

Despite some claims to the contrary, RNA molecules do not exhibit many of the properties that allow proteins to serve as worker molecules in the cell. While RNA has been shown to perform a few roles, there is no evidence that it could perform all necessary cellular functions.[14]

3. The RNA world can't explain the origin of information.

The "Cell Factory" DVD segment referenced earlier provided insights into the great complexities involved in the creation of proteins. But the most fundamental problem with the RNA world hypothesis is its inability to explain the origin of information in the first self-replicating RNA molecule—which experts suggest would have to have been between 200 and 300 nucleotides in length.[15]

How did the nucleotide bases in RNA become properly ordered to produce life? There are no known chemical or physical laws that can do this. To explain the ordering of nucleotides in the first self-replicating RNA molecule, materialists have no explanation other than chance.[16]

The unique RNA sequence required for replication represents complex and specified information (CSI). The odds of specifying 250 nucleotides in an RNA molecule by chance alone are as follows:

$$\frac{1}{4^{250}} \approx \frac{1}{10^{150}}$$

This sequence contains too much CSI—and the probability is too low—to be produced by unguided natural processes.[17] Robert Shapiro puts the problem this way:

> The sudden appearance of a large self-copying molecule such as RNA was exceedingly improbable.... [The probability] is so vanishingly small that its happening even

> once anywhere in the visible universe would count as a piece of exceptional good luck.[18]

ID theorists refer to this obstacle to materialism as the **information sequence problem**.

> **Information Sequence Problem:**
>
> ***Unguided and chance processes could not properly order the sequence in an information-carrying molecule (such as RNA or DNA) to create the first life.***

This is a major obstacle for any materialistic hypothesis of the origin of life.

4. The RNA world can't explain the origin of the genetic code.

Another problem with the RNA world—and any materialistic account of the origin of life—is its inability to explain the origin of the **genetic code**.

> **Genetic Code:**
>
> ***The set of rules used by cells to convert the genetic information in DNA or RNA into proteins.***

There is an important distinction between the genetic code and the information in DNA or RNA: the genetic code is essentially the language in which the genetic information in the DNA or RNA is written.

In order to evolve into the DNA / protein-based life that exists today, the RNA world would need to evolve the ability to convert genetic information into proteins. However, this process of transcription and translation requires a large suite of proteins and molecular machines—which themselves are encoded by genetic information.

Again we see a chicken-or-egg problem, where essential enzymes and molecular machines are needed to perform the very task that constructs them. To appreciate the obstacle this poses to materialistic accounts of the origin of life, consider the following analogy.

If you have ever watched a DVD, you know that it is rich in information. However, without the machinery of a DVD player to read the disk, process its information, and convert it into a picture and sound, the disk would be useless.

But what if the instructions for building the first DVD player were only found encoded on a DVD? You could never play the DVD to learn how to build a DVD player.

So how did the first disk and DVD player system arise? The answer is obvious: intelligent agents designed both the player and the disk at the same time, and purposefully arranged the information on the disk in a language that could be read by the player.

In the same way, genetic information could never be converted into proteins without the proper machinery. Yet the machines required for processing the genetic information in RNA or DNA are encoded by those same genetic molecules—they perform and direct the very task that builds them.

This system cannot exist unless both the genetic information and transcription / translation machinery are present at the same time, and unless both speak the same language. Such a complex system requires an intelligent cause, and is not amenable to a blind and unguided evolutionary explanation.[19]

CRACKING THE CODE

Related to the fourth problem discussed above, the RNA world hypothesis provides no solution to the most crucial step in protein production: matching amino acids to their sequencing instructions according to the genetic code. If this step is not properly performed, the code breaks down and genetic information cannot be used.

As is the case with other machinery involved in creating proteins, this key step in the translation process involves a myriad of specialized enzymes, which of course perform the very process that creates them. A more detailed discussion of the challenge these enzymes pose to an unguided origin of life can be found in Appendix D.

CONCLUSION

Without information, life could not exist. Materialistic explanations for the origin of life's information, like the RNA world hypothesis, fail for a number of reasons.

- RNA can't form without intelligent design.
- RNA can't fulfill the roles of proteins.
- The RNA world can't explain the origin of information.
- The RNA world can't explain the origin of the genetic code or the matching of amino acids to their sequencing instructions.

Other than intelligent design, any model for the origin of life—whether the RNA world hypothesis or some other theory—will stumble over the origin of information. According to materialists, information arose in the first life due to sheer chance. But the odds of producing such ordered information by chance are impossibly low. This refutes the fourth tenet of materialism.

MATERIALISM TENET #4:
The information in life arose from nothing by purposeless, blind processes.

So how does information arise? The past two chapters have shown that life is based upon:

- A vast amount of complex and specified information encoded in a biochemical language.
- A computer-like system of commands and codes that processes the information.
- Molecular machines and multi-machine systems.

Where, in our experience, do language, complex and specified information, programming code, and machines come from? They have only one known source: intelligence.

Even the best efforts of ID critics cannot escape the fact that intelligence is required to solve the information sequence problem. The next three chapters will explore life's complexity in more detail.

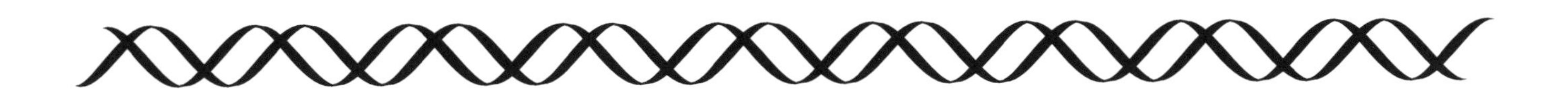

DISCUSSION QUESTIONS

1. Scientists say that DNA functions like a computer. Explain why they make this comparison.

2. A significant amount of information is needed to build a complex structure like a car or a house. How did the information in living cells originate?

3. What is the information sequence problem, and why does it pose an obstacle to the unguided chemical origin of life?

4. One of the largest problems with the RNA world hypothesis is that the first self-replicating RNA must have contained information. If we assume, for the sake of argument, that this information could arise without intelligent design, what other problems remain that materialists must conquer? Do you think they can succeed?

5. Imagine a computer with 450 essential parts scattered around a room. If you couldn't plan ahead, or learn from your mistakes, do you think a computer could ever be built?

6. The genetic code is the set of rules used by cells to convert the information in DNA or RNA into proteins. These rules are abstract concepts, not physical entities. Could they arise in a step-by-step fashion?

7. Can you provide an instance where language-based code, or a machine, arose from a source other than intelligent design?

CHAPTER 9

OPENING THE BLACK BOX

Question:

Can Darwin's theory of evolution be scientifically refuted?

OPENING THE BLACK BOX

"[T]here are presently no detailed Darwinian accounts of the evolution of any biochemical or cellular system, only a variety of wishful speculations."

— Biochemist Franklin Harold[1]

Darwin and his contemporaries had a very limited understanding of the cell and its functions. Biochemist Michael Behe explains, "To Darwin, then, as to every other scientist of the time, the cell was a black box"[2]—something scientists found fascinating but did not understand. By opening this **black box**, modern biochemistry has presented Darwinian theorists with a number of troubling discoveries.

Darwinian evolution is at least a two-step process. First, it requires that random **genetic mutations** cause change in an organism. Next, unguided natural selection must preserve that change in future generations. If mutations do not produce a change, natural selection will never have the opportunity to preserve it.

According to neo-Darwinism, this two-step process is capable of developing complex new traits in populations of organisms. In fact, modern evolutionists claim that this is the driving force behind almost all biological evolution. Is this a realistic explanation? In the last few years, much research has been done, seeking to answer that question.

Let's begin by considering the possible effects of genetic mutations.

EFFECTS OF MUTATIONS

A change in an organism will be beneficial if it provides an advantage that helps the organism survive and reproduce. The process of Darwinian evolution can generally work when only a single mutation is required to gain an advantage. However, when many mutations are necessary before there is a benefit, evolution runs into problems. In *Origin of Species*, Darwin himself recognized this would pose a grave challenge to his theory.

> **Darwin's Test of Evolution**: "If it could be demonstrated that any complex organ existed, which could not possibly have been formed by numerous, successive, slight modifications, my theory would absolutely break down."[3]

Unfortunately for Darwinian theory, the vast majority of mutations are either harmful to an organism or have no effect—and are thus not the type that natural selection tends to preserve.[4] Figure 9-1 shows how natural selection affects four types of mutations.

As seen in the figure, natural selection only favors mutations that are beneficial. The basic problem for Darwinism, therefore, is that a structure must provide a functional benefit *at every step* along its evolutionary path. Evolutionary biologist Jerry Coyne explains:

> It is indeed true that natural selection cannot build any feature in which intermediate steps do not confer a net benefit on the organism.[5]

As committed evolutionists, both Darwin and Coyne claimed they could not envision any organ that could not be built by random mutation and natural selection. However, not all biologists agree. For example, the late Lynn Margulis—an opponent of ID who was a leading biologist—argued that "new mutations don't create new species; they create offspring that are impaired."[6]

Effect of Mutation	Likely Result
1. Beneficial: Increases ability to survive and / or reproduce.	Benefit would tend to be preserved by natural selection.
2. Neutral: No change in ability of organism to survive and / or reproductive.	Natural selection has no effect; it has no reason to preserve the change.
3. Harmful: Decreases ability to survive and/ or reproduce.	Mutation would tend not to be preserved.
4. Deadly: Causes death (or sterility).	Mutation would be eliminated.

Figure 9-1: Four Possible Effects of Mutations.

Of course, if all evolutionary biologists agreed with Dr. Margulis, there would be no debate to investigate. Since there is disagreement, it's important to consider additional evidence. The remainder of this chapter will examine problems that Darwinian evolution would have in creating biological features that cannot be built by "numerous, successive, slight modifications." These problems include:

- The inability of random mutations to produce an amino acid chain that folds into a functional protein.
- The problems encountered in mutating one protein into another.
- The difficulty of producing a hand-in-glove fit between two or more interacting proteins.
- Irreducible complexity as an obstacle to the Darwinian evolution of multi-protein machines.

DARWINISM FOLDS

As indicated in Figure 9-1, mutations will tend to be preserved only if they perform a function that provides a survival advantage to the organism. Evolutionary theory claims that proteins evolve their functions when amino acid sequences are changed by random mutations in the DNA. However, even slight changes to proteins can decrease or even destroy function.

Why is function decreased when proteins are changed slightly? In order to function properly, the chain of amino acids must fold into a specific, stable, three-dimensional shape.

Once proteins adopt their special shape, they can serve as machines or structural components. If they adopt the wrong shape, however, proper function does not occur. As evolutionist and protein biochemist David Goodsell explains, "only a small fraction of the possible combinations of amino acids will fold spontaneously into a stable structure."[7]

Figure 9-2 shows a progression from amino acids to a group of folded protein molecules. The sequential order of the amino acids is critical in allowing a protein to fold into a useful molecule.

For example, the protein **hemoglobin** is a molecule in red blood cells responsible for carrying oxygen to tissues. As seen in Figure 9-3, it is specially shaped to carry molecules containing iron, which in turn gives hemoglobin its oxygen-attracting properties.

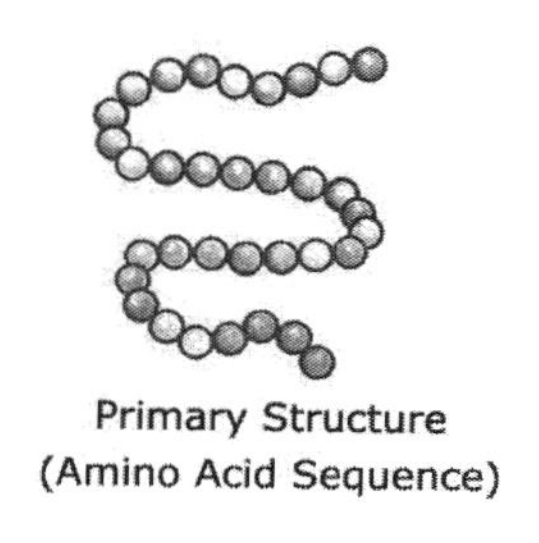

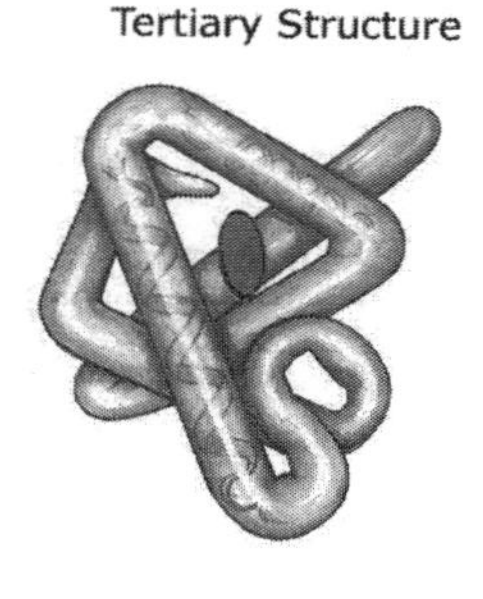

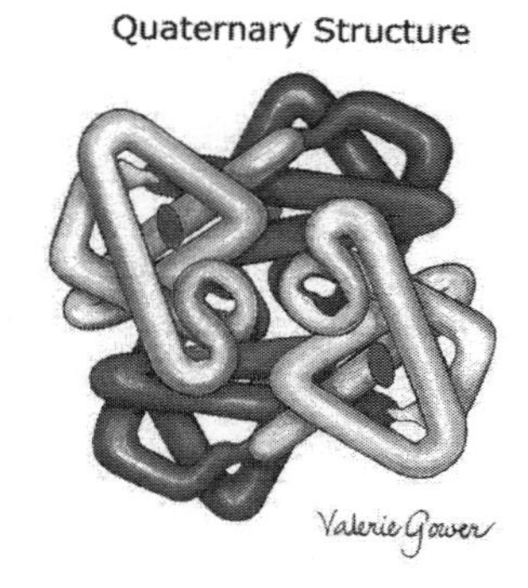

Figure 9-2: Levels of Structure in Proteins.

According to Michael Behe, hemoglobin's ability to perform its function "requires very precise engineering of the shape and amino acid sequence."[8] Behe further notes, "[a] number of genetic diseases are known where a single amino acid change destroys hemoglobin's ability to carry oxygen effectively."[9]

What does this mean for the ability of unguided mutations to produce working proteins? Darwinian evolution must be able to evolve proteins from one stable structure to another, which must remain functional with each successive, slight modification.

Molecular biologist Douglas Axe has rigorously studied how rare properly folded proteins are among possible protein sequences. His research has direct implications for whether functional proteins can be produced by chance DNA mutations.[10]

Axe's experiments mutated the amino acid sequences in known proteins and asked if the resulting three-dimensional shapes were functional—a basic requirement of *any* protein. Axe's study considered a modest protein of only 150 amino acids in length, even though proteins often contain much longer chains.

Published in the *Journal of Molecular Biology*, his research determined that if all possible amino acid sequences (150 in length) are considered, the odds of forming a stable, functional protein fold are 1 in 10^{74}.

To put this number in perspective, it is estimated that there are between 10^{64} and 10^{69} atoms in our galaxy. In other words, the odds of a chain of 150 amino acids yielding a functional protein fold are less than a blindfolded man randomly picking a single, marked atom hidden within our galaxy.

Dr. Axe concludes:

> If you do the experiments and you analyze how much information is required to get a new protein fold, it's just far beyond what you can get by random mutation and natural selection.[11]

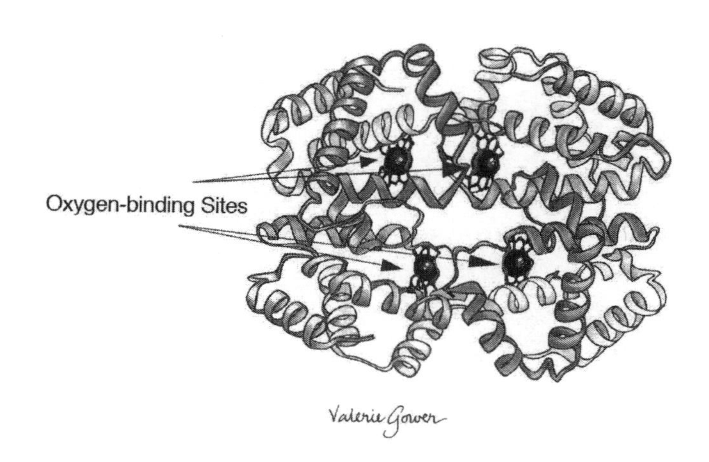

The picture above shows a normal human hemoglobin protein. Arrows indicate the regions where oxygen binds to iron within the hemoglobin protein.

Figure 9-3: Hemoglobin Protein.

DVD SEGMENT 8
"Mount Improbable" (5:22)

From:
Darwin's Dilemma

MULTI-MUTATION FEATURES

Douglas Axe's research looked at creating a new protein with a functional fold, and found that such folds are very rare. This implies that a multitude of mutations would be required—but finding the right sequence just by chance would be next to impossible. But what about slightly mutating an existing protein to create a new one—is this beyond the limits of Darwinism?

In 2010, Axe published another study on protein evolution. He modeled bacteria evolving **multi-mutation features**—those that require multiple mutations before providing any benefit. He found that features requiring more than *a mere six mutations* before providing any beneficial function could not evolve by Darwinian evolution over the entire history of the Earth.[12]

Laboratory experiments suggest that the Darwinian evolution of many new proteins would exceed these limits. In 2011, biologist Ann Gauger published research describing the attempt to convert the function of one protein into that of a closely related, but functionally distinct, protein. Evolutionists claim this kind of transformation happens easily. Gauger found that *at least seven mutations*—and probably many more—would be necessary for this evolutionary conversion.[13]

Single-celled, rapidly reproducing bacteria might be Darwinism's best opportunity for evolving new complex features. But even here there are too few generations, and population sizes are too small. The problem becomes much worse when we look beyond single-celled bacteria to more complex multicellular organisms.

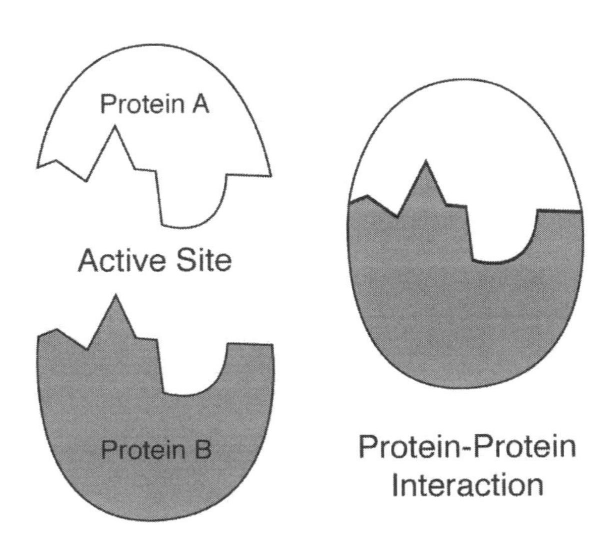

Figure 9-4: The 'Hand-in-Glove' Fit of Proteins.

IF THE GLOVE DOESN'T FIT...

Axe and Gauger's work looked at the ability of Darwinian evolution to create new proteins, or modify existing ones to perform new functions. But many proteins do not work alone.

Proteins must commonly interact or connect through a "hand-in-glove" fit in order to accomplish their cellular functions (see Figure 9-4). Multiple amino acids in each protein must be specifically arranged to give the proper shape for such a fit. Is Darwinian evolution up to the job?

In 2004, Michael Behe and physicist David W. Snoke published a study in the journal *Protein Science*. They started by noting that two (or many more) specific amino acids are often required for building some features of proteins, such as the binding sites necessary for protein-protein interactions. These features therefore require multiple mutations before they can function.

Using a mathematical model, they investigated whether these multi-mutation features could evolve.[14] Their work showed that in many multicellular organisms, evolving even a modestly complex multi-mutation feature would require greater population sizes and more time than would be available over Earth's history.

In 2008, critics tried to refute Behe and Snoke in the journal *Genetics*. But their own investigation found that for complex organisms like humans, obtaining even two specific mutations via Darwinian evolution was "very unlikely to occur on a reasonable timescale." In fact, they admitted that "this type of change would take [more than] 100 million years."[15]

Behe describes the problem such results pose for the Darwinian mechanism:

> [I]f only one mutation is needed to confer some ability, then Darwinian evolution has little problem finding it. But if more than one is needed, the probability of getting all the right ones grows exponentially worse.[16]

The bottom line is this: there is simply too much complex and specified information in many proteins for them to arise by Darwinian mechanisms.

IRREDUCIBLE COMPLEXITY

The studies mentioned in this chapter have shown the difficulties for Darwinian processes to account for beneficial changes at the level of proteins. But that is not the only problem for materialists. Let's move up a level of complexity—from proteins to complex multipart systems.

Many cellular features, such as molecular machines, require multiple interactive parts to function. Behe has further studied the ability of Darwinism to explain these multipart structures.

In his book *Darwin's Black Box*, Behe coined the term **irreducible complexity** to describe a system that fails Darwin's test of evolution:

> What type of biological system could not be formed by "numerous successive slight modifications"? Well, for starters, a system that is irreducibly complex. By *irreducibly complex* I mean a single system which is composed of several interacting parts that contribute to the basic function, and where the removal of any one of the parts causes the system to effectively cease functioning.[17]

As suggested earlier, Darwinism requires that structures remain functional along each small step of their evolution. However, irreducibly complex structures cannot evolve in a step-by-step fashion because they do not function until all of their parts are present and working. Multiple parts requiring numerous mutations would be necessary to get any function at all—an event that is extremely unlikely to occur by chance.

One famous example of an irreducibly complex molecular machine is the **bacterial flagellum** (see Figure 9-5). The flagellum is a micro-molecular propeller assembly driven by a rotary engine that propels bacteria toward food or a hospitable living environment. There are various types of flagella, but all function like a rotary engine made by humans, as found in some car and boat motors.

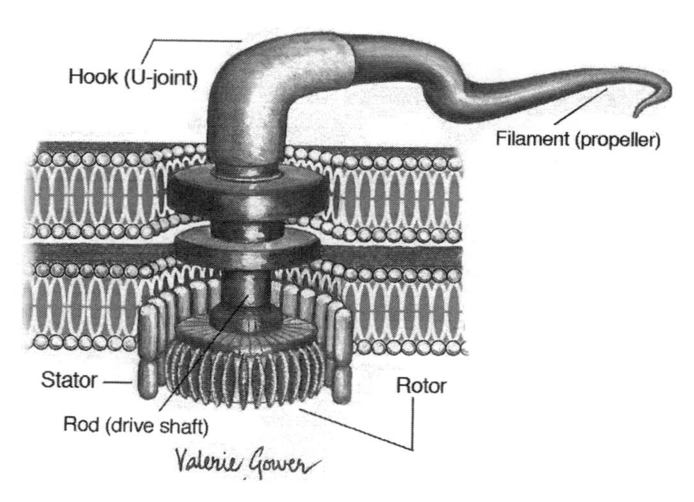

Figure 9-5: Bacterial Flagellum.

Flagella contain many parts that are familiar to human engineers (see Figure 9-6), including a rotor, a stator, a drive shaft, a u-joint, and a propeller. As one molecular biologist wrote in the journal *Cell*, "[m]ore so than other motors,

Figure 9-6: Flagellum Engine Specs.

- **Self-assembled rotary engine**
- **Motor driven by flow of protons**
- **Spins in 2 directions (forward and reverse); can change directions in ¼ turn**
- **Operates around 18,000 rpm; can spin at 100,000+ rpm**
- **Over 30 genes produce over 30 structural parts**
- **According to one study, it might have ~100% energy efficiency[18]**

the flagellum resembles a machine designed by a human."[19]

Genetic knockout experiments by microbiologist Scott Minnich show that the flagellum fails to assemble or function properly if any one of its approximately 35 genes is removed.[20] In this all-or-nothing game, mutations cannot produce the complexity needed to evolve a functional flagellum one step at a time, and the odds are too daunting for it to assemble in one great leap.[21]

The next DVD segment examines the difficulties posed for Darwinian evolution by the flagellum and other complex biological systems.

DVD SEGMENT 9
"Irreducible Complexity" (12:33)

From:
Unlocking the Mystery of Life

SPECULATION WANTED: NO FACTS NECESSARY

Irreducibly complex structures point to design because they contain high levels of specified complexity—i.e., they have unlikely arrangements of parts, all of which are necessary to achieve a specific function.

ID critics counter that such structures can be built by **co-opting** parts from one job in the cell to another.

> **Co-option:**
> ***To take and use for another purpose. In evolutionary biology, it is a highly speculative mechanism where blind and unguided processes cause biological parts to be borrowed and used for another purpose.***

But there are multiple problems co-option can't solve.

First, not all parts are available elsewhere. Many are unique. In fact, *most* flagellar parts are found only in flagella.

Second, machine parts are not necessarily easy to interchange. Grocery carts and motorcycles both have wheels, but one could not be borrowed from the other without significant modification. At the molecular level, where small changes can prevent two proteins from interacting, this problem is severe.

Third, complex structures almost always require a specific order of assembly. When building a house, a foundation must be laid before walls can be added, windows can't be installed until there are walls, and a roof can't be added until the frame is complete. As another example, one could shake a box of computer parts for thousands of years, but a functional computer would never form.

Thus, merely having the necessary parts available is not enough to build a complex system because specific assembly instructions must be followed. Cells use complex assembly instructions in DNA to direct how parts will interact and combine to form molecular machines. Proponents of co-option never explain how those instructions arise.

To attempt to explain irreducible complexity, ID critics often promote wildly speculative stories about co-option. But ID theorists William Dembski and Jonathan Witt observe that in our actual experience, there is only one known cause that can modify and co-opt machine parts into new systems:

> What is the one thing in our experience that co-opts irreducibly complex machines and uses their parts to build a new and more intricate machine? Intelligent agents.[22]

CONCLUSION

Darwinian evolution requires that features arise via "numerous, successive, slight modifications" where each mutation confers a functional benefit. According to the evidence presented in this chapter, many features cannot evolve in that fashion.

- In order to be functional, a protein must be the right shape, but the odds that a random amino acid sequence would yield a stable, functional shape are less than the odds of a blindfolded man randomly selecting a single marked atom within our galaxy. Producing new functional proteins appears beyond the ability of evolution.

- Evolutionists claim closely related proteins can easily evolve from one to another, but such transitions are likely to require multi-mutation steps that are beyond the limits of Darwinism.

- Proteins must be able to join to other molecules with a "hand-in-glove" fit, often requiring two or more specific amino acids. In many complex animals, the probability of even two proteins evolving this way is exceedingly low.

- Biological systems often require many interactive parts in order to function. Such irreducibly complex systems would not be produced by natural selection unless all the components were generated simultaneously—an extraordinarily improbable occurrence under Darwinian evolution.

Reasonable evolutionary explanations for the origin of these complex biochemical features have not been forthcoming. To expand the quotation presented at this beginning of the chapter, after Michael Behe published his book *Darwin's Black Box*, biochemist Franklin Harold admitted:

> We should reject, as a matter of principle, the substitution of intelligent design for the dialogue of chance and necessity; but we must concede that there are presently no detailed Darwinian accounts of the evolution of any biochemical or cellular system, only a variety of wishful speculations.[23]

It would seem that with some biologists, it's not the evidence that drives them to oppose ID, but rather distaste for the conclusion itself. Such philosophical or ideological aversion should have no place in science. If we are willing to follow the evidence where it leads, then biochemical complexity refutes another tenet of materialism.

MATERIALISM TENET #5:
Complex cellular machines and new genetic features developed over time through purposeless, blind processes.

Behe refers to the cell as "the ultimate black box."[24] By opening that box, modern biochemistry has strongly challenged Darwin's mechanism of evolution. Biochemical systems are finely tuned, and thus life contains extremely high levels of complex and specified information. In Chapter 1, we explained that there is only one known cause that can generate such high CSI: intelligence.

DISCUSSION QUESTIONS

1. What are the four potential effects of mutations? Do mutations provide a viable mechanism for generating biological complexity? Why or why not?
2. Irreducibly complex systems require many components to be in place simultaneously in order for the whole system to function. How does this counter Darwin's theory?
3. Does co-option provide a plausible explanation for how irreducible complexity can arise by unguided mechanisms? Why or why not?
4. Though materialists may argue that organisms might still survive some non-advantageous mutations, what would natural selection tend to do in a competitive environment? How would that affect an evolving species?
5. Do you think any explanation given thus far explains the cell's complexity? Elaborate on your answer.

CHAPTER 10

LIFE IS COMPLICATED

Question:

What do you think is the most complicated feature in a living organism?

LIFE IS COMPLICATED

"Biology is the study of complicated things that give the appearance of having been designed for a purpose."

— Evolutionary biologist Richard Dawkins [1]

Many materialists admit the appearance of design but not the reality. However, the deeper scientists peer into biology, the more they become aware of its extraordinary complexity. This chapter will discuss how cells are combined to form complex life forms, like plants and animals. We will also examine the feasibility of another basic tenet of materialism.

MATERIALISM TENET #6:
All species evolved by unguided natural selection acting upon random mutations.

In the previous three chapters, we investigated basic cellular components, like proteins and molecular machines, and the difficulties faced in evolving them. But higher organisms are exponentially more complex—built from billions or trillions of cells, with each individual cell able to produce up to tens of thousands of different proteins.

Consider the changes necessary for an organism as simple as a single-celled bacterium to evolve into a complex multicellular organism like a dog, with trillions of cells. (See Figure 10-1.)

Dogs are just one species. With millions of species of organisms on this planet, life is extraordinarily diverse. Can Darwinian evolution account for the origin of all of this complexity?

FROM CELLS TO SPECIES

Nearly all plant and animal cells contain DNA, which encodes the amino acid sequences of many thousands of different proteins. But in multicellular organisms, not every protein encoded in an organism's DNA is produced in every single cell. The specific set of proteins produced in a cell gives it certain physical properties and the ability to perform certain functions.

Based on their properties and functions, cells are categorized into types. A single multicellular

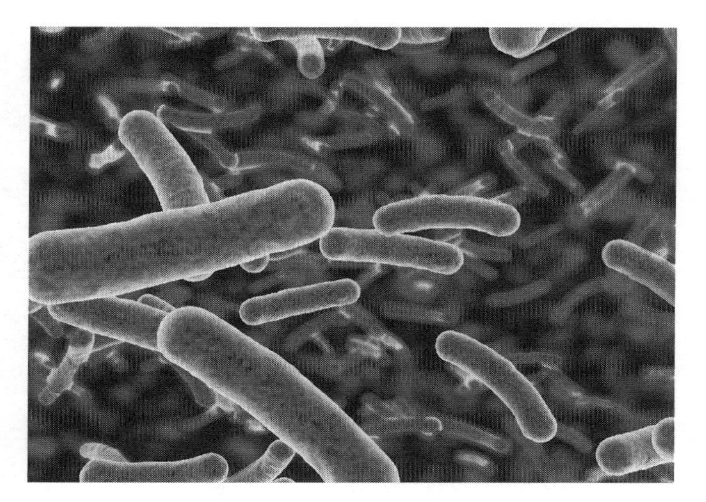

Figure 10-1: A dog is exponentially more complex than a single-celled bacterium.

organism might contain many dozens of different types of cells.

In plants and animals, cells of different types coordinate to perform specific roles, such as the absorption of nutrients, support of other structures, and production of chemicals. These combinations of cells are called **tissues**.

- Plant Tissue Types: **ground**, **dermal**, and **vascular.**
- Animal Tissue Types: **epithelial**, **connective**, **muscle**, and **nervous.**

Tissues, in turn, join together to form organs. In plants, tissues form organs such as roots, stems, or leaves. Animal organs might include the heart, stomach, or skin. Finally, the arrangement of organs forms a complex, living organism with an overall **body plan**. The progression of parts that make up an organism are depicted in Figure 10-2 below.

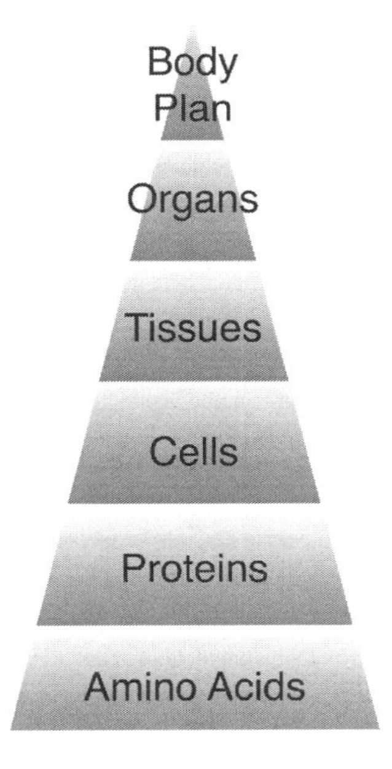

Figure 10-2: Levels of Organization in a Complex Animal.

> **Body Plan:**
> ***An organism's arrangement of organs and other parts that coordinate to perform functions and allow it to survive in its environment.***

Unless coordinated with foresight and planning, the parts at the bottom levels will not produce functional systems at the higher levels. Thus, the information for a body plan comes first, and dictates what other parts are necessary at the lower levels. This is called **top-down** design.

> **Top-Down Design:**
> ***A method of engineering, where the final goal is determined first, and then lower levels of details are identified that would be necessary to achieve that goal.***

Michael Behe explains this concept using the analogy of constructing a tall building with a nice view:

> By top down I mean that the building was planned. Blueprints were followed, supplies ordered, ground purchased, equipment moved in, and so on—all with the final structure of the observation tower in mind.[2]

Top-down design requires forethought and intelligent guidance. Evolutionists disagree, claiming that body plans were produced by an unguided step-by-step process of natural selection. In their view, organisms were built by accumulating **adaptations**.

> **Adaptation:**
> ***A feature that enables an organism to survive and reproduce in its environment.***

Intelligent design certainly acknowledges that minor adaptations (i.e., small-scale microevolutionary changes) can arise by natural selection. However, an enormous amount of new information must be added and coordinated at multiple levels to construct a plant or animal with a new type of body plan. For example, a new class of animals would require different skeletal structures and muscles, as well as new nervous and connective tissues.

Those new tissues may in turn require new cell types, which may require new proteins. Darwinian evolution has no feasible explanation for adding and coordinating this information. As one article in *Scientific American* admitted:

> Although evolutionary theory provides a robust explanation for the appearance of minor variations in the size and shape of creatures and their component parts, it does not yet give as much guidance for understanding the emergence of entirely new structures, including digits, limbs, eyes and feathers.[3]

The following DVD segment provides insight into the development of body plans. There are two references in the segment that should be explained prior to viewing. One is the fossil record—we have not considered it much so far, but will do so extensively in Section IV of this book. The other is a specific portion of the fossil record called the **Cambrian explosion**.

> **Cambrian Explosion:**

A geologically brief period about 530 mya, during which nearly all of the major animal phyla, including many diverse body plans, suddenly appeared.

To examine the validity of Darwinian theory, we will take a closer look at the tightly integrated ways that organisms combine different types of proteins, cells, tissues, and organs to build complex body plans in plants and animals.

DVD SEGMENT 10
"Body Plan"
(5:03)

From:
Darwin's Dilemma

PLANT ONE ON ME

Plants are a diverse group of multicellular organisms, ranging from grasses to ferns, algae to mosses, and strawberries to pine trees. They are generally capable of deriving energy from the sun, through a multi-step chemical reaction called **photosynthesis.**

Chloroplasts and Photosynthesis

Chloroplasts are molecular machines found in green plants and algae that capture energy from the sun using a green pigmentation called **chlorophyll**. During photosynthesis, chloroplasts combine this energy with water (H_2O) and carbon dioxide (CO_2) to create oxygen (O_2) and sugars (e.g., $C_6H_{12}O_6$). Through other biochemical processes, these sugars in turn serve as food for the

plant, helping it grow and, along with minerals extracted from the soil, produce thousands of products like starches, proteins, fats, and vitamins.

Photosynthesis is an intricate and ecologically indispensable cellular process.[4] One scientific paper admitted, "The origin of photosynthesis is a fundamental biological question that has eluded researchers for decades."[5] Indeed, plant biochemistry is so complicated that one botanist noted that "details of the enormously complex chemistry of plants fill many volumes in the scientific literature."[6]

The ecological relationship between plants and animals is mutually beneficial. Plants act as a conduit to transfer energy from the sun to animals. Herbivores eat the plants, and omnivores (like humans) and carnivores then eat the herbivores. Thus, the organisms that animals ingest as food originally derived their energy from the sun. Additionally, the excrement and decaying bodies of animals fertilize the soil, providing nutrients to plants.

There is another beneficial relationship between plants and animals. Plants use carbon dioxide in the air during photosynthesis and produce oxygen that is breathed by animals. In return, animals exhale carbon dioxide, benefiting plants.

Without chloroplasts, virtually all plants would die and the ecosystem would collapse. Higher life forms could not acquire energy or sufficient oxygen, and nothing could survive, let alone evolve.

The Great Wall

While all cells in living organisms have a plasma membrane (see Chapter 7), plant cells are also bounded by a rigid **cell wall** (see Figure 10-3). The wall limits cellular growth and restricts movement, but also provides added protection and structural stability. This allows herbaceous plants like annuals and most grasses to grow, produce seeds, and spread rapidly.

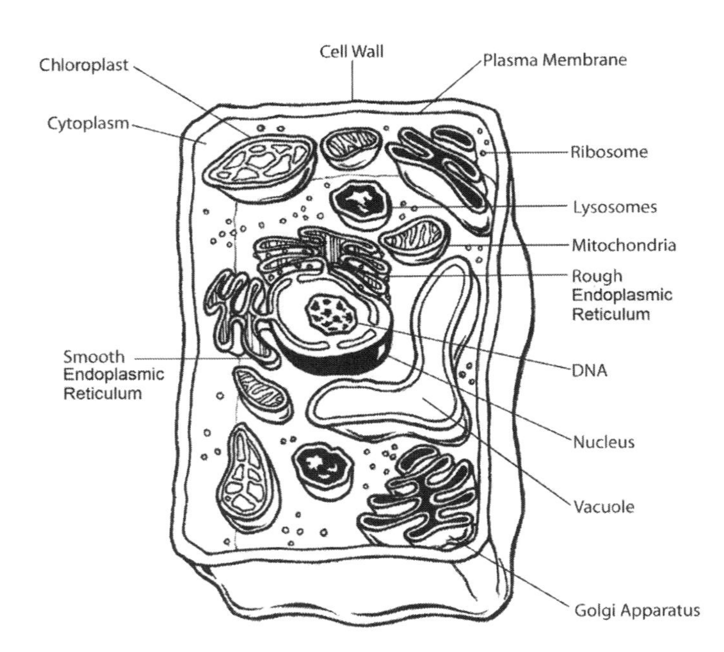

Figure 10-3: Plant Cell.

In larger plants like shrubs and trees, however, the walls of living cells don't offer enough structural support. Instead, woody plants are strengthened by the walls of their older, dead cells. These dead cells form a core in the center of the stem as new cells grow around them. This core allows woody plants to grow taller, better withstand the elements, and live longer.

Plant Tissues

In addition to cell walls and chloroplasts, plants differ from animals in their tissue types. As mentioned earlier, plant tissue types include ground, dermal, and vascular (see Figure 10-3).

Ground tissues are the sites of photosynthesis and food storage. This tissue makes up the bulk of the plant and also helps form the structural foundation of other plant tissues.

Dermal tissue is like the skin of the plant, covering and guarding the outer surfaces. It generates a waxy substance that aids in water conservation and prevents disease by keeping out harmful chemicals or microorganisms. Small, adjustable openings within the derma (called

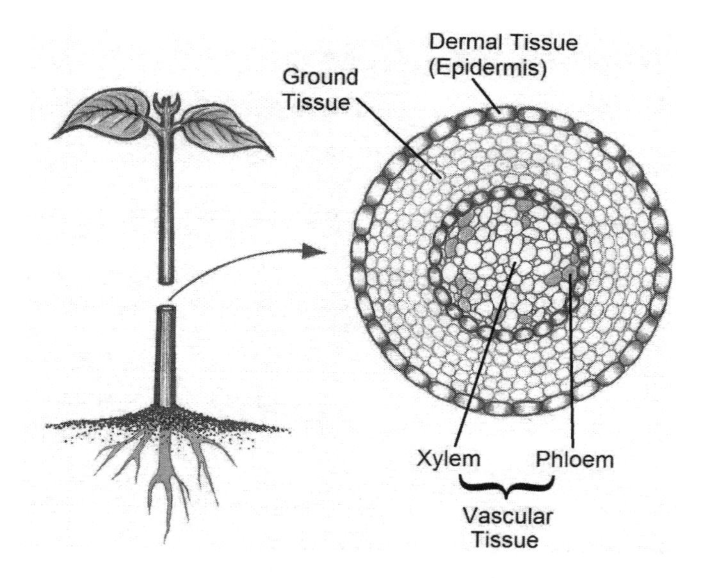

Figure 10-4: Herbaceous Plant.

stomata) allow the exchange of gases, such as oxygen and carbon dioxide, in and out of the plant.

Vascular tissue is made up of the xylem and phloem—channels that carry fluids throughout the plant.

- **Xylem** conducts water and minerals from the roots upward into the plant. The cells making up its channels are dead when fully mature and have rigid, waterproof inner walls.

- **Phloem** conducts sugars and materials down from the leaves to the rest of the plant. Cells in phloem tissue remain alive even when mature.

Without this integrated transportation network, plants could not survive.

Evolutionists confidently assert that they know that plant structures evolved, going so far as to declare that there is overwhelming evidence that all land plants evolved from primitive water-dwelling plants. Evolutionary botanist Brian Capon admits that scientists don't fully understand how plants develop:

> How this process of *cell differentiation* takes place is still not precisely understood. Nor is it known how tissues assume the unique pattern characterizing the anatomy of roots, stems, and leaves. These are only some of the many unresolved mysteries of developmental processes in plants.[7]

If scientists don't understand how plant cells organize into basic plant structures during development, how can evolutionists be sure that blind and unguided mechanisms can account for their origin? In the remainder of this chapter, we will demonstrate how the essential roles of animal tissues also challenge Darwinian explanations.

OH, YOU'RE SUCH AN ANIMAL

Animals are multicellular organisms that feed on other organisms and are generally capable of movement. Many examples of animals—elephants, bats, hummingbirds, and fish—are familiar. But other examples of animals may be unexpected, such as starfish, insects, barnacles, and sponges.

This section will explore a few animal body plans by focusing on the four tissue types (nervous, muscle, connective, and epithelial) that build them.

Nervous Tissue

Nervous tissue is responsible for transmitting electrical signals within the organism.

A single nerve cell, called a **neuron**, is divided into a cell body, which in turn has extensions called **dendrites**, and an **axon** (see Figure 10-5). Dendrites are branches of the cell that receive stimuli and pass messages through the cell body and along the axon to neighboring nerve cells.

A typical nerve cell is a small fraction of the width of a human hair. Depending on the size and composition of an animal, its nervous system can have hundreds to hundreds of billions of neurons.

THEY'VE GOT A LOT OF NERVE

Many bats have especially impressive nervous systems. "Blind as bats," they use their larynx (voice box) to emit a high-pitched sound through their mouth or nose. They listen for the reverberation of the sound off other objects in order to "see" where things are around them. Called **echolocation**, this ability enables bats to navigate in the dark, avoid danger, and find food.

To make echolocation work, the bats also must have ears capable of hearing the echo of sound bouncing off objects as small as a flying insect. In addition, they need a brain capable of interpreting the location of the object by its sound refraction. All of this requires nerves capable of receiving, relaying, and responding to such signals with split-second timing.

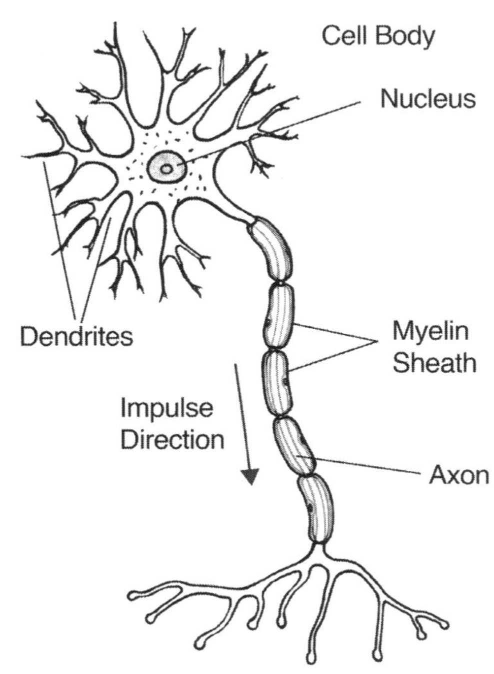

Figure 10-5: Nerve Cell.

Together, the cells in the nervous system facilitate basic biological functions ranging from triggering a heartbeat to detecting heat or cold to signaling hunger, as well as controlling a host of other processes.

Making Evolutionists Nervous

Higher animals like **vertebrates** have a central nervous system that processes information from all body parts to control organ function and behavioral response. Evolving a new organ thus requires much more than merely having the necessary tissues. The organ must be tied into the central nervous system, which must be able to properly interpret the signals.

Much as a piece of computer hardware won't function without the proper software, the central nervous system must be programmed to process signals from the organ and send out appropriate commands in response. Otherwise, the organ is useless to the animal.

Evolutionary explanations for the origin of new organs rarely, if ever, account for the neural circuitry necessary to control and use the feature. Evolutionists probably avoid this discussion because many highly unlikely simultaneous modifications would be necessary to evolve a new, functional organ. Recognizing this problem, mathematician David Berlinski writes:

MUSCLE CAR OF THE BIRD WORLD

Hummingbirds are stunning animals. The bone structure and powerful musculature of their wings allow beats of 75 times per second (or more), creating a humming sound. With flying abilities unlike those of any other bird, hummingbirds can perform maneuvers in more directions than a helicopter: up, down, sideways, diagonal, backwards, hovering, and even upside down.

A hummingbird's heart beats 500–600 times per minute, with the ability to beat up to 1,000 times per minute during periods of high activity. In order to maintain a high energy level, it must eat at least its own weight in nectar and insects daily. This is accomplished in large part due to the muscles in its tongue, which enable it to lick up nectar at a rate of over a dozen times per second.[8]

The hummingbird's body plan enables it to perform acrobatic feats that outstrip the most advanced human technology. Does this point towards an unguided origin or intelligent design?

> If these changes come about simultaneously, it makes no sense to talk of a gradual [evolution]. If they do not come about simultaneously, it is not clear why they should come about at all.[9]

Muscle Tissue

Muscles are composed of large cells called muscle fibers. Many animals have extraordinarily complex combinations of these cells. For example, there are 150,000 muscle units, called "bands," in the trunk of a typical adult elephant.[10]

Muscle fibers do their work via contraction, enabled by molecular machines. **Myosin** is a molecular machine that pulls itself along a track, composed of long proteins called **actin** filaments, to form the basis of muscle contraction. Muscle movement requires the "combined action of trillions of myosin motors."[11]

In addition, muscles are fueled by capillaries full of oxygen-rich blood. The more active the muscles, the more capillaries they have.

There are three types of muscle tissue: smooth, skeletal, and cardiac.

Smooth muscles are involuntary. They are found in places like the gut, blood vessels, and reproductive tract.

Skeletal muscles are voluntary and work together to create movement in conjunction with the skeletal structure of an animal. All the cells in this type of muscle have multiple nuclei.

Cardiac (heart) muscles are involuntary. They are unique in that they start beating in a rhythmic pattern in the embryonic stage and don't stop until the animal dies. This rhythm is achieved through highly conductive "plates" that connect the individual cardiac cells, allowing a swift

SOMETHING'S FISHY

In fish, the tight configuration of epithelial cells forms a protective barrier that prevents excessive amounts of water, minerals, and other chemicals in the environment from entering the skin. The epithelial layer also produces mucus which excludes harmful pathogens and parasites.

Epithelial cells lining the gills of fish are specially designed to absorb oxygen from water. Gills also play an essential role in osmoregulation—the process by which fish balance their absorption of fluids.

For example, in order to balance internal salinity, saltwater fish use molecular pump machines in the gills' epithelial tissue to transport salt out of their bodies, whereas freshwater fish pump salt in. In addition, this system helps maintain a balance within the fish of vital elements like calcium, potassium, and other important minerals.

transmission of the electrical impulses through the heart. The speed of this transmission permits the vital coordination of these cells, allowing the heart to beat uniformly.

Epithelial Tissue

In animals—from worms to lions—cells in epithelial tissue take on a variety of shapes, including flattened, cubical, and elongated. They are solidly packed together in layers that form boundaries and protective barriers.

Epithelial cells are responsible for protecting outer surfaces of the body, keeping out bacteria and other harmful elements in the environment. In addition, epithelial cells absorb and secrete substances.

For example, in the gut they absorb nutrients from food, and in the lungs (or skin of animals without lungs) they absorb oxygen. Glands made of epithelial cells secrete specialized proteins (such as **hormones**) that allow intercellular communication, or digestive enzymes that break down food.

In their protective barrier function, epithelial cells surround internal organs and line interior cavities like the **digestive tract**. Skin is made of epithelial tissue. If the epithelial cells were not in place and functioning, organisms that depend on them would be unable to digest food, breathe, protect themselves, or otherwise survive.

One marine biology textbook asserts that "the most successful and abundant" marine animals "have evolved mechanisms to stabilize" their internal water and salt concentrations.[12] The textbook does not mention that, as one technical paper observes, "osmoregulation in fishes is mediated by a suite of structures" including molecular pumps and transport machines.[13]

No details are given about how these systems arose—typically evolution is simply asserted, as if it requires no scientific justification. Scientists don't understand exactly how these mechanisms work,[14] so how can we know they evolved?

TOO BIG TO FAIL

Connective tissue is the primary component in the sturdy bone structure, trunks and tusks of elephants. But it also performs other special and unexpected roles. For example, within the elephant foot is a thick pad of connective tissue.

Elephants can communicate using **infrasound**—a powerful, low-frequency sound below the human ability to hear.[15]

This sound travels through the ground, like a miniature earthquake. An elephant detects this sound through its feet or by laying its trunk on the ground, using pressure-sensitive pads (encasing nerves) composed of onion-like layers of connective tissue, separated by gel.

Connective Tissue

Just as the name suggests, connective tissue connects parts of the body, providing a framework in which other tissue types are embedded. This tissue forms fat storage cells, blood, bone, cartilage, tendons, and ligaments. Connective tissue cells are found in two general types, loose and dense, and are often isolated from one another by a matrix of non-cellular components.

ANIMAL VERSUS THE MACHINE

While the analogy is far from perfect, the four tissue types in animals bear some resemblance to components we recognize from human technology.

Consider, for instance, the car. In cars, nervous tissue has obvious counterparts: electrical wires transmit electricity from the battery or alternator to the spark plugs, and computer chips control many other functions.

Muscle tissue might be compared to the engine, belts, and drive shaft, all of which generate the car's movement.

While some of the many functions of epithelial tissue might not have analogies in cars, there are similarities. Paint and wax on the car's body fulfill a protective role by preventing rust. Fuel and air filters keep harmful elements out but allow necessary ones to pass through.

Finally there is connective tissue, which is analogous to a car's chassis and body. Like the parts of a car, the organs and tissues in an organism work together to perform a multitude of functions.

These comparisons are readily made between biological systems and machines because *organisms contain machines*.[16] And in all of our experience, machines arise only by intelligence.

When the analogy between biological components and car parts breaks down, it's because biological systems are *more* complicated. For example, the human brain is unimaginably more complex than computer chips in cars. Similarly, our bodies have an immune system and can often heal themselves—abilities many a car owner sitting in a mechanic's waiting room has wished that his or her vehicle had.

If inferior human technology requires design, why can't the design inference be made in the context of biology, which is dramatically more complex?

MY, HOW YOU'VE CHANGED!

One of the most amazing examples of animal complexity is insect growth and development. As they progress to maturity, insects undergo one of three processes: **ametabolism**, **hemimetabolism**, and **holometabolism**.

Ametabolism is the simplest of the three types of development and is only used by a few insects. In this process, young insects (called **nymphs**) have essentially the same body plan as the adult. After emerging from the egg, the insect undergoes only changes in size, but not shape, as it matures. Silverfish, for example, develop through ametabolism.

Ametabolous development:
egg > miniature adult (nymph) > adult

The other two methods of development are more complex, in that they involve **metamorphosis**.

> **Metamorphosis:**

A process of pre-programmed development where an organism changes its body plan.

In hemimetabolism (partial metamorphosis) the insect undergoes gradual, progressive change in form. However in these insects, this change occurs through a process of **instars** (periods of growth and change) and **molts** (the shedding of skin).

After entering the nymph stage, the insect begins to feed. With each subsequent instar and molt, the nymph gradually changes into its adult form until it reaches maturity and is able to breed. Dragonflies, grasshoppers, and crickets develop through hemimetabolism.

Hemimetabolous development:
egg > non-feeding larva > nymph > adult

Holometabolism (complete metamorphosis) is the most common and complicated form of insect maturation. The diverse group that undergoes this type of process includes butterflies, moths, beetles, fleas, bees, ants, and many kinds of flies.

A holometabolous insect emerges from the egg as a hungry **larva**—consuming anything it can while going through multiple instars and molts. The larva may have a simplified body plan, with reduced legs and eyes, and little distinction between major body segments. Its job is to eat and grow. It molts to accommodate its increasing size, but does not change form nor develop adult structures.

At the right time, larval growth slows and stops, and the organism becomes a **pupa**.

- In butterfly species the exterior of the pupa forms a hardened shell called a **chrysalis**.
- Moths and many other holometabolous insects spin a silken case for their pupa called a **cocoon**.
- Other insects develop as a pupa inside the last larval skin.

Metamorphosis now occurs as the pupa undergoes a complete transformation of its body plan. Inside the pupa, the insect *liquefies* itself and restructures most of its body to emerge later as a fully formed adult, often with wings. By breaking down much of its body to supply the building materials for the adult form, the creature is then only alive in the form of a "soup."

The dissolved remnants of its tissues supply raw materials to build the adult form. Additionally, new tissues grow from cells set aside early in development.

After the transformation is complete, developmental hormones signal the insect to emerge from its confinement. For many insects, the force they must exert to emerge and pump fluids into their wings is essential for their development.

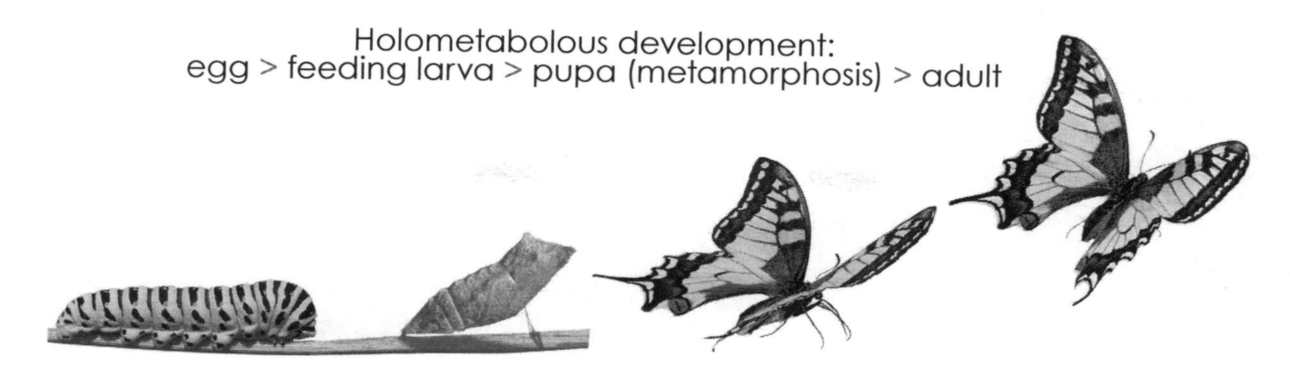

Without expending that effort, they would be deformed.

It is exceedingly difficult to understand the origin of holometabolism in Darwinian evolutionary terms. Neither the larval nor the pupal stage is capable of reproduction—only the adult is. In particular, the pupal stage is an all-or-nothing proposition. It must complete the process and become an adult, or it will die without ever reproducing.

The liquefied organism must be completely rebuilt. For this to occur, large amounts of information—encoding the larval body plan, the mechanisms of transformation during metamorphosis, and the adult body plan—must exist *before* the larva enters this stage.

An organism could not survive complete metamorphosis unless the entire process was fully programmed from the beginning. Such a large jump in complexity requires forethought and planning—things that don't exist in Darwinian evolution. As one evolutionary entomologist acknowledges:

> ... the biggest head-scratcher in evolutionary biology would have to be the origin of the holometabolous insect larva.[17]

But from an ID perspective, metamorphosis is easy to understand: it arose, evidently, by planning and foresight. An intelligent agent could produce the information to program the entire life cycle of such an organism, allowing it to undergo a radical transformation like this. Only a goal-oriented process like intelligent design can explain the mystery of holometabolism.[18]

For a quality documentary presenting the evidence for intelligent design from insect metamorphosis, see the film *Metamorphosis*, available at **www.metamorphosisthefilm.com.**

CONCLUSION

Plants and animals are much more than the mere sum of their proteins. They are extraordinarily complex machines with a diverse range of body plans. What these body plans share is a characteristic common in engineered systems: top-down design. Paul Nelson and Jonathan Wells explain the implications:

> The overall system, not the gene itself, determines its functional role. And such systems, in our experience, can only be intelligently caused.[19]

Many of the complex processes discussed in this chapter, such as photosynthesis and metamorphosis, show the intricate ways that different cell types are used to build tissues and organs, which combine to perform a myriad of functions. The overall body plans require immense amounts of information to purposefully arrange a multitude of parts so that an organism can survive in a particular environment.

Biological organisms are built with an end-goal in mind. In contrast, Darwinian evolution cannot look ahead to see what an organism needs to survive, and provides no explanation for generating the necessary information. The top-down design seen in the blueprints of multicellular organisms had to be planned by an engineer.

Next, we'll look at a few important macro-systems of something more personal—our own bodies.

DISCUSSION QUESTIONS

1. Plants and animals are composed of many different types of tissue, all of which serve important functions. Which do you think are the most significant? Explain why.

2. Describe the various levels of complexity that must arise in order to generate a new body plan. Explain whether you think a blind and undirected process like Darwinian evolution or a goal-oriented process like intelligent design best explains the origin of body plans.

3. The vast majority of animals on this planet use energy originally captured from the sun through photosynthesis. What would happen if the ability to photosynthesize suddenly vanished, or was never present in the first place?

4. Imagine that a new organ develops, but the nervous system is not upgraded to accommodate it. What would happen to the organ? How does the nervous system pose a challenge to Darwinian evolution?

5. Holometabolism is an extremely complicated method of change used by many insects. Could such a process arise in a step-by-step manner, as is required by Darwinian evolution? Explain your answer.

CHAPTER 11

BODY OF EVIDENCE

Question:

Are there any structures in the human body that defy a step-by-step Darwinian explanation?

BODY OF EVIDENCE

"[I]n Man is a three-pound brain which, as far as we know, is the most complex and orderly arrangement of matter in the universe."

— Author and biochemist Isaac Asimov[1]

The human body is a staggeringly complex machine, but we often take it for granted. We rarely spend time amazed by our ability to digest food, breathe, see, hear, feel, walk, talk, and ponder the meaning of life—all simultaneously. Like other animal body plans, humans are built from a myriad of parts and systems which are coordinated to allow us to survive, and do much more.

The more we learn about the human body, the more it appears to defy a Darwinian explanation. Rather, its integrated nature points to an engineering design. In this chapter, we'll consider just three of the many macro-systems of the human body: vision, reproduction, and digestion.

SEEING IS BELIEVING

Optometrists and psychologists have estimated that for most of us, 80% of the information we receive is gained through sight.[2] Because of its importance, we know a great deal about the intricacy of the vertebrate eye (see Figure 11-1).

Just a few of the many essential parts of the eye include:

- A cornea to guard the lens.
- An iris capable of protecting the eye from excessive light.
- Capillaries to provide blood flow.
- Muscles inside and outside of the eye to allow movement and focusing.

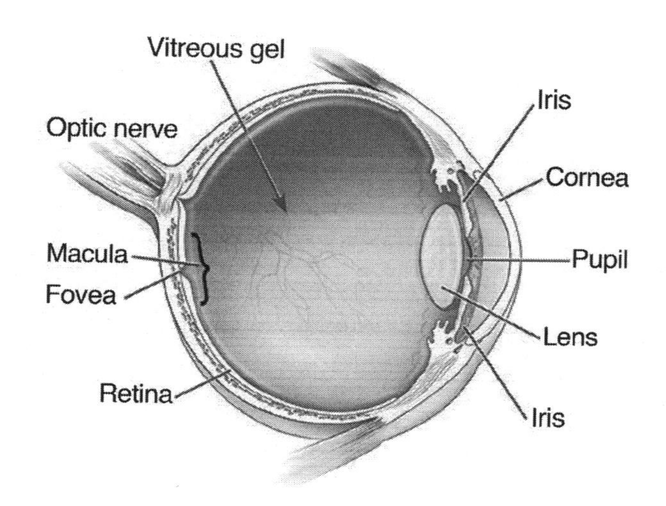

Figure 11-1: Human Eye.

There are parts with familiar names like the retina, the pupil, and the optic nerve, but there are other complex components that aren't so well known. For example, the shape of the eyeball is sustained with the help of a clear gel—the vitreous humor. In the center of the retina is the macula, a region with a high concentration of light-sensitive cells. The fovea, at the center of the macula, provides the sharp vision you need to read this book.

To understand the basics of vision, let's take a ride on a light-ray as it makes its way through the eye.

First, the light hits the cornea, a transparent structure covering the surface of the eye that takes light rays from many different angles and focuses them towards the lens.

Before reaching the lens, however, light must pass through the pupil, a small black circle at the center of the eye. Surrounding the pupil is the iris, a collection of tissues that give your eye its color.

The iris is connected to muscles that work to expand and contract the size of the pupil. Just like the aperture in a camera, this controls the amount of light that reaches the lens and helps you focus on whatever you want to see.

After passing through the pupil, the next stop on our light ray's journey is the lens itself. Just like the lens in a camera, the purpose of the lens in our eyes is to focus light on the back of the eye—the retina.

If we continue the camera analogy, the retina is like the film that collects the light image. But the retina is unimaginably more complex than camera film.

Lining the retina are special cells that convert light patterns into electrical impulses. These cells use proteins that change their shape when struck by light, and chains of interactive proteins that regulate the process.[3]

These electrical impulses are then transmitted to the brain through the optic nerve. The brain then converts those signals into a mental image—*vision*.

EVOLVING THE EYE

The eye is an *extremely* complex system, and many of the components depend on each other to perform their functions. Could this structure have evolved by unguided Darwinian evolution?

After seeing only a fraction of the eye's complexity, Darwin himself wrote:

> To suppose that the eye... could have been formed by natural selection, seems, I freely confess, absurd in the highest possible degree.[4]

Darwin ultimately asserted the eye could have evolved under a scheme of "fine gradations," and began as simple "aggregates of pigment cells."[5]

Modern evolutionary explanations, also lacking detail, still claim that eyes started with a light-sensitive spot that gradually evolved in complexity. But even "simple" eyespots are complex. Evolutionary biologist Sean B. Carroll cautions to "not be fooled by these eyes' simple construction and appearance" since "[t]hey are built with and use many of the ingredients used in fancier eyes."[6] Evolutionists never explain the origin of these light-sensitive spots.

Accounts of the evolution of the eye invoke the abrupt appearance of key features such as the lens, cornea, and iris. Of course each of these features would increase visual acuity, but where did they suddenly come from in the first place? As biologist Scott Gilbert put it, such evolutionary accounts are "good at modeling the survival of the fittest, but not the arrival of the fittest."[7]

Additionally, even a fully functional eye is useless to an organism if there is no nervous system to process the visual signal and trigger an appropriate behavioral response. David Berlinski writes:

> Changes to any part of the eye, if they are to improve vision, must bring about changes throughout the optical system. Without (an) increase in the size and complexity of the optic nerve, an increase in the number of photoreceptive membranes can have no effect. A change in the optic nerve must in turn induce corresponding neurological changes in the brain.[8]

At best, accounts of the evolution of the eye provide a stepwise explanation of one single feature: the increased concavity of eye shape. But the eye is much more than a mere indentation. Michael Behe explains why evolutionary accounts of the eye are superficial:

> Each of the anatomical steps and structures that Darwin thought were so simple actually involves staggeringly complicated biochemical processes that cannot be papered over with rhetoric.[9]

Evolutionary biology has failed to explain the origin of the biochemistry underlying vision. If Darwinism were responsible for sight, our journey on a light ray would go nowhere.

But vision is just one of the many extraordinarily complex systems pointing toward design. Let's explore another.

IT TAKES TWO TO TANGO

The majority of living organisms—including all bacteria—reproduce by cloning themselves, with no mate required to produce offspring. This process of **asexual reproduction** is considered the simplest and earliest form of reproduction, yet even it is a difficult hurdle for Darwinism to overcome.

In asexual reproduction, a cell must first make an identical copy of its DNA. This process requires a multitude of proteins, enzymes and molecular machines. Next, a cell must divide in such a way that its daughter cells contain a full set of the necessary DNA, proteins, cytoplasm, cell membranes, machines, and other parts—all while keeping both the parent and daughter cells operational. One type of cell division is called **mitosis**, a process which requires an enormous amount of complicated and carefully orchestrated cellular machinery.

> **Mitosis:**
>
> ***A process of cell division where a cell clones itself, and the daughter cells have the same chromosomes as the parent cell.***

Reproduction could not have evolved gradually; it had to be present and fully functional for the first life to continue. In Chapter 8, we showed that the first self-replicating organism could not have appeared by blind and unguided chance-based processes. But let's move beyond that issue to examine reproduction from a Darwinian perspective.

On an evolutionary level, asexual reproduction is a highly preferred form of reproduction. Why is that? In a neo-Darwinian world, all that matters is surviving and passing genes on to the next generation. During asexual reproduction, 100% of an organism's genes are passed on to every individual offspring. This maximizes the Darwinian requirement to "survive and reproduce."

However, self-cloning has one drawback: if the parent has bad genes, then, absent an extremely rare beneficial mutation, there's no way to repair the defect. Every single offspring is likely to get the bad genes, resulting in genetic weakness in a population.

Sexual reproduction provides a way around this problem. With two parents, offspring receive two copies of each gene (one from each parent), meaning there's a better chance that at least one copy will work. While harmful mutations can still occur, greater genetic diversity ensures a greater chance of success. As a result, some argue that sexual reproduction provides an evolutionary advantage.

But is an evolutionary transition from asexual to sexual reproduction likely to occur? As paleontologist and *Nature* magazine editor and author Henry Gee has commented, the origin of sex is a "problem for evolutionary biologists."[10] There are at least three reasons to doubt such a transition.

First, reproduction is much more efficient in an asexual world. Just as asexual reproduction is evolutionarily favorable because an organism passes on 100% of its genes, sexual reproduction is *unfavorable* because an organism passes on only 50% of its genes to each offspring. If Darwinian evolution depends on passing on your genes, why would a system evolve that cuts **fitness** in half?

> **Fitness:**
>
> ***The likelihood that a gene, or other trait, will be passed on to the next generation.***

There are other ways that sex can reduce fitness. Since sexually reproducing parents are not identical to their children, lethal biochemical incompatibilities between a mother and her offspring can occur.

Additionally, the increased genetic diversity allowed by sexual reproduction isn't always beneficial. It can work against a parent with good genes if that parent mates with an individual with bad genes. Again, this is evolutionarily unfavorable.

Second, sexually reproducing organisms must expend great amounts of energy to create, preserve, and maintain sex cells (sperm and eggs). Much energy is also spent creating special sexual organs to properly handle sex cells and allow for safe transfer of genetic material.

Most cells replicate through mitosis. But in order to combine genes with another individual to create offspring, sexually reproducing organisms must produce sex cells with half of the normal genetic information. This is called **meiosis**.

> **Meiosis:**
>
> ***A process of cell division used to make sex cells in sexually reproducing organisms, where the daughter cells have half the chromosomes of the original parent cell.***

Third, since it takes "two to tango," a member of one sex cannot reproduce if it is not compatible with the other. Both males and females must evolve sexually compatible body plans simultaneously, and in close proximity.

Could two individual organisms simultaneously develop a new, complex, mutually compatible

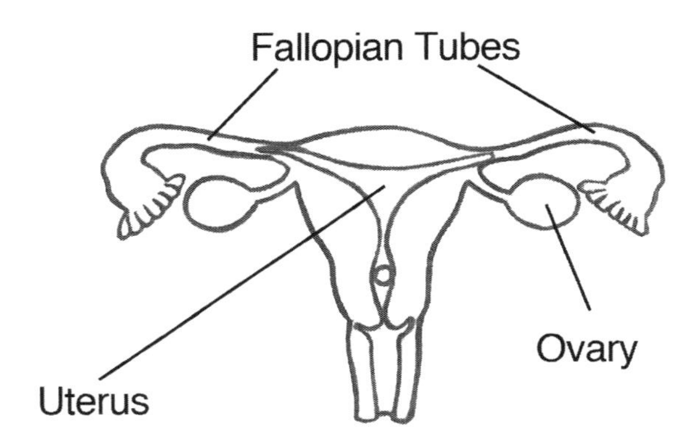

Figure 11-2: Human Female Reproductive System.

reproductive system? To help answer that question, let's consider an example: humans.

In order to reproduce, women need many specialized features, including the following impressive—though partial—list:

- Ovaries for the production and storage of sex cells (eggs) containing half the offspring's genetic inheritance.
- Unique hormones to promote release of eggs (ovulation), and to allow gestation of a baby.
- A vaginal entrance for fertilization.
- Fallopian tubes to: (1) receive the egg released from the ovary, (2) allow a pathway for sperm to reach the egg, and (3) once fertilized, conduct the zygote (fertilized egg) to the uterus (see Figure 11-2).
- A placenta to allow implantation of the zygote and nourishment of the developing baby, keeping it alive.
- Connection to the growing baby through a sturdy umbilical cord—capable of transporting blood rich in oxygen and nutrients, and able to remove cellular byproducts.
- A uterus capable of expanding to over 2000 times its normal size to accommodate the growing offspring (and then shrink back to normal size after delivery).
- Amniotic fluid which helps the lungs develop properly, provides a cushion to protect the baby

from outside forces, and helps the baby move around and retain heat.
- Other hormones to change the female's body in preparation for birth.
- A pelvis and birth canal capable of spreading for delivery of the baby.
- Mammary glands complete with lactation capacity for nourishing the infant.

Men have their own complex features. Besides anatomical differences, which uniquely fit with the female, specific male hormones are essential.

Men have testes for the production and storage of sperm, sex cells that contain the other half of the offspring's genetic inheritance.

Sperm are extraordinary cells. Generated by the millions every day, each is equipped with a flagellating tail capable of propelling it to its target. At its head are enzymes to break down the wall of the egg and start fertilization (see Figure 11-3).

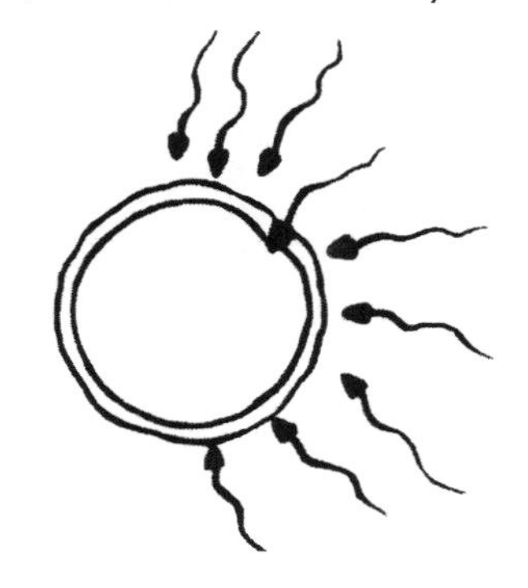

Figure 11-3: Fertilization.

The components of both the female and male reproductive systems are designed to function automatically. During ovulation, the egg leaves the ovary and moves into position in the fallopian tubes while the sperm automatically travels to find it. To aid in the process, tubular pathways and specialized fluids from both the female and male help conduct the egg and sperm toward each other.

Fertilization typically takes place in the fallopian tubes. After fertilization, the zygote makes the long journey (relative to its size) down the fallopian tube with the help of microvilli and cilia—brush-like structures that move the egg along—as well as tubular wall muscles. Eventually, the egg travels into the uterus and is implanted, where it begins to grow.

The amount of fine-tuning that goes into sexual reproduction is impressive. Metabolically, organisms expend much energy on sexual reproduction. Since asexual reproduction is so simple and efficient on an evolutionary level, why is sexual reproduction so prevalent?

NOT NOW, I HAVE A HEADACHE

The complexities, evolutionary costs, and difficulties associated with sex are enough to give an evolutionist a headache.

- Sexually reproducing organisms face a drop in fitness.
- New beneficial mutations may not be passed on.
- Half of the chromosomes must be separated into sex cells.
- Sex cells must be protected and preserved until mating.
- Much energy is spent producing sex organs for mating.
- Individuals risk not finding a mate.
- Reproductive systems in males and females. must develop simultaneously, in close proximity, and have both physical and biochemical compatibility.
- Biochemical incompatibilities between parents and offspring can occur.

No wonder evolutionary zoologist and author Mark Ridley laments, "Sex is a puzzle that has not yet been solved; no one knows why it exists."[11]

DIGESTIVE TRACT

> *"To eat is human. To digest is divine."*
> —*Mark Twain*[12]

Without food, we would die within a matter of days. However, according to Darwinian evolution, our digestive tract (see Figure 11-4) developed through numerous, small, incremental steps. Is that

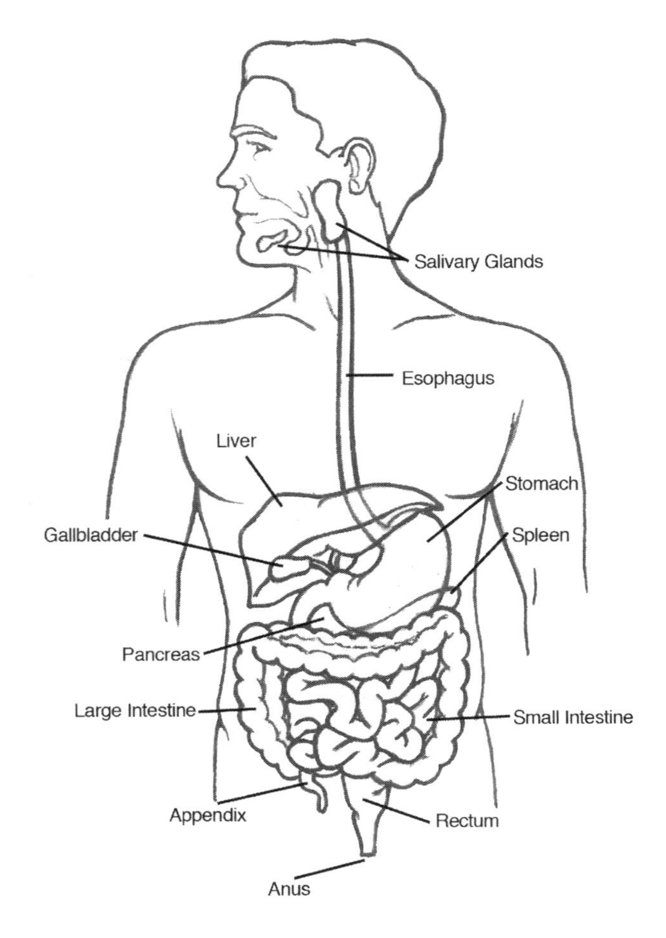

Figure 11-4: Human Digestive System.
a realistic explanation?

The human digestive tract (**alimentary canal**) is nearly 30 feet long in adults, lined along its entire length by a mucous membrane. It is like an assembly line (or more accurately, a *disassembly* line), with various organs arranged in a particular order along the tract to perform specific functions as they break down food and extract nutrients. Let's follow a sandwich as it makes its way down the line—starting with the on-switch.

DOWN THE HATCH

Without hunger, we could forget to feed our bodies and possibly die from malnourishment. The entire process of eating food starts when your digestive system releases hormones that signal your brain to create a desire to eat.

The brain then transmits nerve impulses to the stomach, triggering the release of digestive fluids and contraction of muscles—commonly felt as stomach rumblings. The brain also prompts salivary glands in the mouth to start releasing saliva.

The sandwich enters the digestive tract at the mouth. This specialized organ includes:

- Teeth well suited to bite small pieces off the sandwich, but also strong enough to crush and grind hard foods like nuts and seeds.
- Strong jaw bones and muscles for chewing.
- A tongue equipped with taste buds and nerves to tell whether a food is dangerous or safe, too hot or too cold.

Saliva, which has built up in the mouth, helps break down and moisturize the food, adding the first of many important digestive enzymes.

As the ground-up and increasingly liquefied sandwich moves further into the alimentary canal, it travels down the esophagus. To keep it on the right track, a valve prevents food from going into our lungs—further protected by a coughing response if a single drop finds its way into the wrong passageway.

Different types of muscles inside and outside of the esophagus work together to move the food along. The movement of the muscles—some voluntary (swallowing) and some involuntary (peristaltic action, a wavelike motion)—all coordinate to conduct the meal into the stomach.

Within the stomach, other layers of muscles work together to create peristaltic motion in a different direction. This movement helps to hold the food in the stomach while it is mixed with hydrochloric acid, other digestive enzymes, and mucus. These gastric juices help to break down nutrients and kill harmful bacteria. A protective coating of mucus shields the stomach walls from the acid, lest we digest ourselves.

By this point, the meal isn't looking much like a

sandwich anymore. Past the stomach, the contents move into the small intestine where they are infused with neutralizing enzymes and additional hormones.

At this stage, bile from the liver and gallbladder is also introduced into the mix to break down fatty substances. Pancreatic fluid, an alkaline mixture rich in digestive enzymes, is added to neutralize stomach acid and break down fats, proteins, and carbohydrates.

Throughout the small intestine, the chyme (a mixture of food and juices from the stomach, liver, gallbladder, and pancreas) is met by millions of **villi** (carpet-like fingers) that absorb nutrients. Each villus is covered with an average of 600 microvilli that help increase the surface area and foster the absorption of nutrients.

Within each of the three different sections of the small intestine (duodenum, jejunum, and ileum) the villi have a different shape. Each is formed from multiple types of cells—skin, muscle, and nerve—and equipped with blood vessels and lymph channels. The villi all serve a common purpose—extracting nutrients from the ingested food.

Beneficial bacteria within our intestines also help break down food and produce important elements like vitamins K and B. Next, as the contents move through the large intestines, excess water is removed along with sodium and other salts.

Finally, after the maximum amount of necessary nutrition is absorbed from our food, the leftover waste is transported out of our bodies through another series of muscles and tubes—the rectum and anus.

If any components were missing within our gastrointestinal tract, our bodies would be compromised and weakened—possibly fatally. It is a complex multipart system crucial to our survival.

INTELLIGENT DIGESTION

Very little in the way of details has been proposed about the evolutionary origins of the human digestive system. Even the simplest animal, however, must be able to acquire and intake food, extract and make use of nutrients, and excrete waste. These are not easily acquired traits, yet all must be present in an animal.

Moreover, even "primitive" vertebrate-like animals such as the hagfish have many of the major components of the human digestive system—including a mouth with teeth, an esophagus, stomach, liver, gallbladder, pancreas, intestine, and an anus. Comparative anatomy sheds little light on the evolution of this complex system.

On the other hand, the digestive system parallels engineered systems that, in any other context, we would recognize as having been designed. The entire process must be carefully regulated to make sure each station performs its function at the right time.

Like tributaries joining a river, a network of interconnecting tubes ensure that digestive juices are added to the mixture at the proper stages. Multiple valves at important junctures, and muscles, keep food traveling in the right direction. A complex system of regulatory chemicals ensures that powerful digestive enzymes are activated, released, and neutralized at the proper time. These essential components are coordinated to keep your digestive system focused on digesting food, and not itself.

The engineering design of this system allows multiple, diverse components to work together in careful coordination. The next time you eat a sandwich, ask yourself whether you want unguided Darwinian processes or intelligently engineered design to help your body digest it.

OTHER MACRO-SYSTEMS

We have only discussed three of many complex systems in the human body. There are a number of other vital macro-systems, including:

- Auditory—enables hearing.
- Circulatory—delivers food, oxygen, and other necessary substances to the cells, while taking away waste products.
- Endocrine—glands (such as the thyroid, ovaries, testes and adrenal) that release important hormones.
- Immune—defends against harmful invaders with help from the spleen, bone marrow, thymus and lymphatic glands.
- Musculoskeletal—bones, muscles, joints, and other parts provide support and allow movement.
- Neurological—the body's central information processor.
- Respiratory—takes in oxygen and distributes it into the blood, while expelling carbon dioxide waste.
- Urinary—uses kidneys to filter blood and gets rid of extra water and toxic elements through the urine.

CONCLUSION

Chapters 7 through 9 showed that basic biochemical features such as functional proteins are beyond the reach of Darwinian evolution. This chapter, and Chapter 10, have shown that many macro-systems in higher organisms pose a severe challenge to Darwinism.

The information presented in these chapters on the origin and development of life refutes the sixth tenet of materialism.

MATERIALISM TENET #6:
All species evolved by unguided natural selection acting upon random mutations.

Instead, biological macro-systems show the kind of integrated complexity that, in our everyday experience, we immediately recognize as the result of engineering design.

Despite this compelling evidence for design, some materialists look at the human body—and the body plans of many other organisms—and claim they are too "flawed" to have been designed. Their arguments will be assessed in the next chapter.

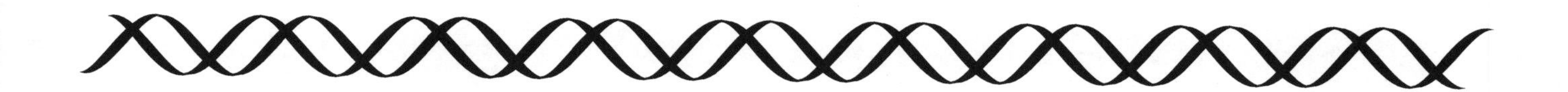

DISCUSSION QUESTIONS

1. Are there any parts of the eye that are unnecessary? Discuss what would happen if your eye was missing its cornea, pupil, vitreous humor, optic nerve, or musculature.

2. Consider what you know about one of the other macrosystems in the human body that were only briefly mentioned in this chapter. Does it require multiple integrated parts to function?

3. Putting pieces of a puzzle together, a person would look for the matching piece to fit with the one she's holding. Until all matching pieces are located and connected, the puzzle cannot be complete. Is the puzzle of reproduction any different? Without the proper anatomy, related hormones and genetic combination, what would happen?

4. Many plants and animals reproduce sexually. Consider the complexity of this type of reproduction, and discuss the feasibility that two organisms in the same place simultaneously developed this mutually compatible reproductive system.

5. Explain some advantages and disadvantages of sexual reproduction over asexual reproduction. Is sexual reproduction likely to have evolved? Why or why not?

6. Can you think of an example of an assembly line that did not arise by intelligent design? If so, describe what it produces.

CHAPTER 12

POORLY DESIGNED ARGUMENTS

Question:

Pop culture would have you believe that our bodies are full of useless or poorly designed parts. Do you agree?

POORLY DESIGNED ARGUMENTS

"Many of the views which have been advanced are highly speculative, and some no doubt will prove erroneous."

— Charles Darwin[1]

Cars break down. Houses deteriorate. Computers crash and printers jam. Organisms die. But despite these apparent imperfections, no one would argue that cars, houses, computers, or printers were not designed by intelligent agents. But what about alleged imperfections in living organisms?

Consider that when humans design an object, each part is expected to have at least a minimal function. What if some biological organs or structures entirely lack function? Would imperfections or lack of function disprove intelligent design?

Philosophers use the term **dysteleology** to describe these attempts to disprove ID based upon considerations of supposedly poor design.

> **Dysteleology:**
>
> ***The view that nature was not intelligently designed, often based on the claim that some natural structures are flawed or functionless.***

Let's examine the issues raised by this view.

POOR DESIGN

The debate over ID becomes confusing when critics **equivocate**.

> **Equivocation:**
>
> ***Changing the meaning of words in the middle of an argument.***

Critics change the meaning of "intelligent," ignoring how ID proponents use the term. They claim that a structure is not "intelligently designed" if its design is not perfect from their perspective. As anti-ID biologist Kenneth R. Miller of Brown University asserts, many biological structures must have evolved "because their structure, their physiology, and even their genetic makeup are all inconsistent with the demands of intelligent design."[2]

However, when ID proponents use the term "intelligent," they simply seek to indicate that a structure has features requiring a mind capable of forethought to design the blueprint. But does intelligent design require *perfect* design?

For that matter, what constitutes biological perfection? Take humans for example. Should our bodies all last 100 years? 200 years? Forever? Should we be impervious to injury and never get sick? These are philosophical or theological questions, having little or nothing to do with science.

Holding biological systems to a vaguely defined standard of "perfect" design is the wrong way to test ID. The examples at the beginning of this chapter—broken machinery, computer failures, and decaying buildings—all show that a structure might be designed by intelligence even if it breaks or has flaws. Intelligent design does not mean perfect design. It doesn't even require optimal design. It means exactly what it says: design by an intelligent agent.

Despite the fact that imperfect design doesn't disprove design, dysteleological arguments about supposed flaws in nature are some of the most popular criticisms of ID. But they share three problems, two of which we have already discussed.

1. Critics' standards of perfection are often arbitrary.
2. An object can have imperfections, but still be designed.
3. Upon closer inspection, alleged design flaws actually turn out to be beneficial.

Three examples—the vertebrate eye, the panda's thumb, and the recurrent laryngeal nerve—show how dysteleology arguments commonly fail.

AN EYE FOR IRRATIONAL ARGUMENTS

Most perceived design flaws are themselves based upon flawed arguments. For example, Kenneth Miller maintains that the vertebrate eye was not designed because the optic nerve extends over the retina instead of going out the back of the eye—an alleged design flaw. According to Miller, "visual quality is degraded because light scatters as it passes through several layers of cellular wiring before reaching the retina."[3]

Similarly, evolutionary biologist Richard Dawkins contends that the retina is "wired in backwards"[4] because light-sensitive cells face away from the incoming light, which is partly blocked by the optic nerve. The result is a small blind spot in each eye, which according to Dawkins shows that the vertebrate eye is "the design of a complete idiot."[5]

Arguments that alleged flaws in nature refute ID should be subjected to two steps of critical scrutiny.

- First, we must determine whether the flaw, if real, would actually be an argument against intelligent design.
- Second, we must investigate whether the flaw is real.

Let's apply these two steps of analysis to the vertebrate eye.

Even if the wiring of the optic nerve is flawed, there is no reason that should refute intelligent design. The fact that some nerves are positioned in a way that Dawkins and Miller find imperfect doesn't negate the purposeful arrangement of parts found in the eye. The vertebrate eye still reflects an integrated system of parts, with high levels of specified complexity. Imperfect design is still design.

Even though this dysteleological argument has already been invalidated by step one of our analysis, we'll continue with step two.

MORE THAN MEETS THE EYE

A closer examination shows that the design of the vertebrate eye works far better than Dawkins and Miller suggest. Dawkins concedes that the optic nerve's impact upon vision is "probably not much,"[6] but the negative effect is even less than he admits.

Only if you cover one eye and stare directly at a fixed point does a small, off-center "blind spot" appear in your peripheral vision as a result of the optic nerve covering part of the retina. When both eyes are functional, your brain meshes the visual fields of both and compensates for the blind spot. Under normal circumstances, the wiring of the nerves does nothing to hinder vision.

Nonetheless, Dawkins is eager to find flaws, and he argues that even if the design works, it would "offend any tidy-minded engineer."[7] But a closer analysis shows that the overall design of the optic nerve actually optimizes visual acuity.

For high quality vision, vertebrate retinal cells require a large blood supply. Because the photoreceptor cells face the back of the retina (with the optic nerve extending over them), they are able to plug directly into the blood vessels that feed the eye, maximizing access to blood.

Biologist George Ayoub suggests a thought experiment in which the optic nerve goes out the back of the retina, the way Miller and Dawkins prefer. Ayoub points out that this design would interfere with blood supply, as the nerve would crowd out blood vessels. In this case, the only means of restoring blood supply would be to place capillaries over the retina—but this would block even more light than the optic nerve presently does.

Ayoub concludes: "In trying to eliminate the blind spot, we have generated a host of new and more severe functional problems to solve."[8]

In 2010, two eye specialists made a remarkable discovery showing that the vertebrate eye compensates for the blockage of light due to the position of the optic nerve. Special "glial cells" sit over the retina and function like fiber optic cables to channel light through the optic nerve wires directly to the photoreceptor cells. According to the journal *New Scientist*, these funnel-shaped cells prevent scattering of light and "act as light filters, keeping images clear."[9]

Miller contends that an intelligent designer "would choose the orientation that produces the highest degree of visual quality."[10] But that seems to be exactly what we find. Researchers who studied glial cells found that the retina is "an optimal structure designed for improving the sharpness of images."[11]

The "flaws" claimed by Dawkins and Miller are not really flaws at all. As mathematician William Dembski observes, their proposals for improving eye function would instead result in "diminishing its visual speed, sensitivity, and resolution."[12]

With the vertebrate eye, we see that the design is optimal. ID critics just protest its design as being untidy or otherwise offensive. This is a common theme in dysteleological arguments: the design might actually function optimally, but evolutionary biologists argue that they could have done better.

TWO THUMBS UP

Another example of an allegedly poor design is the **panda's thumb**. Pandas obtain their nourishment largely from bamboo, which they strip before eating, using an extra appendage on their front paws. This appendage contains a bone but is not an opposable thumb like ours.

In his book *The Panda's Thumb*, evolutionary paleontologist Stephen Jay Gould argued that "odd arrangements and funny solutions are the proof of evolution—paths that a sensible God would never tread."[13] Likewise Miller claims that an intelligent designer would have "been capable of remodeling a complete digit, like the thumb of a primate, to hold the panda's food."[14]

Are Gould and Miller right to contend that the panda's thumb is a "clumsy"[15] feature that is "not necessarily well-designed"[16]?

Figure 12-1 portrays a human hand, and a panda's front paw. Notice that the paw seems to have six digits. The one on the far left, however, is not a true finger but an elongated wrist bone equipped with muscles to hold and strip bark from bamboo.

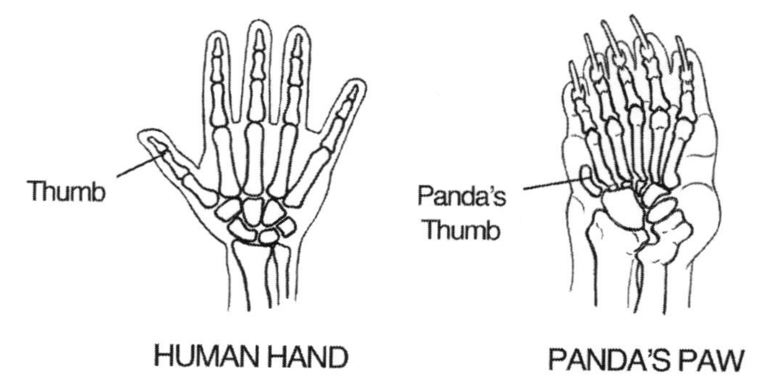

Figure 12-1: Human Hand and Panda's Paw.

It turns out that the panda's thumb is not a clumsy design. A study published in *Nature* used MRI and computer tomography to analyze the thumb and concluded that the bones "form a double pincer-like apparatus" thus "enabling the panda to manipulate objects with great dexterity."[17]

The critics' objection is backed by little more than their subjective opinion about what a "sensible God" should have made.

UNNERVING ARGUMENTS

Another supposed design flaw is the pathway of the **recurrent laryngeal nerve (RLN)**. The RLN provides nerve impulses to the larynx (voice box) of some vertebrates, including humans. But rather than taking a direct route from the brain to the larynx, it loops down below the aorta and comes back up the neck, entering the voice box from below.

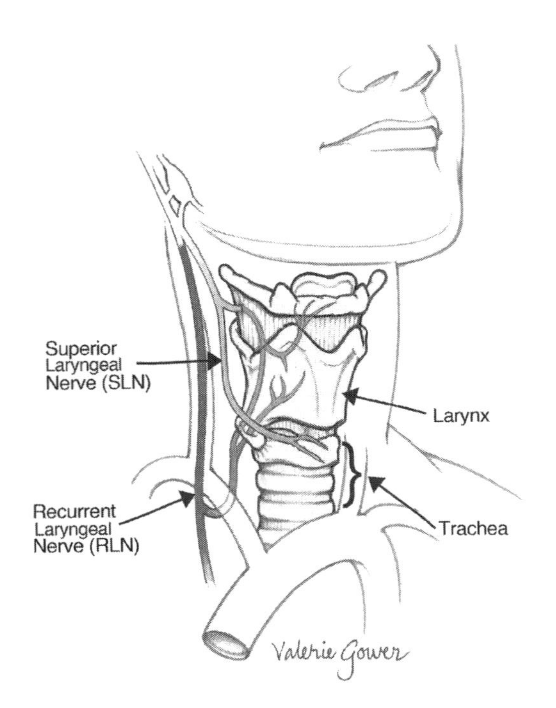

Figure 12-2: Recurrent and Superior Laryngeal Nerves.

Evolutionary biologist Jerry Coyne calls this "[o]ne of nature's worst designs."[18] Richard Dawkins deems it "a ridiculous detour—no engineer would ever make a mistake like that."[19] Coyne and Dawkins go further and argue that the seemingly circuitous route of the nerve is easily explained if one understands that our bodies evolved from a fish body plan and must operate under those constraints.

Case closed? Hardly.

First, as we stated above, even an imperfect design is still a design. Second, there are good reasons to regard the RLN as well-designed.

According to biologist Wolf-Ekkehard Lönnig, about 1% of humans actually have an RLN that goes directly from the brain to the larynx—the route that Coyne and Dawkins claim it ought to follow. However, this condition is associated with health problems, such as breathing difficulties.[20] This implies there are functional reasons for the circuitous route normally taken by the RLN.

Indeed, as the RLN travels from the aorta up to the larynx, it provides nerve filaments to the heart and to the mucous membranes and muscles of the trachea.[21] Clearly, the route of the RLN provides certain benefits.

Moreover, what is rarely disclosed is that there is another nerve that serves the larynx directly from the brain—it's called the **superior laryngeal nerve (SLN)**. Entering the larynx directly from above, the SLN follows the very design demanded by critics like Coyne and Dawkins (see Figure 12-2).

Different medical conditions that occur when either the SLN or RLN is damaged imply that *each* nerve has special functions. This dual-innervation of the voice box—one nerve from above, and one from below—apparently preserves some level of functionality if one nerve gets damaged.[22]

Again, we see that critics reject the possibility of design due to misconceptions about what would work best.

BIOMIMETICS AND POOR DESIGN

Evolutionary biologists like Kenneth Miller argue that organisms were merely "cobbled together,"[23] but since the time of the ancient Chinese, engineers have looked to biological designs for inspiration in devising human technology. A modern field called **biomimetics** has found many cases in which engineers turned to biology to improve technology. For example:

- Birds inspired aircraft design.
- Faster swimwear has been based upon the properties of shark skin.
- Spiny hooks on plant seeds and fruits led to the development of Velcro.
- Mimicking whale flipper shape has improved turbine design.
- Studying hippo sweat promises to lead to better sunscreen.[24]

In another example, researchers from MIT calculated the most efficient layout of mirrors for collecting solar energy only to find out that sunflowers already used this design. As a news article observed, "In matters of clever design, nature has often got there first."[25]

Unsurprisingly, optics engineers also study the eye to improve camera technology, which one tech writer called "one of nature's best designs."[26] It would seem that biological systems like the eye are "cobbled together" so well that engineers regularly mimic them.

As we have shown, alleged examples of "poor design" do not refute intelligent design. However, what about functionless structures? Could they challenge ID?

FUNCTIONLESS ORGANS

Typically, intelligent agents make things that have some function, whether practical or aesthetic. But if a biological feature lacks function, would that necessarily count against intelligent design for that structure?

There are two categories of allegedly functionless organs: those that are said to have lost their function, and those that were supposedly never functional. The former are called **vestigial**.

> **Vestigial Organ:**
>
> ***A biological structure that once had a function but lost part or all of that function through evolution.***

WHAT HAVE YOU DONE FOR ME LATELY?

Looking at vestigial organs, Richard Dawkins frames his dysteleological argument in theological terms: "God wouldn't do it that way." He writes:

> Vestigial eyes, for example, are clear evidence that these cave salamanders must have had ancestors who were different from them—had eyes, in this case. That is evolution. Why on earth would God create a salamander with *vestiges* of eyes?[27]

This is a theological question rather than a scientific one, but Dawkins has a point. Blind cave salamanders probably did descend from salamanders that had functional eyes. Does this refute ID theory?

ID accepts that Darwinian processes can accomplish some types of change. We commonly observe mutations causing deformed individuals that lack certain features. Random mutations certainly have the power to destroy function and cause salamanders to lose functional eyes.

Darwinian mechanisms are probably the best explanation for the functionless state of eyes on blind cave salamanders. This fact doesn't undermine the case for intelligent design. Darwinian evolution might account for loss-of-function, but ID is more concerned with how complex features like functional eyes are gained in the first place.

Moreover, proponents of ID readily acknowledge that natural causes can act upon designed structures. Indeed, we experience the decay of designed objects every day. For example, if a laptop were left exposed to the elements on top of a high mountain for 20 years, it's likely the computer would no longer work. That does not mean that the laptop was not originally designed. It just means that its current state is the result of both intelligent design and subsequent decay.

Thus, if a structure really is vestigial, it seems reasonable to conclude it did not arrive at that state through intelligent design. But that does not mean the structure was not originally designed, nor does it invalidate ID for other complex structures that *do* have function. Scientists must test for design on a case-by-case basis.

This still leaves open the question: *how many structures actually are vestigial?* Let's consider the human body.

WHO ARE YOU CALLING VESTIGIAL?

Many people know someone who experienced severe abdominal pain and soon thereafter was admitted to a hospital to have his appendix removed (see Figure 12-3). About 250,000 appendectomies are performed annually in the U.S.[28] After the surgery, people seem to get along just fine—leading some to suggest the human appendix is a useless organ. As one Darwinian scientist claims, the appendix is a "vestige of our herbivorous ancestry,"[29] and over eons of evolution its function in humans has been lost, or severely reduced.

However, the appendix turns out to have important functions. As one 2009 science news article stated:

> The body's appendix has long been thought of as nothing more than a worthless evolutionary artifact, good for nothing save a potentially lethal case of inflammation. Now researchers suggest the appendix is a lot more than a useless remnant.... In a way, the idea that the appendix is an organ whose time has passed has itself become a concept whose time is over.[30]

One function the appendix serves is to provide a storehouse for beneficial bacteria in our intestines that aid in the breakdown of food. A strong population of these bacteria also helps overcome many harmful bacteria that find their way into the digestive tract.

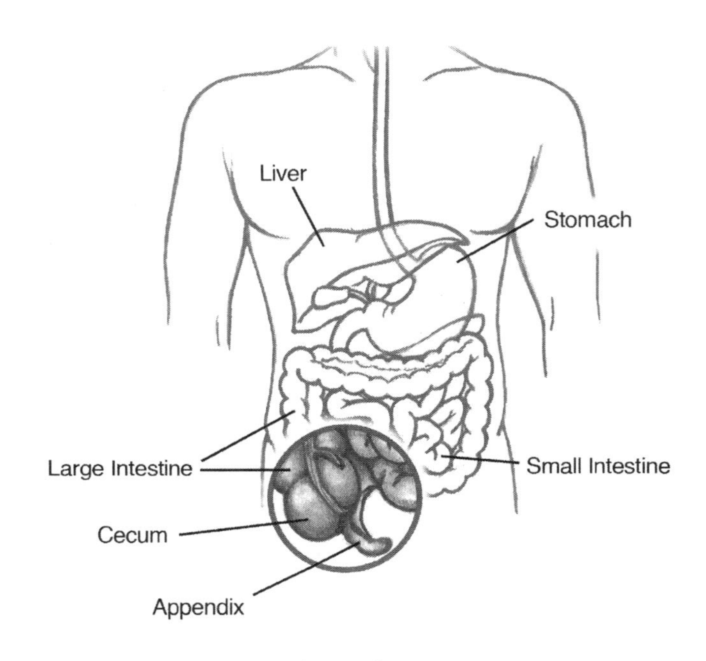

Figure 12-3: Human Appendix.

These are the same kind of **probiotics** that are commonly added to yogurt, drinks, and other foods. The appendix keeps a ready supply of beneficial bacteria that can quickly repopulate our intestines after we suffer from the flu, food poisoning, or other gastrointestinal illnesses.

Additionally, the appendix serves a variety of immune-related functions, helping to produce white blood cells and playing important roles during fetal development.[31]

But if the appendix is so useful, why can people live without it? The answer is that, at the same time humans developed the medical ability to safely remove the appendix, we also developed the knowledge to compensate for its absence.

In modern developed countries our ability to fight disease is aided by antibiotics, antibacterial soap, and cleaner living conditions. In addition, one can replenish beneficial intestinal bacteria with a cup of yogurt or a probiotics capsule from the local grocery store.

But because the appendix is finally considered important, research is being conducted on how to better treat appendicitis.[32] This research is showing that antibiotics can cure many cases of appendicitis. Just as we would treat, rather than remove, an infected kidney, it is now possible to treat—and not just remove—an infected appendix.

In light of this evidence, Duke University immunologist William Parker observed that "Many biology texts today still refer to the appendix as a 'vestigial organ'" but "it's time to correct the textbooks."[33]

THE INCREDIBLE SHRINKING LIST

During the Scopes "monkey" trial in 1925, evolutionary biologist Horatio Hackett Newman suggested that there are over 180 vestigial organs and structures in the human body, "sufficient to make of a man a veritable walking museum of antiquities."[34] In 2008, the pro-evolution journal *New Scientist* reported that since those days the list of vestigial organs "grew, then shrank again" to the point that today "biologists are extremely wary of talking about vestigial organs at all."[35]

Because function and purpose are continually being discovered for so-called vestigial organs, few, if any, are still considered vestigial. In addition to the appendix, structures that were previously—and incorrectly—considered to be vestigial include:

- The tonsils: At one time, they were routinely removed. Now it's known they serve a purpose in the lymph system to help fight infection.[36]

- The coccyx (tailbone): Many evolutionists still claim this is a hold-over from the tails of our supposed primate ancestors,[37] but in fact it's an important structure for the attachment of muscles, tendons, and ligaments that support the bones in our pelvis.

- The thyroid: This gland in the neck was once believed to have no purpose, and was ignored or even destroyed by medical doctors operating under false Darwinian assumptions. Now scientists know that it is vital for regulating metabolism.

Despite the poor track record of identifying vestigial organs, the journal *New Scientist* felt compelled to state, much as professor Newman claimed during the Scopes trial, that "human beings are walking records of their evolutionary past."[38] But the data are pointing in the opposite direction.

Alleged vestigial organs that actually perform important functions don't invalidate ID theory. But what if there are parts that never had function to begin with?

ONE MAN'S JUNK...

For several decades it has been known that a large amount of DNA in multicellular organisms does not code for proteins. In humans, for example, only 1 to 2% of the DNA encodes proteins. According to some Darwinian biologists, much of the rest of the genome, called non-protein-coding DNA, is useless genetic material produced over millions of years of evolution—often referred to as **junk DNA**.

> **Junk DNA:**
> ***Non-coding DNA that allegedly has no function for the organism that carries it.***

Some of this alleged junk (notably **pseudogenes**) is said to be DNA that lost its function, but much of it is said to have never had a function at all. If correct, this would seem to provide an example of biological structures that challenge intelligent design.

> **Pseudogenes:**
> ***Supposedly broken, useless copies of once-functional genes.***

Kenneth Miller makes this argument, claiming that "the human genome is littered with pseudogenes, gene fragments, 'orphaned' genes, 'junk' DNA, and so many repeated copies of pointless DNA sequences that it cannot be attributed to anything that resembles intelligent design."[39]

Likewise, Francis Collins asserts that "45% of the human genome [is] made up of such genetic flotsam and jetsam."[40] Framing the issue theologically, he claims this repetitive junk DNA poses "an overwhelming challenge to those who hold to the idea that all species were created."[41]

Proponents of intelligent design, however, have pointed out that what has been presumed to be "junk" is actually DNA that is poorly understood. Jonathan Wells explains that ID predicted we would discover function for non-coding DNA:

> Since non-coding regions do not produce proteins, Darwinian biologists have been dismissing them for decades as random evolutionary noise or "junk DNA." From an ID perspective, however, it is extremely unlikely that an organism would expend its resources on preserving and transmitting so much "junk."[42]

In fact, ID's prediction has turned out to be exactly correct. A myriad of functions have been discovered for virtually all types of non-coding DNA.[43] Specific functions include:

- Repairing DNA.[44]
- Assisting in DNA replication.[45]
- Regulating gene expression.[46]
- Aiding in folding and maintenance of chromosomes.[47]
- Controlling RNA editing and splicing.[48]
- Helping to fight disease.[49]
- Regulating embryological development.[50]

Indeed, a groundbreaking 2012 study published in *Nature* by an international group of hundreds of research scientists known as the ENCODE Consortium found that "[t]he vast majority (80.4%) of the human genome participates in at least one biochemical [function]."[51]

Since ENCODE only analyzed a minority of human cell types, its lead scientists predicted "it's likely that 80 percent will go to 100 percent," because the evidence suggests "almost every nucleotide is associated with a function of some sort or another."[52] Another *Nature* article stated this evidence "dispatch[es] the widely held view that the human genome is mostly 'junk DNA'."[53]

Even evolution-friendly news media could not ignore the implications of this study. The *New York Times* reported our bodies have a "DNA wiring system that is almost inconceivably intricate."[54] A prominent British newspaper, *The Guardian*, explained the stark implications for junk DNA:

> For years, the vast stretches of DNA between our 20,000 or so protein-coding genes—more than 98% of the genetic sequence inside each of our cells—was written off as "junk" DNA. Already falling out of favour in recent years, this concept will now, with Encode's work, be consigned to the history books.[55]

There's another type of junk DNA worth focusing on because ID critics often talk about it:

pseudogenes. Richard Dawkins writes: "Genomes are littered with nonfunctional pseudogenes, faulty duplicates of functional genes that do nothing."[56] He adds, "creationists might spend some earnest time speculating on why the Creator should bother to litter genomes with untranslated pseudogenes."[57]

Yet a number of studies are detecting function for pseudogenes, raising doubts about the view that they are broken.[58] As two leading biologists report, "pseudogenes that have been suitably investigated often exhibit functional roles."[59] Likewise, a 2011 paper in the journal *RNA* titled "Pseudogenes: Pseudo-functional or key regulators in health and disease?" found that they should no longer be called "junk":

> Pseudogenes have long been labeled as "junk" DNA, failed copies of genes that arise during the evolution of genomes. However, recent results are challenging this moniker; indeed, some pseudogenes appear to harbor the potential to regulate their protein-coding cousins.[60]

A similar 2012 paper in *RNA Biology* observed that "Pseudogenes were long considered as junk genomic DNA," but found that pseudogene function is "widespread," and predicted that "more and more functional pseudogenes will be discovered as novel biological technologies are developed in the future."[61]

The junk DNA paradigm arose due to evolutionary thinking. Its failure has turned out to be a spectacularly bad prediction for Darwinian biology, but a good one for intelligent design.

Unfortunately, this controversy is not a trivial academic exercise. The materialist mindset has had definite negative impacts on science and medicine. Materialists would be well-advised to spend more time studying the details of the biological world, and less time speculating about what God would or wouldn't have done.

STOPPING SCIENCE

The commitment to "junk DNA" and similar thinking has hindered the progress of science, discouraging researchers from discovering functions for non-coding DNA and alleged vestigial organs.

The journal *Science* reported that the junk DNA mindset "repelled mainstream researchers from studying non-coding DNA."[62] Fortunately some bold scientists conducted research—"at the risk of being ridiculed"[63]—that led to the overturning of the junk DNA paradigm.

In 2003, *Scientific American* reported that one type of non-protein-coding DNA called introns was "immediately assumed to be evolutionary junk." According to the article, "[t]he failure to recognize the importance of introns may well go down as one of the biggest mistakes in the history of molecular biology."[64] Junk DNA was discredited in spite of—not because of—Darwinian thinking.

With its commitment to the idea of vestigial organs, materialism did not just hold back scientific progress; it harmed patients. Medical schools steeped in Darwinian thought assumed that organs such as the appendix, tonsils, and thyroid were functionless evolutionary leftovers. As a result, evolutionary thinking led doctors to unwittingly remove important organs.

Though critics often assert that intelligent design would mean the end of scientific investigation, the opposite is actually true. An intelligent design paradigm encourages scientists to seek function for poorly understood biological structures. Guided by such predictions, medicine and biology might have progressed much more rapidly.

As Jonathan Wells concludes in his book *The Myth of Junk DNA*, "assuming that any feature of an organism has no function discourages further investigation. In this respect, the myth of junk DNA has become a science-stopper."[65]

CONCLUSION

As suggested by a *Nature* article titled "What a shoddy piece of work is man," materialists routinely view the human body as having an imperfect design.[66] But intelligent design does not require perfect design, as anyone who has experienced a breakdown in technology will readily attest. Moreover, so-called "vestigial" organs or "junk" DNA have been discovered to have important functions, confirming predictions of ID.

Dysteleological arguments impose critics' personal opinions about how design ought to look, which often turn out to be wrong. From the wiring of the voice box to the vertebrate eye to the panda's thumb, so-called "flaws" in nature turn out to work quite efficiently and exist for good reasons. In fact, the bad dysteleological arguments presented by materialists are proof that something (i.e., their arguments) can be severely flawed, and yet still be designed.

By insisting that much of our DNA and some of our organs are useless evolutionary garbage, opponents of ID have hindered biology from understanding the workings of the cell and the functions of some organs. In contrast, an intelligent design paradigm encourages scientists to explore potential function, offering positive benefits for scientific advancement in medicine and other fields.

DISCUSSION QUESTIONS

1. Describe an intelligently designed item that you tried to use, but then found to be broken or flawed. Do such imperfections indicate that the item was not designed? Explain your answer.

2. Do you agree with claims that the recurrent laryngeal nerve is a flawed design? Is there a rebuttal to that argument?

3. Most people can survive without their appendix. Does that mean it is a functionless, vestigial organ? Discuss your answer.

4. Which is the more scientific approach regarding an unstudied section of non-protein-coding DNA: to assume it has no purpose and ignore it, or to investigate it and test for function? Which approaches would be encouraged by intelligent design and which by neo-Darwinism?

5. In addition to those listed in this chapter, give an example of technology that might have been inspired by a feature of living organisms. Does this point to intelligent design? Why or why not?

6. Is intelligent design a science-stopper? How about Darwinian evolution? Explain your answer.

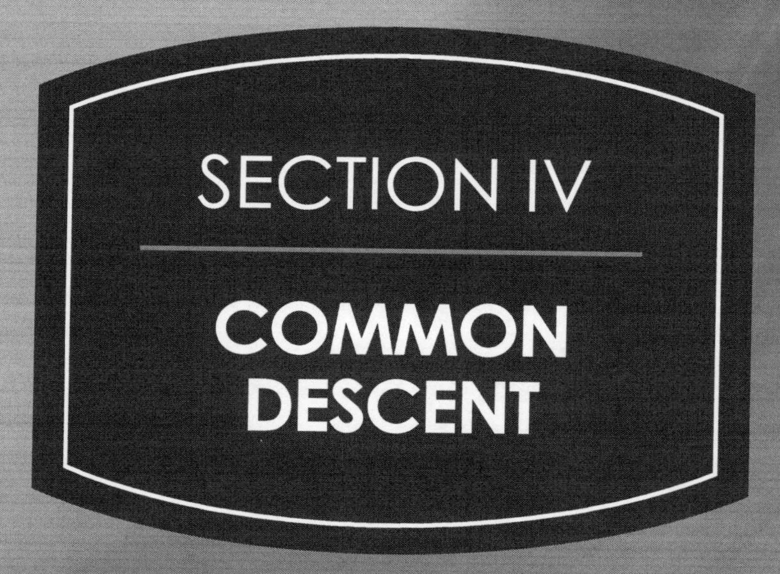
SECTION IV
COMMON DESCENT

CHAPTER 13

TREE HUGGERS

Question:

Do you think all living and extinct species are related through universal common ancestry?

TREE HUGGERS

"Many biologists now argue that the tree concept is obsolete and needs to be discarded. We have no evidence at all that the tree of life is a reality."

—New Scientist, January 21, 2009[1]

Charles Darwin's book *Origin of Species* contains just one illustration: the **tree of life** (see Figure 13-1).[2] While Darwin acknowledged the possibility that there were different progenitors of the various kingdoms of life, he envisioned the history of life forming a grand tree, where "all the organic beings which have ever lived on this earth have descended from some one primordial form."[3]

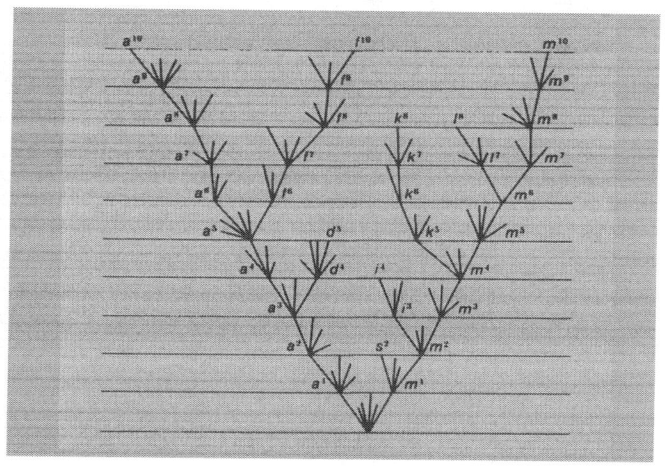

Figure 13-1: Part of Darwin's "Tree of Life" Illustration.

More than 150 years later, the concept of a universal tree of life remains a core aspect of neo-Darwinian theory, and a fundamental tenet of materialism.

MATERIALISM TENET #7:

All living organisms are related through universal common ancestry.[4]

The field of **systematics** studies the relationships between groups of organisms. According to evolutionary systematists, if we go back far enough then you, the fungus growing on the bottom of your foot, and every other living organism on Earth shared a common ancestor.

One opinion article tried to convince readers of common ancestry:

> The evidence that all life, plants and animals, humans and fruit flies, evolved from a common ancestor… is a fact. Anyone who takes the time to read the evidence with an open mind will join scientists and the well-educated.[5]

Likewise, one scientist wrote "Biologists today consider the common ancestry of all life a fact on par with the sphericity of the earth or its motion around the sun."[6] Such strong statements should be supported by strong evidence. To "read the evidence with an open mind" means to follow wherever it leads—even if that risks ridicule or unpopularity. What does the evidence say?

This chapter will assess how scientists construct evolutionary trees in their attempt to demonstrate common ancestry. Before entering this discussion, it's important to reiterate that intelligent design refers to the mechanism of change. It does not claim that species are necessarily unrelated, so it could be compatible with common ancestry. However, as good scientists must, we will test the assumptions of neo-Darwinism.

THE MAIN ASSUMPTION

Evolutionary trees, also called **phylogenetic trees**, are based upon the assumption that *similarity between organisms is the result of inheritance from a common ancestor.*[7]

> **Phylogeny:**
>
> ***A hypothesis describing the ancestral relationships among organisms.***

Evolutionists believe that the degree of similarity between two organisms indicates how closely related they are.

There are two basic types of phylogenetic trees. **Molecule-based trees** are based on comparing DNA, RNA, or protein sequences in different organisms. **Morphology-based trees** are constructed by comparing physical characteristics, such as anatomical and structural similarities of different organisms.

> **Morphology:**
>
> ***The form, structure, and body plan of an organism.***

Let's examine each of those two methods.

MOLECULAR TREES

Molecular trees rely on the assumption that the more similar the DNA, RNA, or protein sequences of two organisms, the closer the evolutionary relationship.

Some evolutionists say that different genes produce phylogenetic trees that are perfectly consistent. For example, when asked about the "strongest, most irrefutable single piece of evidence in support" of evolution, Richard Dawkins replied:

> The most compelling evidence is comparative evidence... You can take any pair of animals you like... and count the number of differences in the letters of a particular gene, and you plot it out, and you find that it forms a perfect branching hierarchy.... [Y]ou then do the same thing for another gene, and another, and another. *You get the same family tree.*[8]

But Dawkins is wrong. Very often, one gene yields one version of the "tree of life," while another gene implies an entirely different tree. Obviously, all trees cannot be right. The basic problem is that biological similarity is constantly appearing in places not predicted by common descent.

The scientific literature is full of papers that report contradictions and failures of molecular data to provide a consistent picture of common descent.[9] In 2009 the journal *New Scientist* published a cover story titled, "Why Darwin was wrong about the tree of life," which summarized this problem:

> For a long time the holy grail was to build a tree of life... But today the project lies in tatters, torn to pieces by an onslaught of negative evidence.[10]

The article explained what happened when one scientist, Michael Syvanen, tried to create a tree showing evolutionary relationships using 2000 genes from a diverse group of animals:

> He failed. The problem was that different genes told contradictory evolutionary stories.... the genes were sending mixed signals.... Roughly 50 per cent of its genes have one evolutionary history and 50 per cent another.[11]

The data were so difficult to resolve into a tree that Syvanen lamented, "We've just annihilated the tree of life."[12]

Other researchers have encountered similar results. *Nature* reported that microRNA molecules "give a totally different tree from what everyone else wants."[13] Another paper titled "Bushes in the Tree of Life" found that "a large fraction of single genes produce phylogenies… at odds with conventional wisdom." The authors suggest that such problems "should force a re-evaluation of several widely held assumptions."[14]

When molecular trees produce such widely conflicting results, they argue against a universal tree of life, not for it.

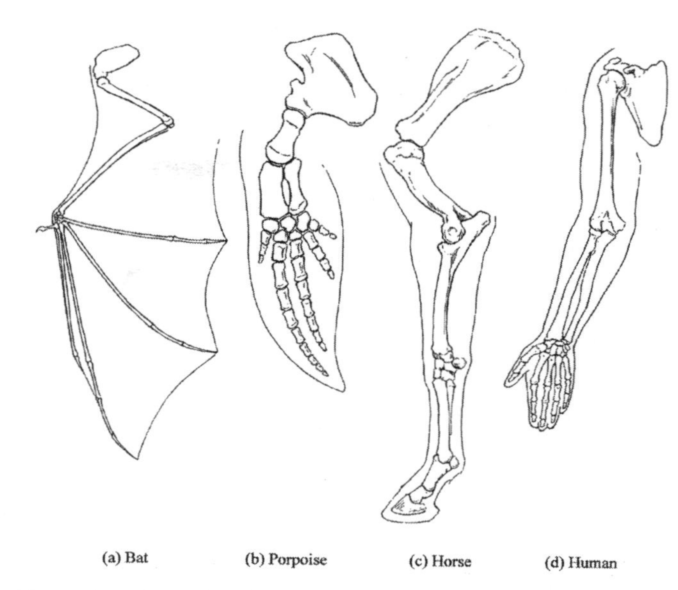

Illustration: Copyright Jody F. Sjogren 2000. Used with permission.

Figure 13-2: Homology in Vertebrate Limbs.

MORPHOLOGY-BASED TREES

Similarities at the molecular level are not the only way evolutionary biologists construct phylogenetic trees. They also compare physical characteristics such as skeletal structure and body plan.

To an evolutionary biologist, a higher degree of anatomical similarity implies a closer evolutionary relationship, leading to the construction of morphology-based trees. These trees rely heavily upon the concept of **homology**.

> **Homology:**

Similarity of structure and position, but not necessarily function. Within neo-Darwinian theory, it means "similarity due to common ancestry."

As seen in Figure 13-2, a common example of homology is the bone structure found in the limbs of various vertebrates.

Evolutionists often cite homology as evidence for common ancestry. However, scientific data has thrown this argument into crisis. The textbook *Explore Evolution* explains:

> [B]iologists have made two discoveries that challenge the argument from anatomical homology. The first is that the development of homologous structures can be governed by different genes and can follow different developmental pathways. The second discovery, conversely, is that sometimes the same gene plays a role in producing different adult structures. Both of these discoveries seem to contradict neo-Darwinian expectations.[15]

There is an alternative to common descent. **Common design** is an entirely reasonable explanation for functional anatomical similarities. After all, why must a designer always start over when a useful design already exists?

MOLECULES VS. MORPHOLOGY

Evolutionists also often say there is widespread agreement between molecule and morphology-based trees, confirming common descent. For example, evolutionary scientist Peter Atkins asserts that "there is not a single instance of the molecular traces of change being inconsistent with our observations of whole organisms."[16]

Once again, the scientific literature tells a different story. Evolutionary trees commonly conflict. As a major review article in *Nature* reported, "disparities between molecular and morphological trees" lead to "evolution wars" because "[e]volutionary trees constructed by studying biological molecules often don't resemble those drawn up from morphology."[17]

To give one example, textbooks often claim common descent is supported because the tree based upon the enzyme *cytochrome* c matches the traditional tree based upon morphology.[18] However, the tree based upon a different enzyme, cytochrome b, can lead to what one study called "an absurd phylogeny of mammals."[19]

There are a number of hypotheses that evolutionists use to explain why phylogenetic trees conflict. We will examine two of them below: horizontal gene transfer and convergent evolution.

TANGLED BUSH OF LIFE

When evolutionary biologists tried to construct a tree showing the evolutionary relationships between the three basic groups of living organisms, they failed. Instead of producing a nice, neat Darwinian tree, they were forced to accept a tangled bush.

The relationships between the three major domains of living organisms, Bacteria, Eukarya, and Archaea, cannot be represented as a tree (see Figure 13-3).[20]

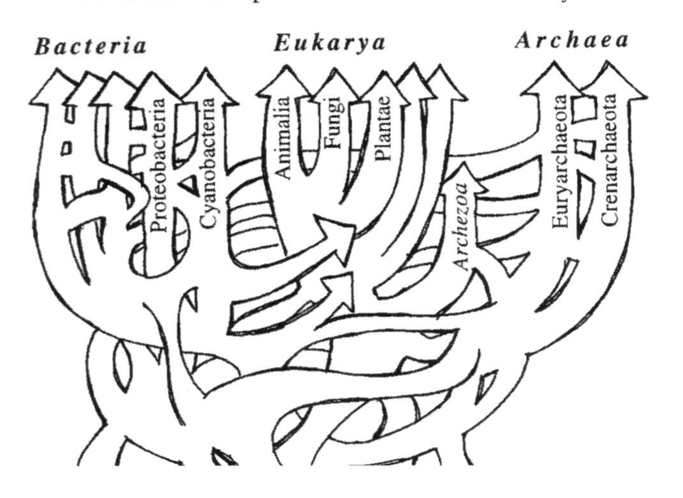

Illustration: From Figure 3, W. Ford Doolittle, "Phylogenetic Classification and the Universal Tree," Science, 284 (June 25, 1999): 2124-2128. Reprinted with permission from AAAS.

Figure 13-3: Tangled Bush of Life.

Because of these problems, leading biochemist W. Ford Doolittle stated, "Molecular phylogeneticists will have failed to find the 'true tree,' not because their methods are inadequate or because they have chosen the wrong genes, but because the history of life cannot properly be represented as a tree."[21]

SHARE WITH YOUR NEIGHBOR

To attempt to explain this non-treelike pattern, materialists like Doolittle appeal to a process called **horizontal gene transfer (HGT)**. In this process, microorganisms obtain genes through mechanisms other than inheritance from a parent, specifically by sharing and swapping genes with their neighbors.[22] For example, they might swap segments of DNA with neighboring cells (see Figure 13-4 on the next page).

The HGT process has been observed in nature—for example, it can spread beneficial traits like antibiotic resistance between bacteria. But evolutionary biologists have extrapolated this process—claiming it applies also to much more complex organisms. This means that when genes appear to invalidate a phylogenetic tree, evolutionists often think they can ignore the conflict, instead blaming it on horizontal gene transfer.[23]

But there are two main reasons why HGT does not validate common descent:

- A number of the tangled pathways at the base of the tree of life (see Figure 13-3) result from genes that encode fundamental cellular machinery. While HGT can occur with *some* bacterial genes, it is highly unlikely that it could occur with all these essential genes.

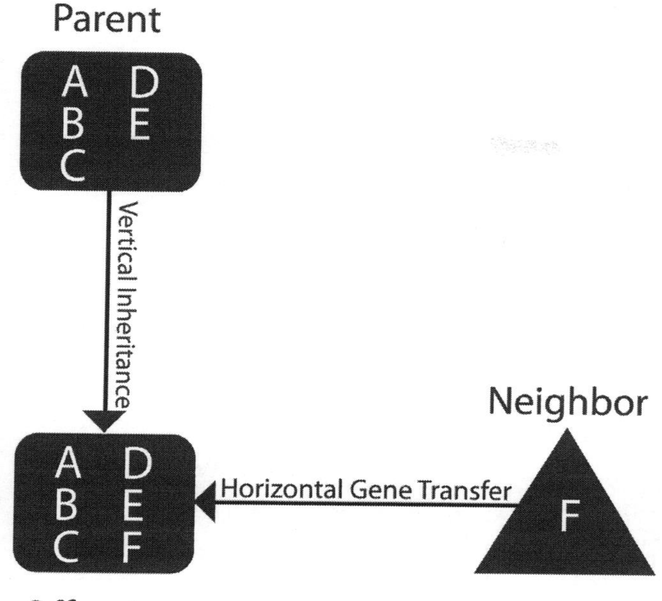

The offspring microorganism at the bottom left obtains most of its genes (A, B, C, D, and E) from its parent through normal processes of inheritance. However, it also obtains one gene (F) from its neighbor. Thus, due to horizontal gene transfer, genes may not always reflect "vertical" ancestry.

Figure 13-4: Horizontal Gene Transfer.

- Conflicts within the tree of life are also prevalent among more complex organisms—such as plants or animals[24]—where such gene-swapping is not directly observed.

Evolutionary biologists claim that HGT can cause conflicts in phylogenetic trees. They then use circular reasoning to claim that conflicts between phylogenetic trees are *evidence* for horizontal gene transfer.[25] They ignore the possibility that the conflicts exist because common descent is false.

Consider this analogy: A store manager tells the owner that the cash register is infested with money-eating microorganisms. "In fact," he insists, "money missing from the cash register is proof that that these bugs exist." Perhaps this is possible, but a responsible owner would not ignore other possible explanations.

The evidence does not support HGT as a solution for discrepancies in the tree of life. Scientific conclusions should be based upon evidence, not unfounded extrapolations.

There is another approach used by evolutionists to explain away data that conflicts with common ancestry.

CONVERGENT EVOLUTION

You may recall the "main assumption," mentioned at the beginning of the chapter, that Darwinian evolutionists use when constructing trees: *similarity implies inheritance from a common ancestor*. But what about situations where that assumption is clearly untrue—i.e., when two organisms share a trait that their supposed common ancestor could not have possessed?

A striking example is the similarity between marsupial and placental "saber-toothed cats," which are classified as very different types of mammals due to their two distinctly different ways of bearing young.

According to current evolutionary theory, the common ancestor of these two cats was a small

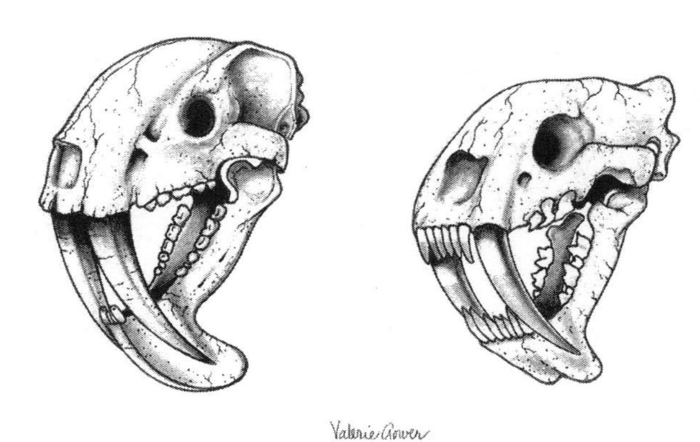

The marsupial saber-toothed cat *Thylacosmilus atrox* (left) has a skull that is very similar to the placental saber-toothed cat *Eusmilus sicarius* (right). Not to scale.

Figure 13-5: "Convergent Evolution" in Saber-Toothed Cats.

rodent-like mammal with a very different body plan. Thus, their highly similar skull structures (see Figure 13-5) had to develop independently and could not have been inherited from a common ancestor.

Common descent does not explain these similarities. Evolutionists try to explain this evidence by claiming these distinctly different cats evolved the same traits independently through **convergent evolution**.

> **Convergent Evolution:**
> ***Two or more species independently acquiring the same trait, supposedly through Darwinian evolution.***

Convergent evolution isn't cited only to explain skeletal morphology. Evolutionists are also often forced to invoke it at the genetic level, where they say different species evolved the same DNA sequences independently.

For example, organisms as diverse as jellyfish, insects, and mammals use the same genes to govern the development of their very different types of eyes, even though their presumed common ancestor couldn't have had anything like a modern eye, if it had eyes at all.[26] Likewise, bats and whales use similar genes for echolocation, but their presumed common ancestor didn't have this capability.[27]

Even evolutionists commonly call these findings "surprising."[28] Why is this? It's difficult enough to evolve a complex structure once. But the odds of evolving a similar feature multiple times, independently, in different lineages, are very low.

Richard Dawkins acknowledges this point, stating "it is vanishingly improbable that exactly the same evolutionary pathway should ever be travelled twice."[29] Yet he admits there are "numerous examples... in which independent lines of evolution appear to have converged, from very different starting points, on what looks very like the same endpoint."[30]

Rather than admitting that convergent evolution provides a challenge to Darwinian claims, Dawkins simply believes "[i]t is all the more striking a testimony to the power of natural selection."[31]

Beyond the fact that it is highly unlikely, convergent evolution presents an even more difficult problem for neo-Darwinian theory. It indicates that biological similarity *does not* necessarily imply common ancestry, challenging the heart of the methodology used to infer common descent.[32] The textbook *Explore Evolution* explains this problem:

> Convergence, by definition, affirms that similar structures do not necessarily point to common ancestry. Even neo-Darwinists acknowledge this. But if similar features can point to having a common ancestor—and to not having a common ancestor—how much does "homology" really tell us about the history of life?[33]

Perhaps the main assumption of phylogenetic trees should be rewritten as: "similarity implies inheritance from a common ancestor, *except when it does not*." Rather than being a helpful solution for neo-Darwinists, convergent evolution undermines the reasoning used to construct phylogenetic trees.

Darwinian evolution is supposed to have no goal, yet convergent evolution implies species are evolving the same complex traits over and over again. Is there a better explanation?

COMING DOWN OUT OF THE TREES

Biology provides numerous examples of highly similar structures shared by different organisms which cannot be explained by common ancestry, but are unlikely to be generated repeatedly by natural selection. What other mechanism could generate similar complex features in diverse species?

ID theorists Jonathan Wells and Paul Nelson think they have an answer to that question:

> An intelligent cause may reuse or redeploy the same module in different systems, without there necessarily being any material or physical connection between those systems. Even more simply, intelligent causes can generate identical patterns independently.[34]

Common design is often the best explanation for cases of supposed convergent evolution. Even when functional biological similarity does fit the "tree" predicted by common ancestry, common design is also a viable explanation.[35]

In the context of human technology, designers regularly reuse parts, programs, or components that work in different designs. For example, engineers use wheels on both cars and airplanes, while technology designers put keyboards on both computers and cell-phones. Why not re-use common blueprints for body plans in different organisms?

However, evolutionary biologists continue to turn a blind eye to the possibility of common design. Attempting to provide an analogy for common descent, evolutionary biologist Tim Berra shows this thinking:

> [I]f you compare a 1953 and a 1954 Corvette, side by side, then a 1954 and a 1955 model, and so on, the descent with modification is overwhelmingly obvious.[36]

If anything here is "overwhelmingly obvious," it's that Corvettes did not evolve by unguided natural processes, but were intelligently designed (see Figure 13-6). The type of error that Professor Berra made is so common that the term **Berra's blunder** now describes taking evidence for common design, and mistakenly calling it evidence for common descent.

Illustration: Copyright Jody F. Sjogren 2000. Used with permission.

Figure 13-6: Berra's Blunder and Corvette "Evolution."

CONCLUSION

While textbooks and the media often present evolutionary trees as fact, they are based entirely upon assumptions. Moreover, those assumptions commonly fail, as one gene may produce one version of the tree of life, while another gene yields a totally different tree. There are a multitude of biological similarities among different species that cannot be explained by common descent.

A good scientific hypothesis is one that can be tested. But under the thinking that evolutionary biologists use to build phylogenetic trees, there is essentially no way to conclusively test universal common ancestry.

First, evolutionists use circular logic by *assuming* that similarity implies inheritance from a common ancestor. This thinking assumes the truth of the argument without objectively demonstrating common ancestry was the cause.

Second, when the main assumption fails, evolutionists fall back on explanations like convergent evolution or horizontal gene transfer.[37] But these claims often push materialist explanations beyond belief.

The truth is that common ancestry *is* testable, but perhaps evolutionary biologists prefer not to test it for fear it might fail. Proponents of common descent are forced into reasoning that similarity implies common ancestry, except for when it doesn't. The one assumption most are not willing to question is universal common ancestry itself.

Perhaps different genes are telling different evolutionary stories because all organisms are not ancestrally related. For those open-minded enough to consider it, common design may be a superior explanation to common descent.

In the coming chapters, we will continue analyzing the most prominent arguments offered to support common ancestry.

DISCUSSION QUESTIONS

1. What is the main assumption used when building phylogenetic trees? Is this assumption valid?

2. Give an example of a component that is re-used in different human designs. Why is that component used repeatedly?

3. What is convergent evolution? Why does it cause difficulties for the construction of phylogenetic trees?

4. Is common design a better explanation than natural selection for convergent evolution? Elaborate.

5. When Professor Berra used the changing Corvettes to represent descent with modification, what was his blunder?

6. After reviewing the evidence in this chapter, what do you think about universal common ancestry versus common design? Discuss.

7. If *universal* common ancestry is not valid, is it possible that some lower taxonomic groups are still related? (For example, are all felines related to one another?) Justify your explanation with examples.

CHAPTER 14

FAKES AND MISTAKES

Question:

Have you ever found a false or misleading claim made by the media or in a textbook?

FAKES AND MISTAKES

"False facts are highly injurious to the progress of science, for they often endure long."

—Charles Darwin[1]

One of the authors of this book grew up next-door to a curmudgeonly but friendly retired dentist named Max. Before studying dentistry, Max studied biology at UC Berkeley in the earlier part of the twentieth century. He was also a strong believer in Darwinian evolution. When asked why, Max exclaimed: "Because ontogeny recapitulates phylogeny!"

ONTOGENY RECAPITA-WHAT?

Max was referring to **recapitulation theory**, which was popularized in 1866 by the influential German biologist Ernst Haeckel—a follower of the ideas of Charles Darwin. The theory holds that the development of an organism (ontogeny) replays (recapitulates) its evolutionary history (phylogeny).

The theory is based on the belief that, over eons of time, certain fish evolved into amphibians, some of which evolved into reptiles. In turn, some reptiles evolved into birds, and others into mammals. Recapitulation theory holds that each **embryo** of higher life forms replays its presumed ancestral history as it develops. In other words, at one point between your conception and birth, you resembled a fish.

For decades, this concept was taught to biology students as fact, although it has been known to be false for well over a hundred years. The evidence shows that vertebrate embryos do not replay their supposed earlier evolutionary stages.[2] Nonetheless, there are many people like Max who were taught recapitulation theory in school—and still believe it's true.

Anyone can inadvertently pass along a bad idea, right? However, there is a darker side to recapitulation theory. Not only is the concept wrong, but it has been promoted through fraudulent embryo drawings concocted by Haeckel (see Figure 14-1). According to the journal *Science*, "[g]enerations of biology students may have been misled" by **Haeckel's embryo drawings**.[3]

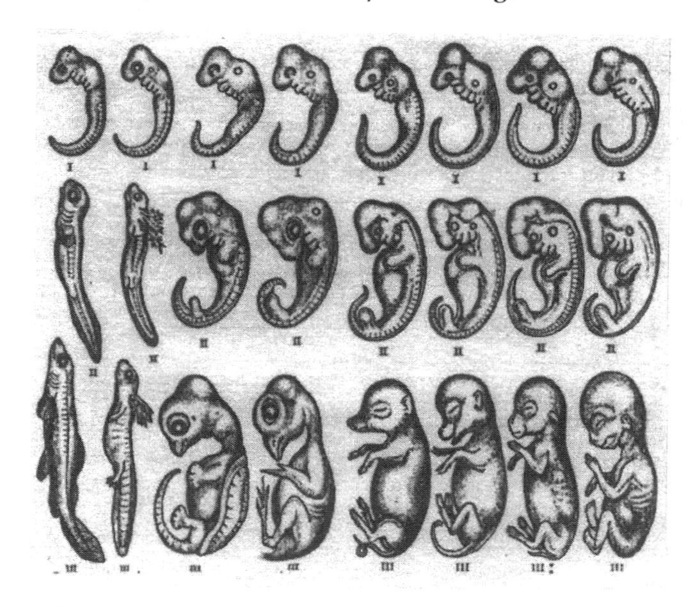

Figure 14-1: Haeckel's Embryo Drawings.

Haeckel's drawings obscured the differences between the earliest stages of embryonic development, making the embryos look more similar than they actually are. A leading developmental biologist has called the drawings "one of the most famous fakes in biology."[4]

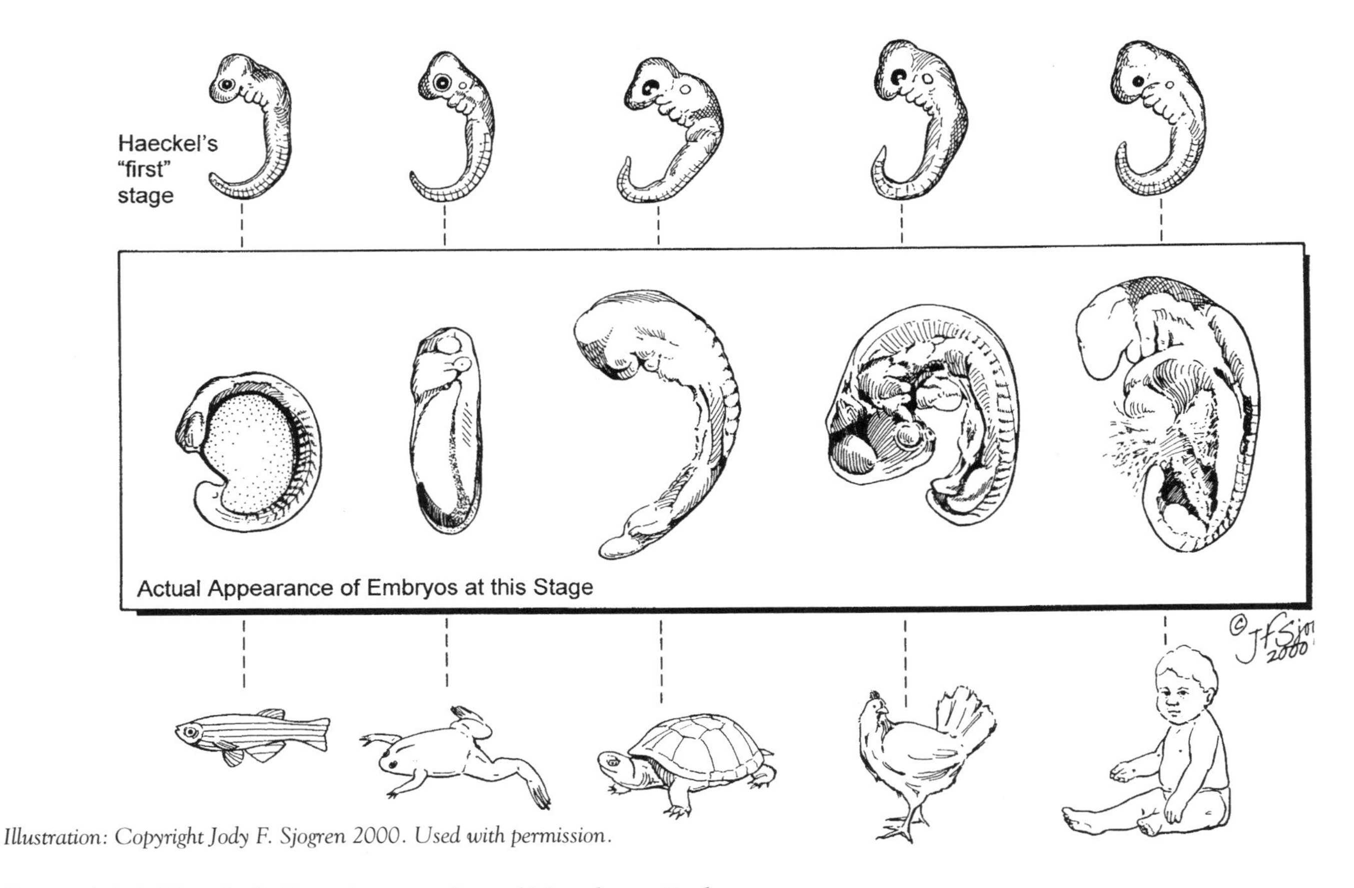

Illustration: Copyright Jody F. Sjogren 2000. Used with permission.

Figure 14-2: Haeckel's Drawings vs. Actual Vertebrate Embryos.

Figure 14-2 compares Haeckel's drawings with embryos that have been accurately depicted.

Recapitulation theory has been removed from most (though not all[5]) modern biology textbooks, and it no longer serves as an important concept in evolutionary biology. But what about Haeckel's drawings?

IMPLAUSIBLE DENIABILITY

In 2008, the *New York Times* reprinted material from the National Center for Science Education, admitting that Haeckel's drawings were "long-discredited" but claiming they have not been used in textbooks since "20 years ago."[6] The reality is that multiple recent biology textbooks have used Haeckel's drawings to promote evolution.[7]

Embryologist Michael K. Richardson acknowledged that there are "at least fifty recent biology textbooks which use the drawings uncritically"[8] and admitted the drawings "are still widely reproduced in textbooks and review articles, and continue to exert a significant influence on the development of ideas in this field."[9] This led Stephen Jay Gould to exclaim:

> [W]e do, I think, have the right to be both astonished and ashamed by the century of mindless recycling that has led to the persistence of these drawings in a large number, if not a majority, of modern textbooks![10]

Haeckel's drawings are often replaced by other drawings or misleading photographs, but the effect is still the same: students are not being told the whole truth about the evidence.

Textbooks still commonly make two inaccurate

DVD SEGMENT 11
"Haeckel's Embryos"
(2:05)

From:
Icons of Evolution

claims about vertebrate embryos.

- To support common ancestry, they overstate similarities between embryos during their earliest stages of development.

- They allege that vertebrate embryos contain gill slits reflecting fish ancestry.

GILL SLITS

One feature exaggerated in Haeckel's illustrations was folds of skin on the neck of the first depicted stage of embryos. Modern textbooks continue to use inaccurate drawings and claim that the folds are actually gill slits, citing them as evidence of our alleged fish ancestry. But biologist Jonathan Wells explains in *The American Biology Teacher* that this is not correct:

> [H]uman embryos do not really have gills or gill slits: like all vertebrate embryos at one stage in their development, they possess a series of "pharyngeal pouches," or tiny ridges in the neck region. In fish embryos these actually go on to form gills, but in other vertebrates they develop into unrelated structures such as the inner ear and parathyroid gland. The embryos of reptiles, birds and mammals never possess gills.[11]

OVERSTATING SIMILARITIES

Textbooks use Haeckel's inaccurate illustrations or other misleading illustrations or photos to claim that vertebrate embryos are highly similar in their early stages, and that these similarities demonstrate the embryos' common ancestry.[12] But this too is incorrect, for two reasons:

- As shown earlier, Haeckel's drawings and similar illustrations exaggerate or invent similarities.

- The misleading drawings and photos do not uniformly depict an early stage of development. Embryologists recognize that vertebrate embryos actually start development quite differently.[13]

As a paper in *Nature* stated, "Counter to the expectations of early embryonic [similarities], many studies have shown that there is often remarkable divergence between related species both early and late in development."[14] Or, as one article in *Trends in Ecology and Evolution* stated, "despite repeated assertions of the uniformity of early embryos within members of a phylum, development before the phylotypic stage is very varied."[15] But what is this midpoint phylotypic stage, and does it provide evidence for common descent?

EMBRYONIC HOURGLASS

Accepting that Haeckel was wrong, a new theory emerged in the twentieth century called the **developmental hourglass theory**. This model of vertebrate development acknowledges that vertebrate embryos start differently, but claims that they pass through a highly similar stage midway through development. Called the **phylotypic**, or **pharyngula stage**, its similarities are said to provide evidence for common ancestry.

Again, the data points in a different direction. As shown Figure 14-3, there are minor similarities but overall the embryos are quite different. In recent years some embryologists have come to doubt that the pharyngula stage of vertebrate

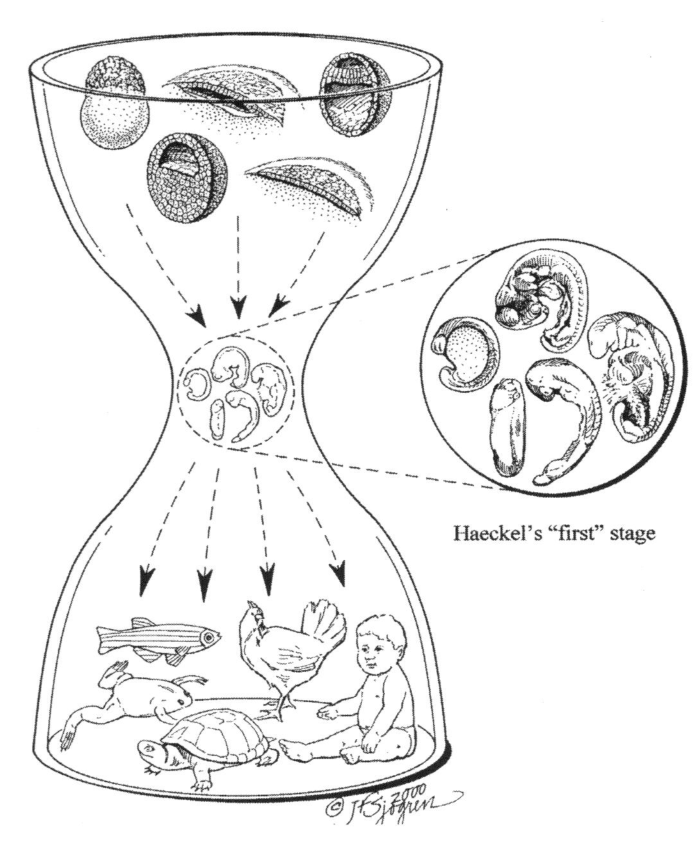

Illustration: Copyright Jody F. Sjogren 2000. Used with permission.

Figure 14-3: Embryonic Hourglass.

development even exists. One paper found that during this purportedly similar stage, vertebrate embryos show diversity in:

- Body size.
- Body plan.
- Growth patterns.
- Timing of development.[16]

The researchers conclude that the evidence is thus "[c]ontrary to the evolutionary hourglass model" and "difficult to reconcile" with the existence of a pharyngula stage.[17]

THE PEPPERED MYTH

Unfortunately, Haeckel's embryo drawings are not the only example of biology textbooks making incorrect claims about the evidence for evolution.

In England (and many other parts of the world) there is a species of **peppered moths,** which have black and white speckled wing spots (Figure 14-4). Some members of the species have more black scales and are darker than their "sister" moths, which have more white scales and a lighter color. Whether we're talking about the lighter or the darker form, they all belong to the same species.

Figure 14-4: A Peppered Moth.

According to the evolutionary story taught in many textbooks, originally there was a predominance of lighter-colored moths because they blended in with the light-colored lichen on tree trunks. This camouflage allowed the lighter moths to avoid being seen and eaten by birds.

However, during Britain's industrial revolution, airborne pollutants killed the lichens and darkened the tree trunks with soot. The darker forms of the moth now had a selective advantage, as they could blend into the tree trunks and avoid becoming a tasty bird snack.

In Chapter 2, we made a distinction between micro- and macroevolution. The change in moth coloration is an example of small-scale change—microevolution. All that changed was the relative frequency of light- and dark-colored moths, both of which existed before the Industrial Revolution began. If the peppered moth story is true, all it demonstrates is microevolution.

We'll say more about microevolution later in this chapter, but there are other problems with the conclusions drawn from the moth studies.

For example, there is controversy over whether peppered moths normally rest on tree trunks in the wild. Many textbook photographs showing peppered moths on tree trunks are staged. In what biologist Jonathan Wells called "unnatural selection,"[18] experimenters artificially glued moths to tree trunks to determine if they would be eaten by birds. Some experts have suggested that the moths typically rest in hidden locations where they are out of sight from hungry birds.

While camouflage to avoid predation by birds may still be a factor in determining moth coloration frequencies, Wells explains that the evidence is inconclusive and textbooks are inaccurate: "In particular, it is misleading to illustrate the story with photographs showing moths on tree trunks where they do not rest in the wild. Our students deserve better."[19]

According to researchers, environmental laws reduced the amount of pollution, and selection now favors the light-colored moths, which have once again come to dominate the population. Therefore, the peppered moth is now becoming, at best, an example of oscillating selection, where there is no net evolutionary change over time.

SON OF A FINCH

Another popular icon of evolution is the beak size in Galápagos finches.[20] Biology textbooks commonly cite these finches as an example of **adaptive radiation**.

> **Adaptive Radiation:**
> ***The supposed rapid diversification of species after entering an empty habitat, or niche.***

According to the standard finch story, about 14 million years ago a group of finches found their way from the South American mainland to the **Galápagos Islands**. Over time, they diversified into about a dozen species that inhabit various habitats in the island group. One textbook boasts that the finches provide one of "the most important studies of natural selection in action."[21]

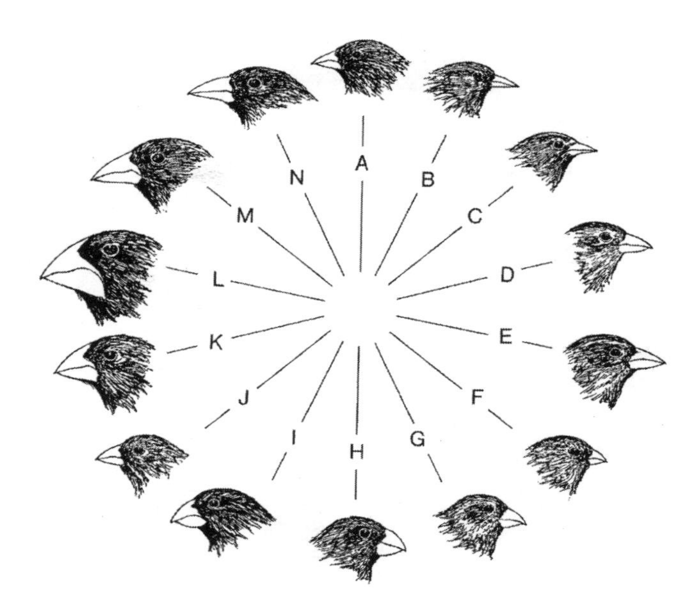

Illustration: Copyright Jody F. Sjogren 2000. Used with permission.

Figure 14-5: Darwin's Finches.

But what exactly do the finches show natural selection can accomplish? Figure 14-5 shows sketches of the various Galápagos finch species. Aside from small differences in beak shape, size, and feeding habits, the finches are highly similar. In fact, it can be difficult to tell them apart.

In his Pulitzer Prize-winning book *The Beak of the Finch*, Jonathan Weiner compares the largest and smallest species of Galápagos ground finch and remarks that they are "almost indistinguishable."[22] A paper in *BioScience* noted that after a full 14 million years of evolution, the finches remain highly similar and even "retain the ability to interbreed and produce viable, fertile hybrids."[23]

The small-scale differences among the finch species do not demonstrate that natural selection can cause large-scale evolutionary changes, such as the origin of new body plans. Rather, the finches show that after millions of years of evolution, very little has changed.

If anything, the Galápagos finches demonstrate the limits of natural selection. A famous study by the field biologists Peter and Rosemary Grant found finches with larger beaks survived better during a drought, as they were able to crack the tougher seeds that remained. When the drought ended, the average beak size in the finch population returned to normal.[24]

Just like the peppered moths, Galápagos finches only provide an example of oscillating selection, with no net evolutionary change.

DVD SEGMENT 12

"Darwin's Finches" (3:10)

From:
Icons of Evolution

THE EVOLUTION BAIT-AND-SWITCH

In Chapter 2, we introduced the concepts of microevolution and macroevolution. Microevolution generally refers to small-scale changes within an existing species. Macroevolution refers to large-scale changes, including the evolution of higher taxa and fundamentally new biological features.

Microevolution is not a subject of debate because there is much evidence of small-scale changes within species over time. However, proponents of Darwinism commonly pull an evolutionary "bait-and-switch," citing small-scale changes in the colors of moths or the sizes of finch beaks, and then extrapolating to claim that fundamentally new types of organism can evolve.

Some scientists doubt that such extrapolations are warranted. For example, even if the peppered moth story turns out to be true, at best it would document microevolution. This led evolutionary biologist Robert L. Carroll to ask:

> Can changes in individual characters, such as the relative frequency of genes for light and dark wing color in moths adapting to industrial pollution, simply be multiplied over time to account for the origin of moths...?[25]

Many scientists think the answer to that question is "no." They believe that Darwinian scientists are greatly over-extrapolating from the available data. According to the prominent ID thinker Phillip Johnson, "When our leading scientists have to resort to the sort of distortion that would land a stock promoter in jail, you know they are in trouble."[26]

Studies about natural selection consistently show only small-scale changes. For example, in 2009, the journal *Nature* celebrated Darwin's 200th birthday by publishing a collection of "evidence for evolution by natural selection."[27] Yet the studies cited showed trivial degrees of change, such as a minimal variation in the size of a fish species, a slight increase in the number of eggs for a bird population, or small changes in the coloration of spots on guppies—all clearly examples of microevolution.

Evolutionists may respond by saying that macroevolution simply occurs after many repeated rounds of microevolutionary change. However, science must be based upon **empirical** data, not unwarranted extrapolation. Observed examples of natural selection in action provide evidence of small-scale—not large-scale—evolutionary change.

Some leading scientists doubt that microevolution can be extrapolated to explain macroevolutionary change. One paper in *International Journal of Developmental Biology* asked if macroevolution is "just cumulative microevolutionary changes?" The paper concluded, "microevolution does not provide a satisfactory explanation for the extraordinary burst of novelty during the Cambrian Explosion"[28]—a topic we will cover in more detail in the next chapter.

CONCLUSION

Regardless of exactly how we define micro- or macroevolution, Darwinian evolution has limits. As Darwin himself admitted, "if any complex organ existed which could not possibly have been formed by numerous, successive, slight modifications, my theory would absolutely break down."[29]

Modern evolutionists don't seem interested in testing Darwin's challenge. The empirical examples of natural selection offered by textbooks and science journals represent unimpressive, small-scale microevolutionary change. They wrongly extrapolate from these examples to claim that natural selection can produce vast biological diversity.

ID proponents take a more rigorous approach that seeks to test what can, or cannot, be produced by Darwinian mechanisms. As we have seen, there are many structures that meet Darwin's challenge and defy a Darwinian explanation.

In his book *Icons of Evolution*, Jonathan Wells asks, "If there is such overwhelming evidence for Darwinian evolution, why do our biology textbooks, science magazines, and television nature documentaries keep recycling the same old tired myths?"[30]

The fact that these icons of evolution—finch beak sizes, moth colors, or exaggerated embryo drawings—are so weak, and yet so commonly used, should tell us something. If evolutionary biologists had better evidence, we would know about it.

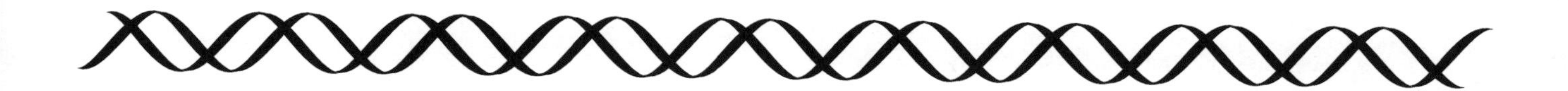

DISCUSSION QUESTIONS

1. Materialists often try to show that microevolution demonstrates the validity of macroevolution. Do you think they accomplish their objective? Discuss why or why not.

2. The embryo drawings by Ernst Haeckel have been known to be erroneous for many years. Why do you think they are still sometimes used? Do you think they help or hurt the case for common ancestry?

3. Suppose that the peppered moth story turns out to be correct, and birds in fact eat moths off tree trunks, preferring those without camouflage. Would this demonstrate macroevolution?

4. The Galápagos finches have long been used as examples of natural selection. Do they provide good support for Darwin's theory? What are the strengths and weaknesses of the finch evidence?

CHAPTER 15

SUDDEN, GRADUAL CHANGE

Question:

Do you think the fossil evidence supports Darwinian evolution?

SUDDEN, GRADUAL CHANGE

"Sit down before a fact as a little child, be prepared to give up every preconceived notion, follow humbly wherever and to whatever abysses nature leads, or you shall learn nothing."

—Evolutionary biologist Thomas Henry Huxley[1]

The hallmark of Darwinian evolution is gradual change (Figure 15-1). In his *Origin of Species* Darwin proposed that all organisms evolved via descent with modification, and predicted that "[t]he number of intermediate varieties, which have formerly existed on the earth, [must] be truly enormous."[2]

However, there was a problem with the evidence for common descent. Darwin himself acknowledged that the fossil record did not contain these intermediate forms of life:

> Why then is not every geological formation and every stratum full of such intermediate links? Geology assuredly does not reveal any such finely graduated organic chain; and this, perhaps, is the most obvious and gravest objection which can be urged against my theory.[3]

Fossils are the preserved remains of dead organisms. A lack of fossil evidence for evolutionary transitions presents a critical problem for materialists. It bears directly on one of their fundamental tenets.

MATERIALISM TENET #7:

All living organisms are related through universal common ancestry.

Although Darwin acknowledged the missing intermediate links, he considered the problem temporary and anticipated that such links would eventually be found.

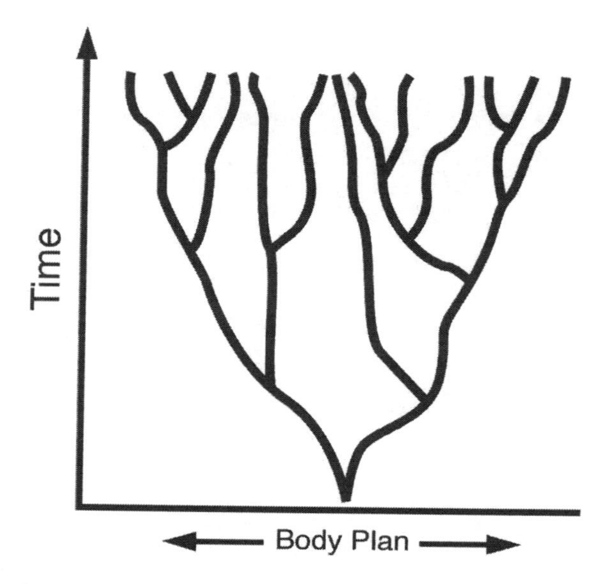

Under Darwin's theory of evolution, one species turns into another through slow, smooth, gradual changes. Therefore, many transitional organisms should have existed.

Figure 15-1: Darwinian Gradual Evolution.

IMPERFECT EXCUSE

Darwin attempted to save his theory of gradual evolution by claiming that intermediate fossils are not found because of "the extreme imperfection of the geological record."[4] In other words, if the record were simply more complete, we would not lack transitional fossils. Paleontologists today call this the **artifact hypothesis**.

In recent decades, however, leading paleontologists have recognized that gaps between major taxonomic categories are not simply the result of an incomplete fossil record. Rather, many of the gaps are real and are unlikely to significantly change based on new discoveries.[5]

A study published in *Nature* found that among the higher taxonomic categories (family and above), "the past 540 million years of the fossil record provide uniformly good documentation of the life of the past."[6] Likewise, an article in the journal *Science* warned that "[e]volutionary biologists can no longer ignore the fossil record on the ground that it is imperfect."[7]

Despite the completeness of the fossil record, only a tiny fraction of known fossil species are claimed to be **transitional forms**. As paleontologist and committed evolutionist Stephen Jay Gould stated:

> The absence of fossil evidence for intermediary stages between major transitions in organic design... has been a persistent and nagging problem for gradualistic accounts of evolution.[8]

Occasionally the media goes into a frenzy claiming that some newly discovered fossil was an intermediate form. Inevitably, questions arise about the validity of the claims, and alleged transitional forms are often later disproven. In the next chapter, we will discuss some of the more widely-publicized examples of presumed transitional forms.

The scarcity of transitional fossils is problematic for Darwinian theory, but it points to another aspect of the fossil record that is very difficult for evolutionary thinking to explain. The record consistently shows a pattern where new forms come into existence abruptly, which many have called "explosions" in the history of life.

EXPLOSIONS OF LIFE

As explained in Chapter 4, current scientific beliefs hold that after the formation of our solar system there was no life on Earth for several hundred million years. For about the next 3 billion years, single-celled organisms dominated our planet. Then suddenly, during the Cambrian period about 530 million years ago, nearly all of the major living animal phyla appear in the fossil record (see Figure 15-2). This landmark event in the history of life is called the Cambrian explosion, and it took place within a geological eye blink—5 to 10 million years (or possibly much less).

Figure 15-2: History of Time.

Event	*Billions of Years Ago*
Big Bang	**13.7**
Formation of our solar system	**4.5**
Earth becomes habitable	**3.8**
Cambrian explosion	**0.53**

Before the Cambrian, very few fossils having anything to do with modern phyla are found in the fossil record (see Figure 15-3). As one invertebrate zoology textbook states:

> Most of the animal phyla that are represented in the fossil record first appear, "fully formed" and identifiable as to their phylum, in the Cambrian some 550 million years ago.... The fossil record is therefore of no help with respect to understanding the origin and early diversification of the various animal phyla...[9]

Even though problems with the artifact hypothesis are widely known, evolutionary scientists still commonly invoke it in trying to make sense of the Cambrian explosion. They

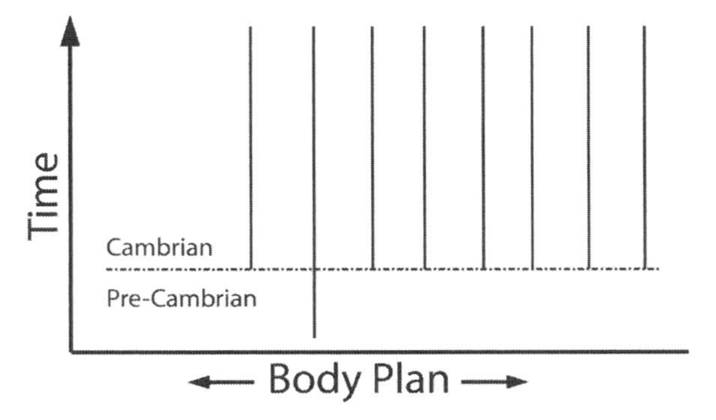

Vertical lines indicate the actual fossil record of the phyla. The dotted line marks the Cambrian / Precambrian boundary.

Figure 15-3: The Cambrian Explosion.

postulate that the Precambrian fossils were simply too small, or too soft-bodied, to be preserved.

This explanation has severe deficiencies. In particular, small and soft-bodied organisms were commonly fossilized, and in fact the Cambrian explosion is full of them.

Additionally, scientists *have* discovered a few fossils in the Precambrian—but they are not considered to be ancestors to the complex animal phyla that appear in the Cambrian. Indeed, Precambrian sponge embryos have been discovered, demonstrating that if small, soft-bodied transitional organisms existed, they could have been fossilized.

The Precambrian fossil record and its failure to document the evolution of the Cambrian fauna are discussed in more detail in Appendix E. The Cambrian explosion is explained further in the following DVD segment.

In addition to the Cambrian, there are other examples of explosions in the history of life. For example:

- Most major fish groups appear abruptly.[10]
- Plant biologists observe that the initial appearance of many land plants "is the terrestrial equivalent of the much-debated Cambrian 'explosion' of marine faunas."[11]
- Later in the fossil record there is an explosion of flowering plants, sometimes referred to as the "big bloom."[12]
- Vertebrate paleontologists believe there was a mammal explosion with the abrupt appearance of many orders of mammals. Niles Eldredge, an evolutionary paleontologist and curator at the American Museum of Natural History, explains that "there are all sorts of gaps: absence of gradationally intermediate 'transitional' forms between species, but also between larger groups—between, say, families of carnivores, or the orders of mammals."[13]
- There is also a bird explosion, with major bird groups appearing in a short time period.[14]
- As we will discuss in Chapter 17, some have even described the abrupt origin of our own genus, *Homo*, as an explosion.

The sudden appearance of new forms tends to be the rule, rather than the exception, in the fossil record. As one zoology textbook explains:

> Many species remain virtually unchanged for millions of years, then suddenly disappear to be replaced by a quite different, but related, form. Moreover, most major groups of animals appear abruptly in the fossil record, fully formed, and with no fossils yet discovered that form a transition from their parent group.[15]

Because the fossil record has not confirmed Darwin's predictions of gradual evolution, some scientists have sought to find a model where "changes in populations might occur too rapidly to leave many transitional fossils."[16]

PUNC EQ

In 1972, Stephen Jay Gould and his colleague Niles Eldredge proposed the theory of **punctuated equilibrium** (often referred to as "punc eq"). According to this theory, most evolution takes place in small populations over relatively short geological time periods[17] (see Figure 15-4). These hypothetical periods of rapid change are

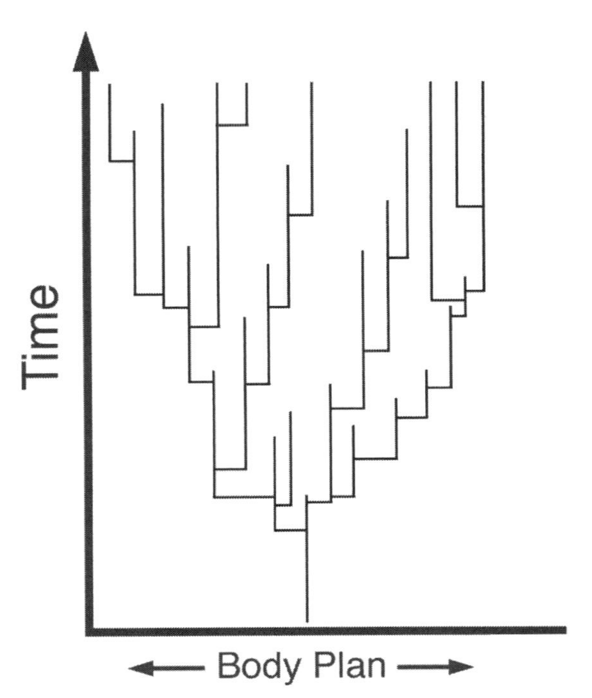

Under the punc eq model, evolution is said to occur over short time periods in small populations, so transitional forms are unlikely to be fossilized.

Figure 15-4: Punctuated Equilibrium.

interspersed between long time spans without much change, called **stasis**.

By proposing reduced transitional population sizes and time spans, punc eq drastically limits the expected number of transitional organisms. Since most individuals do not fossilize, the likelihood of finding transitional forms is decreased.

It was hoped the theory might account for the abrupt appearance of new species, as in the Cambrian explosion, and the lack of transitional forms in the fossil record. However, we must ask whether the punc eq model requires too much genetic change too fast.

Two important aspects of Darwin's theory include:

- Evolving populations might be quite large, increasing the chances that favorable variations will occur.[18]
- Long time periods would be available for the origin of biological complexity.

A basic problem with the punc eq model is that it deprives Darwinism of these two advantages. It compresses the vast majority of evolutionary change into small populations that lived during a small segment of time, allowing too few rolls of the dice for beneficial mutations to arise.

Debates over punc eq have led to bitter disputes within the evolutionary community. Paleontologists, because they are acutely aware of problems with the fossil evidence, have favored this model. But some influential biologists oppose it because it directly conflicts with gradualism—a basic premise of Darwinian theory.

Critics of punc eq have observed that the "genetic mechanisms that have been proposed to explain the abrupt appearance and prolonged stasis of many species are conspicuously lacking in empirical support."[19] Less civil critics called the model "evolution by jerks."[20]

Others, however, have sought a magic bullet mechanism at the genetic level to account for rapid biological change. Many have hitched their hopes to a field called **evolutionary developmental biology** (often called "**evo-devo**"), which claims that evolution proceeds by mutations in the genes controlling the development of an organism.

EVO-DEVO TO THE RESCUE?

Proponents of evo-devo contend that changes to the master genes that control the development of an organism, such as *Hox* or *homeobox* genes, can cause large, abrupt changes in body plans. But there are at least three major problems with this hypothesis.

First, developmental genes are tightly interconnected. Changes in one will affect many others, making most mutations lethal. As one writer in the journal *Nature* cautioned, "macromutations of this sort are probably frequently maladaptive."[21]

Second, ***Hox* genes** do not encode proteins that build body parts—they merely *direct* the genes that encode body parts. At most, *Hox* mutations can only rearrange parts that are already there; they cannot create truly novel structures:

> [H]omeobox genes are selector genes. They can do nothing if the genes regulated by them are not there.... It is totally wrong to imply that an eye could be produced by a macromutation when no eye was ever present in the lineage before.[22]

Third, despite years of research, the best examples of evolutionary change produced by evo-devo mechanisms are meager. They often entail loss, rather than gain of function.[23] Some of the most impressive evidence for evo-devo amounts to changes in coloration spots on the wings of flying insects.[24] Obviously such small-scale changes will not produce new body plans.

Those who seek to explain how large changes in an organism evolve by genetic mutations are faced with a problem: major mutations are not viable, but viable mutations are not major. The textbook *Explore Evolution* explains:

> Small, limited mutations (like those that produce antibiotic resistance) can be beneficial in certain environments, but they don't produce *enough* change to produce fundamentally new forms of life. Major mutations can fundamentally alter an animal's anatomy and structure, but these mutations are always harmful or outright lethal.[25]

It seems that instead of firing magic bullets, evo-devo is shooting blanks.

DEMAND EVIDENCE

The failure of evo-devo creates major difficulties for punc eq. Although punc eq has devotees among paleontologists (and some biologists), it requires too much biological change in too little time. No known genetic mechanism can account for this rapid change.

What is more, punc eq can appear to be simply an excuse for why transitional fossils are missing.[26] Would you believe someone who claimed to capture fairies and Leprechauns on video, but when asked to produce the film, declares "well, they are on camera but they are too small or too fast to be seen"?

Molecular biologist Michael Denton critiques punc eq by arguing:

> [U]nless we are to believe in miracles... such transitions must have involved long lineages including many collateral lines of hundreds or probably thousands of transitional species.[27]

Evidence supporting the predictions of Darwinian theory—gradual change and universal common ancestry—is lacking in the fossil record. How does intelligent design explain this data?

INTELLIGENT DESIGN AND THE FOSSIL RECORD

As this chapter has discussed, the history of life shows a pattern of explosions where new fossil forms come into existence without clear evolutionary precursors. The abrupt appearance of fully formed body plans in the fossil record requires a cause that can rapidly infuse massive amounts of information to encode and integrate the necessary proteins, cell types, tissues, and organs.

There is only one known cause that can accomplish these tasks: intelligence. Specifically:

- Design theorists have observed that intelligent agents are uniquely capable of rapidly infusing large amounts of information into the biosphere.[28]
- Only intelligence can produce the complex and specified information necessary to

coordinate many levels of organization into a functional body plan.[29]

- Only intelligence can conceive of a fully formed blueprint before implementing the design in the world.[30]

CONCLUSION

As this chapter has shown, the fossil record does not bear out the predictions and expectations of neo-Darwinian evolution. Additionally, there are severe problems with punc eq and evo-devo models that indicate that evolution lacks a mechanism to rapidly generate new biological information.

Instead, the history of life shows a pattern of explosions where new fossil forms come into existence without clear evolutionary precursors. The fossil record supports ID's prediction that species might appear abruptly, indicating the rapid infusion of new information into the natural world.

Design theorists have observed that intelligent agents are capable of doing just that—using large amounts of information to create blueprints, which lead to fully functional machines and new types of organism.

The quotation with which we opened this chapter, from the famous evolutionary biologist, Thomas Henry Huxley, suggested that good scientists throw away preconceptions and follow the evidence where it leads. What would scientists discover if they simply took the fossil record at face value and did not try to force-fit it into a model of neo-Darwinian evolution?

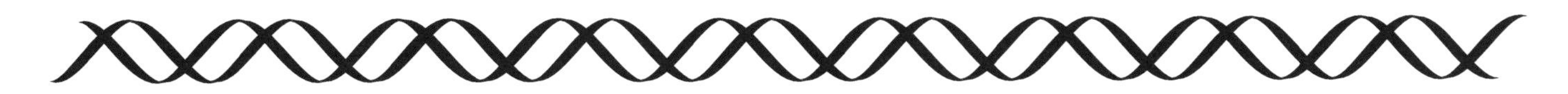

DISCUSSION QUESTIONS

1. What is the predominant pattern of change observed in the fossil record? How does it affect Darwin's theory of slow, gradual change?

2. When the fossil record did not produce the transitional forms predicted by Darwinian theory, many evolutionists claimed the reason was due to the incompleteness of the fossil record. Do you agree with their explanation? Why or why not?

3. What is the significance of the dramatic appearance of Cambrian fossils for the debate over intelligent design and evolution? Is this the only instance of the explosive appearance of new life-forms?

4. Punctuated equilibrium and evo-devo are models used to explain the abrupt appearance of new life-forms in the fossil record. Evaluate the strengths and weaknesses of these explanations.

5. A good scientific theory is one that predicts that confirming evidence will be discovered. Does punc eq predict such evidence in the fossil record?

6. How can intelligent design explain the abrupt appearance of body plans in the fossil record?

CHAPTER 16

SKELETONS IN THEIR CLOSET

Question:

When you hear a media story about the new discovery of a "missing link," do you tend to believe what you hear, or are you skeptical?

SKELETONS IN THEIR CLOSET

"To take a line of fossils and claim that they represent a lineage is not a scientific hypothesis that can be tested, but an assertion that carries the same validity as a bedtime story—amusing, perhaps even instructive, but not scientific." **—Nature Magazine senior editor Henry Gee**[1]

May 2009 saw a carefully orchestrated public relations campaign involving leading paleontologists, top TV networks, some of the Internet's most popular websites, and numerous other media outlets in a coordinated effort to promote evolution to the public. A fossil primate, dubbed "**Ida**" by her discoverers, was introduced by the media as the "eighth wonder of the world"[2] whose "impact on the world of palaeontology" would be like "an asteroid falling down to Earth."[3]

Famed BBC broadcaster Sir David Attenborough got involved in the campaign, making a documentary titled *Uncovering Our Earliest Ancestor: The Link*, which said Ida was "the link that connects us directly with the rest of the animal kingdom."[4]

Why is such hype necessary to promote so-called "missing links" to the public? Because, as we learned in Chapter 15, the fossil record has not been kind to gradualistic theories of Darwinian evolution by natural selection. Evolutionists would welcome some good news for a change.

Was Ida the "link" that finally connected us to distant ancestors? We'll answer that question at the end of the chapter, but first we must review a number of other presumed "missing links"[5] that have been widely discussed since Darwin wrote his famous book. This chapter will investigate the alleged transitions from:

- Fish to amphibians.
- Reptiles to birds.
- Land mammals to whales.

We will also look at the hypothesized horse lineage and the manner in which some fossils have been manipulated in an attempt to support an invalid evolutionary hypothesis.

A FISH WITH A WRIST?

The first four-legged vertebrates (i.e., **tetrapods**) found in the fossil record were amphibians—animals that spend their lives both walking on land and swimming in the water.

Amphibians, such as frogs, newts and salamanders, share a few traits with other tetrapods, as well as with fish. Likewise, some living fish can walk out of the water on their fins and breathe air. Does that mean they are transitional forms? Not in the least.

These organisms have lived a land-plus-water lifestyle for eons without evolving into anything other than more fish, frogs, newts, and salamanders—indicating that an organism can share traits with multiple groups and yet not necessarily be part of an evolutionary transition.

But what about a possible transition between fish and amphibians? Is there any evidence that such an animal has existed?

If you pay attention to bumper stickers, you have undoubtedly encountered the "Darwin fish"—a hypothetical fish with legs and feet. In 2004, some paleontologists discovered a fossil in the Canadian Arctic they said was an actual Darwin fish, named ***Tiktaalik roseae***.

Much like Ida, *Tiktaalik* has been heavily promoted to the public. It was featured in a PBS / NOVA documentary, multiple *New York Times* articles, and *Time* magazine, which called it the evolutionists' "Exhibit A in their long-running debate with creationists."[6] Evolutionary biologist Jerry Coyne praised the fossil as "[o]ne of the greatest fulfilled predictions of evolutionary biology."[7]

If true, such a transitional form would have to bridge the significant differences between fish and amphibians. The most obvious difference is that fish have fins, whereas tetrapods do not. Nearly all tetrapods (including amphibians) have legs.[8]

What caused Coyne to praise *Tiktaalik* as such a success story for evolution? There are two arguments—both of which fail on closer inspection.

First, evolutionary biologist Neil Shubin, who co-discovered *Tiktaalik*, called the fossil "a fish with a wrist."[9] However, a closer inspection shows that it had fishlike fins, a very different skeletal structure from a tetrapod leg, and no wrist bones.[10] Two leading experts acknowledged these large differences, admitting there "remains a large morphological gap"[11] between the fins of *Tiktaalik* and the limbs in tetrapods.

Second, Coyne believed that *Tiktaalik* (dated at about 375 million years ago) was found between the supposed fish ancestors of tetrapods and the first true tetrapods in the fossil record. As Shubin put it, the fossil was purportedly found in "rocks of just the right age."[12] (See Figure 16-1.) However, in 2010 this claim fell apart when tetrapod tracks were discovered in Poland that predated *Tiktaalik* by about 20 million years.[13]

The problem is this: *Tiktaalik's* presumed tetrapod descendants now appear 20 million years before *Tiktaalik* itself appears—far from "just the right age." As one leading scientist put it, "If—as the Polish footprints show—tetrapods already existed" millions of years before *Tiktaalik*, "then an enormous evolutionary void has opened beneath our feet."[14]

Contrary to the massive public relations campaign for *Tiktaalik*, the fossil evidence does not support the Darwinian claim that it was a transitional intermediate between fish and tetrapods.

DINO-BIRD?

Most evolutionists claim birds descended from reptiles—specifically a group of dinosaurs called **theropods**.

> **Theropod:**
> ***A type of carnivorous, bipedal dinosaur with small forelimbs.***

Figure 16-1: Approximate Geological Timeline of Early Tetrapods.

Fossil	Millions of years ago[15]
Tetrapod tracks (reported in 2010)	397
Hypothetical fish ancestors of tetrapods	390
Tiktaalik roseae	375
Tetrapod body fossils	360

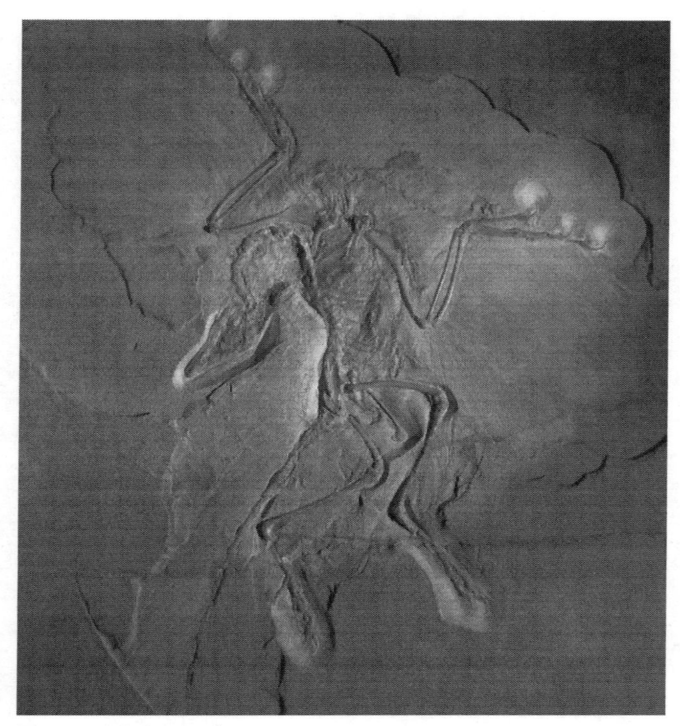

Figure 16-2: Archaeopteryx.

Long before *Tiktaalik* became an icon of evolution, one of the most celebrated presumed examples of a transitional fossil was ***Archaeopteryx***—said to be an intermediate form between theropods and birds (see Figure 16-2).

First discovered in Germany in 1861, *Archaeopteryx* is the earliest known fossil of a bird thought to have been able to fly. Initially, it surprised and excited scientists because it had feathers like birds, but also features generally associated with reptiles—claws on its forelimbs, teeth, and a bony tail. Darwin's *Origin of Species* had been published only two years earlier, and his followers argued that this mix of traits showed a predicted transitional intermediate between reptiles and birds.

However, in the past hundred years, scientists have learned a great deal about reptiles and birds, and the case for *Archaeopteryx* as a transitional form has weakened significantly. While there is no doubt that *Archaeopteryx* represents a bird species with a mosaic of reptilian and bird traits, that mixture does not necessarily mean it descended from reptiles. Let's consider its "reptilian" traits.

Wing claws are functional appendages found on some living birds.[16] For example, in the marshland surrounding the Amazon River lives the **hoatzin**, a bird that resembles *Archaeopteryx* in many respects. Both have a similar body size, small head, long neck, long tail, and shallow breastbone—making them poor flyers. Additionally, like *Archaeopteryx*, the hoatzin is born with claws on its wings.

Far from being useless holdovers from reptilian ancestors, the hoatzin's claws serve an important purpose. Hoatzins nest above water, and if a predator invades the nest, the juvenile will instinctively drop into the water and then use its claws to climb back up to the nest. There are good reasons to suspect that the wing claws of *Archaeopteryx* served a similar climbing purpose.

As for its teeth and bony tail, these traits were not unique to *Archaeopteryx*. Many species of extinct birds known from the fossil record had teeth and/or bony tails. But the fact that prehistoric birds had this mix of traits does not necessarily mean that they were descended from reptiles.

William Dembski and Jonathan Wells liken *Archaeopteryx* to the platypus, a mammal that lays eggs and has a bill like a duck, but "has never been considered a transitional form between birds and mammals."[17] The argument could just as easily be made that such mosaics were designed to include a mix of specific traits necessary for survival.

There are also fossil considerations that challenge the status of *Archaeopteryx* as an evolutionary link. According to vertebrate paleontologist Robert A. Martin, the theropods "all occur in the fossil record after *Archaeopteryx* and so cannot be directly ancestral."[18] Martin also acknowledges a much larger problem, since within the line of reptiles that supposedly led to birds, "no [fossils]... are known that could be a common

Figure 16-3: Reptiles vs. Birds.

REPTILES	BIRDS
Cold-blooded (ectothermic)	**Warm-blooded (endothermic)**
Slow metabolism	**Fast metabolism**
Three-chambered heart[20]	**Four-chambered heart**
No feathers	**Feathers**
Typically abandon young[21]	**Instincts to care for young**
Breathe using diaphragm	**Breathe using air sacs**

ancestor to both dinosaurs and birds."[19]

This doesn't bother evolutionists, who, as we saw in Chapter 15, respond by saying the fossil record is incomplete. Yet nothing in the fossil record indicates an ancestor—theropod or otherwise—to *Archaeopteryx*. So, from what, if anything, did birds evolve?

To explore that question, we need to examine structural differences.

FROM REPTILES TO BIRDS

Proponents of the reptile-to-bird hypothesis must explain transitions between some major differences in the two classes (see Figure 16-3).

A fossil cannot be a true intermediate form unless the larger overall story of a supposed Darwinian evolutionary transition makes sense. For example:

- The alleged ancestors ought to appear *before* the descendants in the fossil record.
- There must be sufficient time allowed in the fossil record for all the necessary changes to evolve.
- The transitional organisms must be able to survive, and even gain some advantage, while evolving from one body plan to another.

The third criterion prompts the question: *Could such features as the avian respiratory system, the four-chambered heart, and feathers have evolved from the reptile body plan?* Let's examine this issue.

BIRD BREATHS

Like all other vertebrates, reptiles use a diaphragm to breathe—inflating and deflating the lung. However, the design of the bird respiratory system is completely different from that of reptiles. In fact, it's unlike that of any other animal.

Birds do not expand and contract their breathing cavity. Instead, they use multiple air sacs, which act like bellows, to force air through a continuous circuit in the lungs. Since there is no mixing of oxygen-rich and oxygen-poor air, respiration is more efficient in birds than in other vertebrates. This enables birds to maintain the high rate of metabolic activity needed for flying, as well as for breathing at high altitudes.

Bird air sacs are not rigid and require a protected space in which to function. The largest and most important air sacs for birds are in the abdominal cavity. Evolutionary zoologist John Ruben explains why the skeletal structure of theropod dinosaurs would not allow sufficient space for abdominal air sacs, and could not be a precursor to the bird body plan:

> [T]heropod dinosaurs had a moving femur and therefore could not have had a lung that

> worked like that in birds. Their abdominal air sac, if they had one, would have collapsed. That undercuts a critical piece of supporting evidence for the dinosaur-bird link.[22]

The origin of the avian respiratory system thus raises the problem of non-functional intermediate stages.

DINOSAURS OF A FEATHER?

Despite forceful pronouncements in the media reporting the discovery of feathered dinosaurs, there is more here than meets the eye. In particular, the evolutionary origin of feathers faces two major obstacles: the fossil record and the structure of feathers themselves.

The Fossil Record

The problem with alleged feathered dinosaur fossils is that they generally aren't feathered, aren't dinosaurs, or aren't fossils. They fit into one of three categories: Dinofuzz, Secondarily Flightless Birds, and Frauds.

Dinofuzz: Some alleged feathered dinosaurs are covered in a down-like material, but it isn't feathers. Labeled "dinofuzz" by critics, one scientific article reported that these "wispy hair-like structures are so unlike modern feathers that skeptics have all but dismissed the possibility that the two could be related."[23]

Secondarily Flightless Birds: Some feathered dinosaurs are not dinosaurs at all, but are best viewed as secondarily flightless birds—birds that lost their ability to fly. As paleornithologist Alan Feduccia argues, alleged feathered dinosaurs "are replete with features of secondarily flightless... birds."[24] Modern-day examples of secondarily flightless birds likely include the ostrich or the emu.[25]

Frauds: In 1999, *National Geographic* rushed to print a major story about a fossil named ***Archaeoraptor***. The magazine claimed it was a "missing link between terrestrial dinosaurs and birds"[26] and "exactly what scientists would expect to find in dinosaurs experimenting with flight."[27]

Shortly after the article was published, however, *Archaeoraptor* was shown to be a forgery—a skillful fusion of a dinosaur fossil and a bird fossil. Though the original story was a full-length article in *National Geographic*, the retraction was buried in a short letter to the editor months later.[28]

This episode led Storrs Olson, a bird expert at the Smithsonian Institution, to issue a harsh critique of the dinosaur-to-bird hypothesis:

> The idea of feathered dinosaurs and the theropod origin of birds is being actively promulgated by a cadre of zealous scientists acting in concert with certain editors at *Nature* and *National Geographic* who themselves have become outspoken and highly biased proselytizers of the faith. Truth and careful scientific weighing of evidence have been among the first casualties in their program, which is now fast becoming one of the grander scientific hoaxes of our age...[29]

Despite these criticisms, popular science media depictions commonly portray feathered dinosaurs as established fact. Alan Feduccia cites multiple examples of such drawings which he observes are based upon "no evidence."[30] Intelligent design would not be refuted by the finding of a feathered dinosaur fossil. But at this stage, claims of their existence are not supported by the data.

The Intelligently Designed Structure of Feathers

There are many different types of feathers. When considering the origin of these structures, authorities have asked, "How did these incredibly strong, wonderfully lightweight, amazingly intricate appendages evolve?"[31]

The traditional theory taught that feathers

evolved when scales on reptiles mutated to become frayed. Somehow these frayed scales gave the reptile an advantage of increased lift, eventually leading to the evolution of feathers for flight.

The main problem with this hypothesis is that feathers and scales are very different structures. Feathers essentially develop as hollow tubes that grow out of special follicles in the skin, whereas scales are flat, folded skin that develop quite differently. Critics also argue that feathers are so well-suited for flight that there would have to be many transitional stages between scales and fully functional flight feathers.[32]

In 2003, a cover story for *Scientific American* stated that the lack of evidence for the "scale" hypothesis meant the "long-cherished view of how and why feathers evolved has now been overturned."[33] The explanation offered in place of the scale hypothesis, however, is no better.

According to the latest theories, feathers and other advanced features that allow birds to fly did not originally evolve from scales for the purpose of flight.[34] Instead, the *Scientific American* article vaguely proposed that feathers provided "some kind of survival advantage,"[35] perhaps providing insulation for the organism.

Failing to explain the evolution of flight, materialists now propose that feathers and other complex features that are finely tuned for flight are accidental byproducts that evolved for entirely different purposes. Yet they admit that "[t]he entire avian body is structured for flight."[36] Birds, in this evolutionary account, were favored by some extraordinarily good luck.

Additionally, proponents of the gradual evolution of feathers must explain a striking fact: the feathers of the earliest bird, *Archaeopteryx*, are essentially identical to those of today's birds.[37] Apparently little change has taken place over eons of time, even though a great deal of change would have been necessary to produce feathers initially.

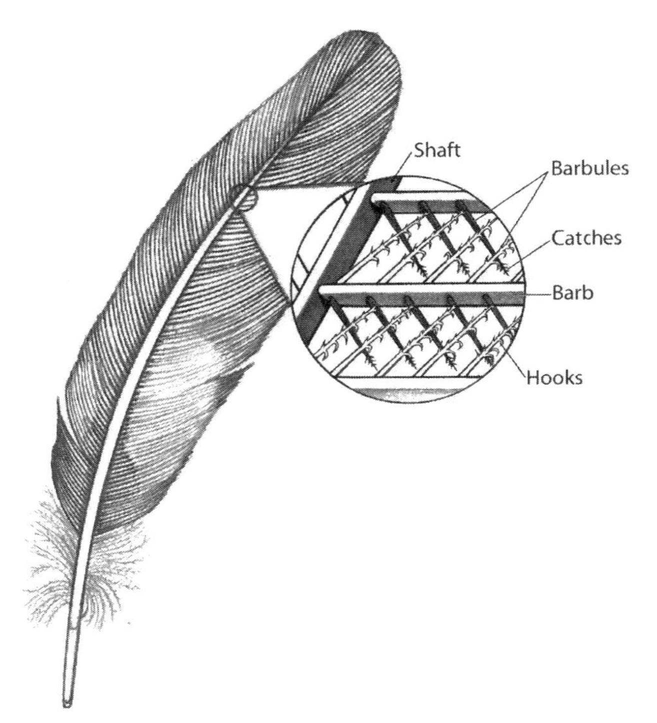

Figure 16-4: Irreducibly Complex Structure of a Flight Feather.

Feathers branch somewhat like a tree with a trunk, branches, and twigs. Figure 16-4 portrays a flight feather, showing "branches" called barbs, and "twigs" called barbules.

Barbules facing upward have hooks, and those that face downward have catches. The irreducibly complex interaction of hooks and catches keeps the barbs in line, giving the feather the rigid shape needed for lift.

An ideal structure for flight would be both rigid and lightweight. The layout of barbs and barbules not only optimizes these qualities but points strongly to design.

When the barbs become misaligned, birds instinctively preen their feathers, "zipping" the hooks and catches together and restoring proper shape. Space between the barbs allows air to pass through, making feathers lightweight.

Leading evolutionary authorities have marveled that feathers "have no clear antecedents in ancestral animals and no clear related structures" in other living species.[38] Arguably irreducibly complex, flight feathers could not function unless all of their features are present. The degree to which they optimize lift, rigidity, and weight would make the most brilliant aerospace engineer jealous.

HEARTFELT MOMENT

Some organs are not essential for survival. The heart is not one of them. Loss of heart function results in immediate death. Thus, function must be maintained during every step of heart evolution.

As described in Figure 16-5, reptiles hearts have three chambers, while bird hearts have four.

To test the feasibility of maintaining functionality during a gradual change from one type of heart to the other, consider some of the necessary modifications:

- The forked abdominal aorta must be replaced with a single aorta.
- The septum must be extended to create a separate ventricle chamber.
- Veins and arteries must be extensively reorganized.
- Various other structural changes must be made to walls and valves.[39]

Evolving a four-chambered heart would require severe rerouting of blood flow through the system. Anytime you remodel the plumbing in a house, the water has to be turned off. In living organisms, however, if you turn off the blood flow, the organism dies.

Clearly, there are many obstacles to the theory that birds evolved from reptiles.

A HORSE IS A HORSE, OF COURSE

Horses are another group of fossils commonly cited in support of macroevolution[40]—as one evolutionary biologist claimed, "horses are a poster child of evolution."[41] Textbooks and museums often display charts showing horse fossils.[42] They typically start with smaller four-toed creatures, which grow larger and lose toes until the series

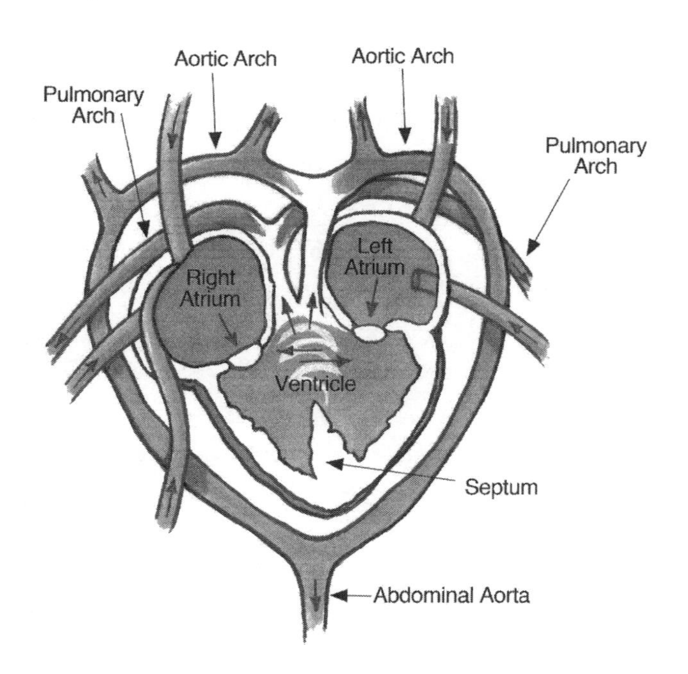

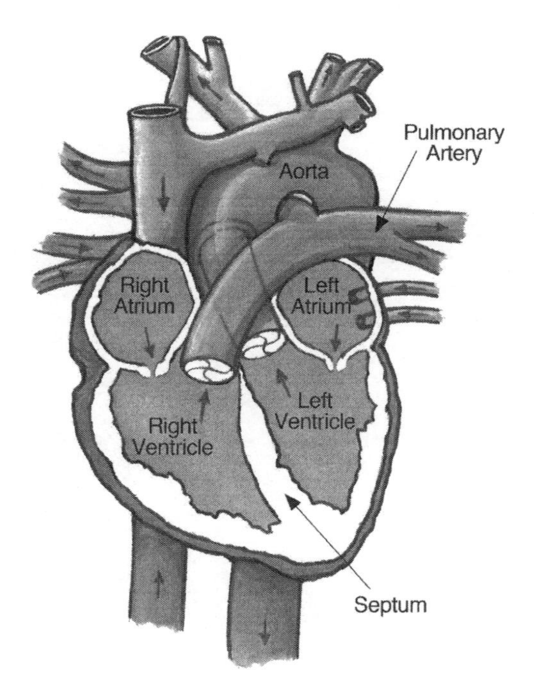

Figure 16-5: Three-Chambered vs. Four-Chambered Heart.

ends with modern horses, which have hooves.

Aside from the fact that the horse body plan does not significantly evolve throughout the series, there is a significant problem with this evolutionary story: it's an imaginary lineage.

In the actual fossil record, some of the earlier horses are larger than later ones. Additionally, the fossils portrayed in the supposed lineage span different continents, separated by vast expanses of ocean.

Both location and sequence conflict with the evolutionary series as it is often portrayed.[43] Even a prominent evolutionary website admits that the horse fossils merely "represent branches on the tree and not a direct line of descent leading to modern horses."[44]

Some defenders of Darwinism will say in response that a messy, branching evolutionary tree nevertheless demonstrates evolution. Perhaps, but then why do textbooks and museums feel the need to manipulate the horse fossil record?

Niles Eldredge called the series "speculative," and finds it "lamentable" that it "has been presented as the literal truth in textbook after textbook."[45] Likewise, one technical book on evolution called classical portrayals of horse evolution "not accurate," and admitted they "put the chart before the horse."[46]

A WHALE OF A TALE

According to evolutionary thinking, certain groups of fish left the water and evolved into land-dwelling animals with legs, some of which eventually evolved into mammals. However, as the theory goes, some mammals later returned to the sea (e.g., whales), losing the very hind limbs that once walked their ancestors out of the water.

As with the horse, evolutionists also claim that fossils documenting the presumed transition from land mammals to whales provide a "poster child for macroevolution."[47] In particular, there are some fossil land mammals that have ear-bones similar to whales, and there is evidence of whale-like aquatic mammals with hind limbs.

Is this enough evidence to demonstrate a transition? Hardly. Even if there are fossils that look like potential intermediate forms, if the overall evolutionary story does not make sense, then the fossil cannot be transitional. To put it another way, if a puzzle piece has the right picture on it, but the wrong shape, it can't be a part of the puzzle.

A great many changes would have been necessary to convert a land mammal into a whale. Here are just a few of them:

- Emergence of a blowhole, with musculature and nerve control.
- Modification of the eye for permanent underwater vision.
- Ability to drink sea water.
- Forelimbs transformed into flippers.
- Modification of skeletal structure.
- Ability to nurse young underwater.
- Origin of tail flukes and musculature.
- Blubber for temperature insulation.[48]

Each of these changes would necessarily involve many mutations. At this point, whale evolution runs into a time problem. The fossil record requires that the evolution of whales from small land mammals took place in less than 10 million years.[49] That may sound like a long time, but it actually falls dramatically short.[50] Biologist Richard Sternberg has examined the requirements of this transition mathematically and concludes: "Too many genetic re-wirings, too little time."[51]

Whale origins provide an interesting case for studying problems with evolution. On the rare occasions when there actually are fossils that show potentially intermediate traits, unguided neo-

Darwinian evolution is invalidated by the short amount of time allowed by the fossil record. If the poster child of macroevolution doesn't hold up to scrutiny, what does this tell us about other cases where evolutionists tout supposed transitional fossils?

DARWIN ON TRIAL

Perhaps ironically, evolutionary paleontologist J. G. M. Thewissen has encouraged scientists to treat whale evolution "like a legal trial, in which all the evidence is carefully scrutinized to arrive at the best interpretation" and there must be "no evidence that is incompatible"[52] with the final verdict. In other words, scientists should not ignore unfavorable evidence.

To continue the legal analogy, many of these alleged transitional fossils bring to mind a murder trial where a witness testifies he saw a suspect pull the trigger and shoot the victim. This sounds compelling, until the suspect is brought into the courtroom in a wheelchair, and the jury learns that he has been paralyzed from the neck down since birth. Regardless of what the witness thinks he saw, the suspect is physically incapable of firing a gun.

In the same way, many of these supposed evolutionary transitions face grave obstacles.

- Birds have structures that could not evolve from reptiles in a series of functional intermediates.
- The alleged horse lineage indicated in textbooks and museums is invalidated by the fossil record.
- Whales arose too quickly for neo-Darwinian processes.

Despite the contrary evidence, evolutionists stridently defend the validity of these transitions. In these and other cases, we might ask: *Is fossil interpretation guided by an objective examination of the evidence, or by the constraints of evolutionary thinking?*

CONCLUSION

Remember the beginning of this chapter where Darwin's promoters in the media touted Ida as "the link that connects us directly with the rest of the animal kingdom"?[53] A few months later, Ida's prestige came crashing down after scientists inspected the fossil and determined that "[m]any lines of evidence indicate that [Ida] has nothing at all to do with human evolution."[54]

In promoting evolution to the public, scientists and the media alike have shown a tendency to rush to judgment. Scientists bear a responsibility to avoid overstating the results of their discoveries and investigations. Journalists have a responsibility to investigate scientific claims and to do their utmost to present realistic assessments—including multiple credible viewpoints, when they exist—to the public.

The lesson seems clear: When the scientific and popular media excitedly announce the discovery of another "missing link," the public is well-advised to maintain a healthy skepticism.

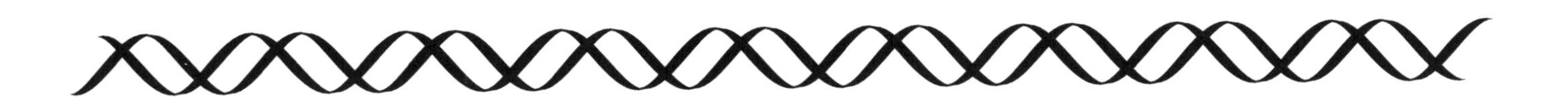

DISCUSSION QUESTIONS

1. Consider the fossil evidence that evolutionists cite to support common descent. What do you think are their strongest and weakest examples? Explain.

2. List the complex parts in flight feathers. Could they evolve in a stepwise fashion, where there is an advantage gained at each small step? Explain your reasoning.

3. Do you think there is conclusive fossil evidence for dinosaurs with feathers? Why or why not?

4. According to the short timescale allowed by the fossil record, the Darwinian evolution of whales is extremely improbable—yet evolutionists consider it to be a "poster child" for their theory. What does this say about other instances where evolutionists claim to have supposed transitional forms?

5. Imagine you are a reporter covering the discovery of a new fossil that is being promoted by evolutionary scientists as a "transitional form." If you want to remain objective and print a fair and accurate story, what questions would you ask about the fossil?

CHAPTER 17

WHO'S YOUR DADDY?

Question:

Which side of your family is most likely descended from ape-like ancestors?

WHO'S YOUR DADDY?

"Each Neanderthal expert thought the last one I talked to was an idiot, if not an actual Neanderthal."

—PBS producer Mark Davis, on interviewing paleoanthropologists[1]

The origin of humans—studied by **paleoanthropologists**—is perhaps the most hotly debated evolutionary field. An article in the journal *Science* explained how biases and bitter disputes between researchers permeate the study of human evolution:

> The field of paleoanthropology naturally excites interest because of our own interest in our origins. And, because conclusions of emotional significance to many must be drawn from extremely paltry evidence, it is often difficult to separate the personal from the scientific disputes raging in the field.[2]

Raging disputes? Some might wonder how scientists can look at the same fossils and develop notably different conclusions. It's because much of paleoanthropology is not based on objective criteria, but on subjective interpretations of scant fossil evidence.

The one point that most paleoanthropologists seem to agree upon is their belief that humans evolved from ape-like ancestors. Beyond that, the details get very sketchy and inconsistent.

Illustration: Copyright Jody F. Sjogren 2000. Used with permission.

Figure 17-1. Ape-to-Man Icon.

The basic idea is reflected in the classic ape-to-human progression (see Figure 17-1), familiar to anyone who reads science textbooks or newspapers. But is this idea based on evidence from the fossil record or is it another misleading icon of evolution?

Paleoanthropologists confidently offer accounts of human evolution, but their narratives often conflict with one another and with the evidence. In order to investigate human origins in more detail, we should become familiar with a few key individuals who appear prominently in these stories.

WILL THE REAL HOMINIDS PLEASE STAND UP?

Members of the family Hominidae are called **hominids**. While there has been much disagreement among paleoanthropologists, the most prominent view at present is that hominids include humans, great apes (chimpanzees, gorillas, and orangutans), and all extinct species leading back to their most recent alleged common ancestor. A smaller group, the **hominins**, includes just humans and chimpanzees, and any extinct organisms leading back to their presumed most recent common ancestor.

The taxonomic definitions of these groups have evolved in recent years, but today much of the focus on human origins centers upon hominins. As we will see, the hominin fossil record is fragmented, and does not document the evolution of humans from ape-like precursors.

Let's take a closer look at some of the key hominin species.

Australopithecus

Australopithecus is an ape-like genus thought to have arisen a little over 4 million years ago (mya). In Latin, the name means "southern ape," since the first australopithecine fossils were found in southern Africa. These small-brained hominins went extinct about 2 mya.

There are two major groups of australopithecines, the larger "robust" forms and the smaller "gracile" forms. Most evolutionary paleoanthropologists would claim we are descended from the gracile forms, such as *Australopithecus afarensis*, which includes the famous fossil "**Lucy**."

One important member of *Australopithecus* that will receive attention is ***habilis***. As we'll see, *habilis* has been classified under *Homo*, but recent reanalyses have shown it is much more like the australopithecines and modern apes than like humans.

Homo

Homo, which means "man" in Latin, is a genus that includes *Homo erectus*, Neanderthals, our own species *Homo sapiens*, and, according to some opinions, possibly several other extinct species. Scientists date our genus from about 1.9 mya to the present day.[3]

Homo erectus is an extinct member of the genus *Homo* that is generally believed to have lived from about 1.9 mya until less than 100,000 years ago.[4] The name means "upright walking man."

The name **Neanderthal** comes from the Neander Valley in Germany where some specimens have been found. Their fossils date from 30,000 to 200,000 years ago. They look nearly identical to us, except for slightly larger and thicker skulls.

The taxonomic name for modern humans is ***Homo sapiens***, which in Latin means "wise man." Due to their high level of anatomical similarity, some would classify both *Homo erectus* and Neanderthals as members of *Homo sapiens*. Generally speaking, however, modern humans and their culture are thought to have appeared some 50,000 years ago.

Now that we've made the introductions, let's take a closer look at our genus.

ALL IN THE FAMILY

Most members of the genus *Homo* are very similar to modern humans. In fact, the differences between the human-like members of the genus *Homo* can be explained as the result of micro-evolutionary change.[5]

For example, below the neck, *Homo erectus* is extremely similar to modern humans.[6] Further, as seen in Figure 17-2, *erectus* skulls are within the range of modern human variation.

Neanderthals are commonly portrayed as bungling, primitive relatives of modern humans. But recent anatomical, genetic, and cultural data have led many scientists to believe Neanderthals were a sub-race that was part of our own species.[7] Not only is their body shape within the range of modern human variation,[8] but anatomical evidence suggests they were capable of speech.[9] Further, they have been found associated with signs of art and culture—including burial of the dead.[10]

Whether they went extinct is debated—it's possible that we are them, and they are us. Paleoanthropologist Erik Trinkaus of Washington University has argued, "They may have had heavier brows or broader noses or stockier builds, but behaviorally, socially and reproductively they were all just people."[11]

Figure 17-2. Cranial Capacities.[12]

Taxon:	Cranial Capacities (in cubic centimeters):	Species Resembles:
Gorilla (*Gorilla gorilla*)	340-752 cc	Modern Apes
Chimpanzee (*Pan troglodytes*)	275-500 cc	
Australopithecus	370-515 cc (avg. 457 cc)	
Homo habilis	Avg. 552 cc	
Homo erectus	850-1250 cc (avg. 1016 cc)	Modern Humans
Neanderthals	1100-1700 cc (avg. 1450 cc)	
Homo sapiens (modern man)	800-2200 cc (avg. 1345 cc)	

AUSTRALOPITHECUS VS. *HOMO*

The Holy Grail for evolutionary paleoanthropologists has been to find a fossil species that links *Homo* back to some kind of ape-like ancestors. For a time, *Australopithecus* seemed like a good candidate.

By far, the most famous australopithecine fossil is Lucy because 40% of her bones were uncovered, making her one of the most complete known fossils among early hominins. However, the fossil evidence doesn't support the claim that Lucy's species was ancestral to *Homo* (Figure 17-3).

While australopithecines were capable of some upright walking, their long forelimbs and other traits also indicate they spent significant amounts of time in trees.[13]

Even when traveling on the ground, australopithecines had a mode of locomotion different from *Homo*.[14] For example, one study in *Nature* found that Lucy had the hands of a knuckle-walking ape,[15] leading commentators to observe that her body was "quite ape-like" with "relatively long and curved fingers, relatively long arms, and funnel-shaped chest."[16]

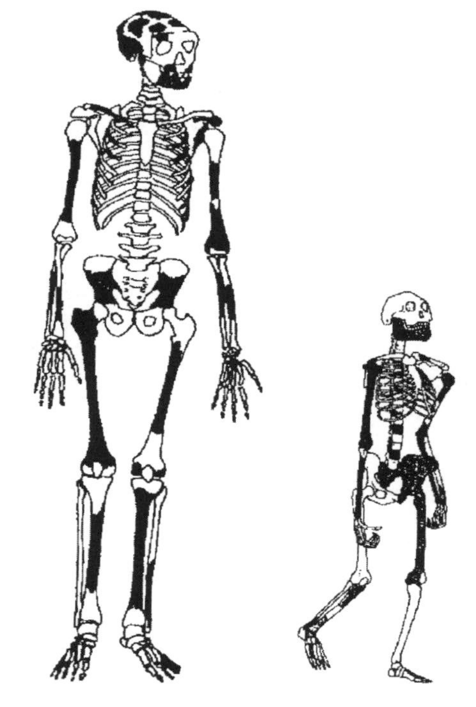

Illustration: From Figure 1, John Hawks et al., "Population Bottlenecks and Pleistocene Human Evolution," Molecular Biology and Evolution, copyright 2000, 17 (1): 2–22, by permission of the Society for Molecular Biology and Evolution.

A comparison of *Australopithecus* (right) to early *Homo* (left). Black bones indicate those which have been discovered.

Figure 17-3: Skeletons of Homo vs. Australopithecus.

Additionally, australopithecines' inner ear canals—responsible for balance and related to locomotion—were different from those of *Homo* but similar to those of great apes.[17] There are also sharp differences between the pelvic bones of australopithecines and *Homo*,[18] prompting one expert to say that "Lucy's erect posture is unlike that seen in modern humans and is still a mystery."[19]

Paleoanthropologist Leslie Aiello states that when it comes to locomotion, "Australopithecines are like apes, and the *Homo* group are like humans. Something major occurred when *Homo* evolved, and it wasn't just in the brain."[20]

There is another extinct species that some leading paleoanthropologists have claimed is a link between our genus and more primitive hominins. *Habilis*, dated at about 1.9 mya,[21] was placed within *Homo* to imply an ancestral link between *Homo* and the australopithecines. But multiple studies have challenged this claim.

A study in the *Journal of Human Evolution* examined habiline remains and found the species was not only similar to modern apes, but *more* similar to living apes than were other australopithecines.[22] An author of the paper called those results "unexpected in view of previous accounts of *Homo habilis* as a link between australopithecines and humans."[23]

Similarly, an analysis by leading paleoanthropologists published in *Science* found that *habilis* is very different from *Homo*, and should be reclassified within *Australopithecus*, since its characteristics fit overwhelmingly into that group.[24] Additionally, since *habilis* does not predate the earliest true members of *Homo*, it could not have been a precursor of our genus.[25]

In the journal *Science*, leading paleontologists identified significant differences between *Australopithecus* and *Homo* (see Figure 17-4).[26] It is worth noting that all of the differences also apply to *habilis* compared to *Homo*.

According to the paper, brain size is the only trait for which there are any fossils with an intermediate state between *Australopithecus* and *Homo*. But this is not necessarily strong evidence that *Homo* evolved from less intelligent hominins (see Figure 17-5). Intelligence is determined largely by internal brain organization, and thus involves much more than just brain size.[27] Finding a few skulls of intermediate size does little to bolster the case for human evolution.

Figure 17-4: Significant Differences Between Australopithecus and Homo.

- **Body size.**
- **Body shape.**
- **Mode of locomotion.**
- **Jaws and teeth.**
- **Development.**
- **Brain size.**

Textbooks, news stories, and TV documentaries sometimes display skulls lined up from small sizes to larger ones, trying to imply an evolutionary progression of primitive species to more intelligent ones. But brain size is not always a good indicator of intelligence. Case in point: Neanderthals had an average skull size larger than modern humans, but few would claim we are less intelligent than they were.

Figure 17-5: How Important are Skulls of "Intermediate" Size?

NO MONKEY IN THE MIDDLE

Hominid fossils generally fall into one of two categories: human-like fossils and ape-like fossils, with major differences between them. In 2004, evolutionary zoologist Ernst Mayr recognized this problem:

> The earliest fossils of *Homo*... are separated from *Australopithecus* by a large, unbridged gap.[28]

Similarly, a paper in the *Journal of Molecular Biology and Evolution* recognized that going from the australopithecines to *Homo* would require a "genetic revolution" since early *Homo* "was significantly and dramatically different from... australopithecines in virtually every element of its skeleton and every remnant of its behavior."[29]

Likewise, two scientists wrote in the journal *Nature*, "Not for nothing has [early *Homo*] been described as a hominin without an ancestor, without a clear past."[30] Given a lack of transitional evidence, the sudden appearance of the *Homo* genus has been called the "big bang theory of human evolution."[31]

Despite the fanfare in the media, paleoanthropologists have yet to find fossil evidence demonstrating that humans evolved from more primitive hominins. According to the evidence, the genus *Homo* appears abruptly.[32]

"MISSING LINK" CELEBRITIES

Scientists are dependent on grants issued to those judged to be able to make significant contributions to their field. In paleoanthropology, this often means finding fossils linking humans to ape-like ancestors.

Of course, scientists have other motivations such as fame, scientific curiosity, or the validation of a personal philosophy. Combined, these various lures have created a mad scramble to find and publicize transitional fossils—which the media commonly calls "missing links." This in turn, has resulted in a long series of news articles citing "breakthroughs" that are "mind-blowing"—discoveries that often turn out to be vastly overstated.[33]

History should teach us caution.

In 1912, a fossil was discovered in a gravel pit near Piltdown, England. This purported half-ape, half-man skull was touted as proof of evolution. *The New York Times* published the headline "Paleolithic Skull Is a Missing Link," and the fossil was featured as court-approved evidence for human evolution during the nationally-followed Scopes Trial, which became the subject of national news.[34]

But there was a problem: "**Piltdown Man**" was a forgery pieced together from the jaw of an orangutan and the skull of a human. Unfortunately, the hoax wasn't exposed until 1953. In the meantime, countless people took it as factual evidence of human-ape common ancestry.

Over the decades, there have been many other cases of mistaken anthropological identity.

In the 1960s, an ape-like fossil named *Ramapithecus* was promoted as an ancestor of humans solely on the basis of teeth and skull fragments. As two paleoanthropologists lamented, "an entire *Ramapithecus* walking upright has been 'reconstructed' from only jaws and teeth" with "his legitimacy sanctified by millions of textbooks and Time-Life volumes on human evolution."[35]

But those claims were overturned in the 1970s and 1980s when more complete specimens were discovered, showing the species was unlike humans, and probably an extinct orangutan-like ape.[36]

Recent years have seen further examples of widely publicized, alleged human ancestors being overturned by the evidence. In the previous chapter we discussed the rise and fall of Ida, but she was not the only "missing link" erroneously promoted in 2009. The media also declared ***Ardipithecus ramidus*** the most recent common ancestor of apes and humans.

The journal *Science* called "Ardi" the "breakthrough of the year," yet admitted that her bones were originally "crushed" and required over a decade of "reconstruction."[37] Indeed, other paleoanthropologists noted that Ardi was initially "crushed nearly to smithereens" and its pelvis resembled "Irish stew."[38]

Later, when cooler heads prevailed, multiple studies found that Ardi was more similar to apes than humans, and concluded she was not a human ancestor.[39] As *Time* Magazine reported, critical scientists regard "the hype around Ardi to have been overblown."[40]

Textbooks—and even some technical books—often use fictional drawings of hominins to diminish the differences between humans and apes in the minds of readers and influence their thinking to accept common ancestry. For example, australopithecine apes might be portrayed as highly intelligent, whereas early members of the genus *Homo*, such as Neanderthals, are shown as bungling, primitive brutes.[41]

Regarding such inaccuracies, the famed physical anthropologist Earnest A. Hooton wisely cautioned in 1931 that "alleged restorations of ancient types of man have very little, if any, scientific value and are likely only to mislead the public."[42]

WILL DNA SAVE THE DAY?

Lacking clear fossil evidence to make their case for ape-human common ancestry, some evolutionists have turned to DNA.

The Myth of 1%

A popular argument for evolution is that human and chimp DNA are only 1% different, which shows that we share a common ancestor with apes.[43] Such a claim raises two important questions.

1. Is the 99% human/chimp DNA-similarity statistic correct?

In 2007 the journal *Science* published an article subtitled "The Myth of 1%," observing that human and chimp DNA often differ by much more than 1%. According to the article, the claim that human and chimp DNA are just 1% different "reflects only base substitutions, not the many stretches of DNA that have been inserted or deleted in the genomes."[44] The paper explained that the actual differences amount to at least "35 million base-pair changes, 5 million [insertions and deletions] in each species, and 689 extra genes in humans."[45]

There are at least two other reasons to disbelieve the 99% claim:

- Genetic comparisons typically focus only on differences in protein-coding DNA, and ignore differences in noncoding DNA. However, since scientists have realized the importance of non-coding ("junk") DNA for building body plans, some have begun to suggest that "it's the junk that makes us human."[46]

- Our genome is in some ways very different from that of chimps. One study in *Nature* compared human and chimp y-chromosomes, and found that they "differ radically in sequence structure and gene content."[47]

Geneticist Richard Buggs has suggested that depending on how you do the analysis, human and chimp genome similarity might be as low as 70%.[48]

But there is a deeper question to be asked.

2. If humans and chimps are genetically similar, would that demonstrate common ancestry?

Evolutionists assume that functional genetic similarity indicates common ancestry. But as we pointed out in Chapter 13, intelligent agents often reuse functional components in different systems (e.g., wheels for cars and wheels for shopping carts). Even the evolutionary geneticist Francis Collins admits that such similarity alone "does not, of course, prove a common ancestor," because a designer could have "used successful design principles over and over again."[49]

So, even if the 1% figure were true, functional genetic similarities between humans and chimps could easily be explained by common design.

Chromosomal Con-Fusion

A final piece of DNA evidence often cited as demonstrating human-ape common ancestry involves the fact that humans have 23 pairs of chromosomes, while apes have 24. Evolutionists claim that human chromosome 2 has a structure similar to what one would expect if two chimpanzee chromosomes were fused together at their ends (called **telomeres**).

No one is claiming that this alleged fusion event somehow turned an ape-like creature into a human.[50] However, as the story goes, chromosomal fusion shows that humans are descended from species with chimp-like genetics.

Let's assume, for the sake of argument, that there was a chromosomal fusion event within humans. What would this show? All the fusion would show is that, at some point within our lineage, two chromosomes became fused.

If we think outside the neo-Darwinian box, then the following scenario becomes possible:

1. The human lineage was designed separately from that of apes,
2. A chromosomal-fusion event occurred, and
3. The trait spread throughout the human population.[51]

In such a scenario, the evidence would appear precisely as we find it, without any common ancestry between humans and apes. Human chromosomal fusion tells us nothing about whether our human lineage leads back to a common ancestor with apes.

But there's more to this story, as the evidence for chromosomal fusion isn't as neat and tidy as is often claimed. In fact, the alleged fusion point contains much less telomeric DNA than it should if two chromosomes were fused end-to-end. A paper in *Genome Research* found that the alleged telomeric sequences are "highly diverged" from what would be expected if humans and apes shared a recent common ancestor.[52]

At most, all the chromosomal fusion evidence does is reconfirm something we already knew: humans and apes share similar functional genetics. But this similarity could be explained by common design just as well as by common descent.

HUMAN EXCEPTIONALISM

Citing evidence from fossils and genetics, this chapter has pointed to distinct gaps between humans and apes. But according to the field of evolutionary psychology, it isn't just the human body and DNA that evolved. Our *behavior* is also said to be programmed by natural selection.

Could our behavior and mental abilities have evolved?

It should be obvious that there are significant differences between humans and apes. For one, humans are the only primates that always walk upright, have relatively hairless bodies, and wear clothing. But the differences go far beyond physical traits and appearances.

Humans are the only species that uses fire and technology. Humans are the only species that composes music, writes poetry, and practices religion. When it comes to morality, bioethicist Wesley J. Smith observes that:

> [W]e *are* unquestionably a unique species—the only species capable of even contemplating ethical issues and assuming responsibilities—*we uniquely are capable of apprehending the difference between right and wrong, good and evil, proper and improper conduct...*[53]

Humans are also the only species that seeks to investigate the natural world through science. Additionally, humans are distinguished by their use of complex language. As MIT professor and linguist Noam Chomsky observes:

> Human language appears to be a unique phenomenon, without significant analogue in the animal world.... There is no reason to suppose that the "gaps" are bridgeable.[54]

Other linguists have suggested that this finding would imply "a cognitive equivalent of the Big Bang."[55]

Because of this evidence, some scholars have argued that humans are *exceptional.*[56]

> **Human Exceptionalism:**
> ***A view holding that the human race has unique and unparalleled moral, intellectual, and creative abilities.***

Materialists often oppose **human exceptionalism** because it challenges their belief that we are little more than just another animal. The next time someone tries to minimize the differences between humans and apes, remind him that it's humans who write scientific papers studying apes, not the other way around.

EVOLUTION GOES PSYCHO

While evolutionary accounts of human language face great obstacles, there is no doubt that complex language would provide a great survival advantage once it came to exist. But some of humanity's most cherished activities don't appear to offer any evolutionary benefit at all.

The requirements of Darwinian selection are simple: organisms must survive and spread their genes. Michael Ruse and E. O. Wilson thus explain that under Darwinism, "ethics... is an illusion fobbed off on us by our genes to get us to cooperate."[57] In other words, in a strictly Darwinian world, there is no such thing as objective morality, true selflessness, or pure **altruism**.

> **Altruism:**
> ***Unselfish regard for the welfare of others.***

Altruistic behavior appears to run counter to natural selection and should have been eliminated long ago. But here we are, and humans exhibit astounding examples of altruism. The field of evolutionary psychology purports to solve this conundrum by claiming that seemingly unselfish behavior actually gives kickbacks to your selfish genes.

For example, if I share food with my neighbor, perhaps later he'll return the favor. This is called *reciprocal altruism*. Similarly, if I neglect my own reproductive success to help my sister raise her kids, some of my genes might still be passed on. This is called **kin selection**.[58]

In recent years, such theories have captured the minds of journalists. In 2006, *The New York Times* gave a glowing review to the book *Moral Minds: How Nature Designed Our Universal Sense of Right and Wrong*, promoting the hypothesis that "people are born with a moral grammar wired into their neural circuits by evolution."[59]

Humans do appear hard-wired for morality, but were we programmed by unguided evolutionary processes?

Natural selection cannot explain extreme acts of human kindness. Regardless of background or beliefs, upon finding strangers trapped inside a burning vehicle, people will risk their own lives to help them escape—with no evolutionary benefit to themselves.

Evolutionary biologist Jeffrey Schloss explains that Holocaust rescuers took great risks that offered no personal biological benefits:

> The rescuer's family, extended family and friends were all in jeopardy, and they were recognized to be in jeopardy by the rescuer. Moreover, even if the family escaped death, they often experienced deprivation of food, space and social commerce; extreme emotional distress; and forfeiture of the rescuer's attention.[60]

Francis Collins elaborates on this theme, using the example of Oskar Schindler, who risked his life "to save more than a thousand Jews from the gas chambers. That's the opposite of saving his genes."[61] Schloss adds other examples of "radically sacrificial" behavior that "reduces reproductive success" and offers no evolutionary benefit, such as voluntary poverty, celibacy, and martyrdom.[62]

In spite of the claims of evolutionary psychologists, many of humanity's most impressive charitable, artistic, and intellectual abilities outstrip the basic requirements of natural selection. If life is simply about survival and reproduction, why do humans compose symphonies, investigate quantum mechanics, and build cathedrals?

Natural Academy of Sciences member Philip Skell explained why evolutionary psychology does not adequately predict human behavior:

> Darwinian explanations for such things are often too supple: Natural selection makes humans self-centered and aggressive—except when it makes them altruistic and peaceable. Or natural selection produces virile men who eagerly spread their seed—except when it prefers men who are faithful protectors and providers. When an explanation is so supple that it can explain any behavior, it is difficult to test it experimentally, much less use it as a catalyst for scientific discovery.[63]

Contrary to Darwinism, the evidence indicates that human life isn't about mere survival and reproduction.

CONCLUSION

ID is compatible with common descent, and we have kept an open mind about the possibility of human/ape common ancestry. But the evidence discussed in this chapter has shown many reasons to be skeptical:

- Hominin fossils fall into two groups: ape-like species and human-like species, with a distinct gap between them.
- Humans and apes have many genetic similarities, but this evidence could be explained by common design, and does not dictate inheritance from a common ancestor.
- Humans exhibit many unique intellectual abilities and behaviors that appear beyond explanation by natural selection.

Whether we're investigating fossils, DNA evidence, or stories about the evolution of the human mind, the media have a long history of advancing dubious evidence that humans evolved from ape-like creatures. The lesson to be learned is to keep an open mind and think critically when reading about the latest "proof" of human evolution.

DISCUSSION QUESTIONS

1. Materialists use the fossil record in an attempt to show that humans evolved from ape-like ancestors. Is the evidence for their case strong or weak? Explain your answer.

2. Does *Australopithecus* provide a convincing ancestor for *Homo*? Why or why not?

3. Comparisons between human and chimpanzee DNA have led evolutionists to claim we evolved from a common ancestor. What do you think? Discuss this question in detail.

4. This chapter listed various human behaviors that provide no apparent evolutionary benefits. Can you think of examples of humans who displayed these behaviors? Are there other behaviors that cannot be explained by natural selection?

5. Discuss the differences between apes and humans, and the likelihood that these differences could have arisen through unguided mechanisms.

6. Recall the last time you heard the media declaring the discovery of a "missing link." Did you find the story credible? Why or why not?

SECTION V

SUMMARIZING THE EVIDENCE

CHAPTER 18

TAKING INVENTORY

Question:

Based on the evidence, which do you think offers the best explanation for the origin of the universe and life: materialism or intelligent design?

TAKING INVENTORY

"[T]here is ample reason to believe that Darwinism is sustained not by an impartial interpretation of the evidence, but by dogmatic adherence to a philosophy even in the teeth of the evidence."

—Law professor and author Phillip E. Johnson[1]

Look at the image to the right. No one would conclude that the treehouse grew naturally as part of the tree. The construction of such a structure would require high levels of complex and specified information. As we have demonstrated throughout this book, there are numerous structures in the universe, especially within living organisms, which are far more complex than this treehouse.

The more we look at nature, the more we find fine-tuning. From the life-friendly architecture of the universe, to the information in the DNA of the simplest living organism, specified complexity is everywhere. Functional proteins, molecular machines, and exquisitely engineered body plans contribute immensely to the complexity of nature.

We must now ask, *what cause is capable of generating these unlikely structures?* With respect to the origin of biological features, we will consider two processes: Darwinian evolution, and intelligent design.

SEARCHING FOR THE ANSWER

Intelligence is a goal-directed process that is capable of thinking with will, forethought, and intentionality to achieve some end-goal. This stands in contrast to neo-Darwinian evolution, which essentially is a blind, trial-and-error process.

The difference between the two processes is analogous to a random Internet search engine versus an intelligently programmed one. To understand this difference, let's say our goal is to find information about jelly beans.

Under Darwinian evolution, mutations arise in a manner that is random, or blind, with respect to the needs of the organism. In our analogy, this would be like going to a search engine, typing in "jelly beans," but then a random page on the Internet is returned. Perhaps you get a page about septic tanks. Or maybe dung beetles, or Lady Gaga. Who knows? Darwinian evolution is blindly picking pages at random, without considering your needs.

If the random page you are given by the Darwinian search has information about jelly beans, then you keep it. If not, you discard it, and try again. Needless to say, it would take innumerable tries using this random search process to learn about jelly beans.

How about an intelligently designed search? The key difference is that ID does not rely upon random mutations that are blind to the needs of the organism. A rational intelligent agent, with foresight, can think ahead to see what needs exist. In our analogy, this would mean that the search engine would be *smart*—programmed to return only information that deals specifically with jelly beans.

As Stephen Meyer explains:

> Intelligent agents have foresight. Such agents can select functional goals before they exist.... The causal powers that natural selection lacks—almost by definition—are associated with the attributes of consciousness and rationality—with purposive intelligence.[2]

Based upon our studies of intelligence, we can understand its causal abilities to generate new complex and specified information. Only intelligence can find or generate the rare amino acid sequences necessary to produce functional biological structures, or the highly unlikely universal laws and constants necessary for life.

DVD SEGMENT 14
"Design Inference"
(11:44)

From:
Unlocking the Mystery of Life

REVIEWING THE EVIDENCE

We have presented a great deal of positive evidence for intelligent design. To summarize this evidence, let's review what we've learned, and along the way we'll evaluate the philosophy of materialism.

The first chapter listed seven tenets of materialism and stated that if any could be proven wrong, then the philosophy would be negated. Let's investigate how many of the seven are still standing.

MATERIALISM TENET #1:

Either the universe is infinitely old, or it appeared by chance, without cause.

The Evidence Shows:

The universe had a beginning and is not infinitely old. The Big Bang theory, which requires that the universe had a beginning, is supported by:

1. The detection of red-shifts in the frequency of light from all visible galaxies, meaning that the universe is expanding.

2. The detection of background radiation that matches what would be expected from the Big Bang event.

Adding to this claim is the philosophical *kalam* argument which states that anything that has a beginning also has a cause.

To explain the existence of the universe, materialists have resorted to speculative arguments such as:

1. Self-creation. This logically absurd concept requires that the universe existed before it was created.
2. Chance. Chance is not a cause, but merely a descriptive and predictive tool related to probability. Chance explains nothing.
3. Quantum theory. The claim that the universe began as just another quantum particle does not explain where the quantum vacuum (a sea of energy) came from. Quantum theory provides no explanation as to why any quantum outcome occurs, or how the universe came to exist.

There is no evidence to support any of these hypotheses. Both science and philosophy refute the first tenet of materialism.

MATERIALISM TENET #2:

The physical laws and constants of the universe were produced by purposeless, chance processes.

The Evidence Shows:

The universal constants and laws are finely tuned for complex life. This shows that exceedingly high levels of complex and specified information are built into the architecture of the universe.

The degree of fine-tuning is so extraordinary that, to cope with the data, materialists have resorted to wildly speculative hypotheses such as the oscillating universe model and the multiverse. However, the evidence does not support these ideas. In fact, multiverse-type thinking—which invents vast numbers of unobservable universes to avoid solving difficult problems—undermines the ability of science to investigate nature.

We live in a finely tuned universe and on a rare and especially privileged planet, with all the factors necessary for complex life. The probability that the cosmological constant, initial entropy, and other factors arose by chance is so extremely remote as to be unworthy of scientific consideration—refuting Tenet #2.

MATERIALISM TENET #3:

Life originated from inorganic material through blind, chance-driven processes.

The Evidence Shows:

The most basic living cell is a miniature factory that runs on machines built by information. Two of the many highly complex machines found in most living cells include

- Ribosome. Called an "incredibly beautiful complex entity"[3] that is "ingeniously designed for [its] functions,"[4] this molecular machine requires a minimum of 53 proteins. Even bacterial cells contain thousands of ribosomes.

- ATP synthase. This protein-based molecular machine uses a rotary motor to create ATP energy molecules. It has been described as "one of the wonders of the biological world."[5]

Cells use molecular machines to perform basic functions, such as protein production, transportation within the cell, energy production, disposal of waste, replication, and protection from outside elements.

Conditions on the early Earth could not have created even the preliminary building blocks of life, much less other basic requirements such as a protective cell membrane. Life is too complex to have originated from inorganic material by unguided, chance-based processes.

MATERIALISM TENET #4:

The information in life arose from unguided, blind processes.

The Evidence Shows:

DNA is an information-carrying molecule that is essential to all life. There are no known chemical or physical laws that can order the nucleotide bases in DNA. The cell membrane, DNA, and proteins form an interdependent, complex system in which all three must have been present at the start for a cell to survive. The DNA/protein basis of life is too complex to arise by chance.

In the face of the complexity of DNA/protein-based life, the primary materialistic explanation is the RNA world. That hypothesis is invalid for four reasons:

1. RNA production requires intelligent design.
2. RNA can't fulfill the necessary roles of proteins.
3. The RNA world can't solve the information sequence problem and explain the origin of genetic information.
4. The RNA world can't explain the genetic code.

The probability of unguided causes generating the information in life is so low that it renders such materialistic theories unworthy of scientific consideration. A goal-directed search process like intelligent design is required.

MATERIALISM TENET #5:

Complex cellular machines and new genetic features developed over time through purposeless, blind processes.

The Evidence Shows:

Darwinian evolution requires that structures evolve by "numerous, successive slight modifications." But research by scientists like Douglas Axe shows that amino acid sequences that yield stable, functional proteins are too rare to be produced incrementally by random mutations.

Additionally, research by Michael Behe and David Snoke shows that in multicellular organisms, the Darwinian model cannot even produce many simple protein-protein interactions. Critics who attempted to refute their research actually determined that the development of new features by Darwinian means could not be achieved within a reasonable time.

Darwinian evolution requires that random mutations arise before natural selection can act to preserve them. But some biological systems require many parts to be present all at once before there is any function. These irreducibly complex structures cannot evolve in a stepwise fashion. Rather, in our experience, complex, information-rich structures like machines have only one known cause: intelligence.

MATERIALISM TENET #6:

All species evolved by unguided natural selection acting upon random mutations.

The Evidence Shows:

Irreducible complexity is found not just in biochemical machines, but also in larger-scale biological systems like vision or digestion. The rationale applied to Tenet #5 applies even more strongly in light of the variety and complexity of animal body plans.

The transition from one type of organism to another would require far more than merely developing a new feature. A myriad of coordinated changes would be necessary. Organisms are composed of integrated systems of organs, which are in turn built from various tissue types. Tissues are composed of many different cell types that are determined by the production of thousands of different proteins. Such a top-down coordination of parts requires intelligent planning to design a blueprint.

Intelligent design does not require optimal design. Nevertheless, alleged examples of bad design often turn out to be very good designs. Classical examples of "poor" design like the wiring of the vertebrate eye or the panda's "thumb" are elegant designs that function efficiently.

Noncoding DNA is not "junk" but in fact performs many vital cellular functions. Likewise, so-called vestigial organs like the appendix, tonsils, and coccyx play important roles in our bodies. The Darwinian assumption that poorly understood parts are useless evolutionary leftovers has held back the progress of science.

Whether we study insect metamorphosis, plant photosynthesis, or hummingbird acrobatics, animal and plant complexity offer breathtaking examples of biological fine-tuning. In our own species, the integrated complexity of systems such as the eye, reproduction, and digestion strain unguided Darwinian evolution beyond believability.

Unguided Darwinian processes do not present a realistic option to create complex organisms. This effectively refutes Tenet #6.

MATERIALISM TENET #7:

All living organisms are related through universal common ancestry.

The Evidence Shows:

The theory of intelligent design can accommodate common ancestry, yet the tree of life hypothesis conflicts with much of the evidence. Attempts to construct such a tree through molecular and anatomical homology have failed. One gene or feature produces one evolutionary tree, while another gene or feature yields an entirely different tree. Materialist attempts to explain away this conflicting data through "convergent evolution" are actually better accounted for by common design.

Classical lines of evidence used to support common descent—such as presumed similarities in vertebrate embryos—are weak because vertebrate embryos actually begin development quite differently from one another. Common textbook examples of natural selection, such as the Galápagos finches or peppered moths, cannot be extrapolated to support macroevolution or universal common ancestry. They only demonstrate small-scale change—microevolution.

The fossil record is full of explosions of new life-forms that lack evolutionary precursors. This abrupt appearance of species cannot be explained by "punc eq" or "evo-devo," as neither provides a viable mechanism for rapidly producing new body plans. Examples of supposed transitional forms often fall apart on closer inspection.

The fossil record shows that our genus *Homo* appears abruptly and does not imply an evolutionary connection to ape-like species such as the australopithecines. While humans do share functional similarities with apes, these can be explained by common design. Instead, humans exhibit many behaviors and intellectual abilities that natural selection cannot explain. Human common ancestry with apes is not supported by the evidence.

Many materialists still claim the tree of life is a valid concept—but maintain that we just haven't yet discovered the correct means of constructing it. However, if universal common ancestry cannot be demonstrated by the fossil record, the genetic data, or anatomical homology, why must we believe it is correct?

EPIC FAIL

Let's do the math. For the refutation of materialism and naturalism, only one of the seven tenets had to fail. But six have been refuted by the evidence, and serious doubts have been presented regarding the seventh (common descent). Materialism should not be considered a viable position, and as we'll discuss in the next two chapters, materialists should not be given veto power over scientific inquiry.

This book has presented evidence that nature is full of structures rich in specified and complex information. From the origin and fine-tuning of the universe to the origin and development of life's complexity, the evidence provides a powerful, positive scientific argument that an intelligent agent has been at work.

Despite the failure of materialism, intelligent design is not merely a negative argument against that philosophy. Throughout this book, we have presented the positive case for design. In the next section, we'll discuss the conflict over ID within our culture and answer some common objections.

SECTION VI

ACADEMIC FREEDOM

CHAPTER 19

MATERIALISM OF THE GAPS

Question:

Why do you think some scientists fight so strongly against intelligent design?

MATERIALISM OF THE GAPS

"The struggle is always between the individual and his sacred right to express himself and... the power structure that seeks conformity, suppression, and obedience."

—Former U.S. Supreme Court Justice William O. Douglas[1]

The first chapter of this book opened with the advice of Socrates, who challenged us from across the millennia to "follow the argument wherever it leads." Throughout the book, we have endeavored to do just that as we explored the scientific evidence.

At this point, you might feel as we do—that the evidence strongly supports intelligent design. However, you may also wonder why so many in the scientific community and elsewhere oppose ID. That question brings us to this chapter, which explores the relationship between ID, materialism, and our culture.

Over the last century, materialists have gained pervasive influence over the scientific establishment, as well as other important facets of society such as education, the media, and the legal system. Many scientists believe that materialism is essential for scientific endeavors, and they fight to stifle the debate over intelligent design.

The intensity of this opposition raises two significant questions:

1. If the evidence indicates that intelligent design is the best explanation for many aspects of nature, why is ID opposed by so many scientists?
2. If materialist philosophy is in great doubt, how does it maintain its cultural power?

The answers to these questions will provide a broader perspective on this debate, and on the scientific enterprise in general.

OPPOSITION FROM THE SCIENTIFIC ESTABLISHMENT

When trying to understand scientific opposition to ID, people make two common mistakes.

One mistake is to believe that scientists universally oppose ID. That's not true. While ID is currently a minority scientific viewpoint, there is a growing community of scientists who are sympathetic to intelligent design. This expanding community pursues a vibrant research program and publishes data supporting design in peer-reviewed scientific journals.

Another mistake is to assume that if the scientific establishment opposes ID, it always does so based on objective scientific investigations. Some scientists do oppose ID because of their interpretations of the evidence—and of course, there might be strong rebuttals from the ID viewpoint to their claims. But there are other reasons for opposition to ID which have little or nothing to do with evidence. Four prominent reasons are:

- Saving science from the ignorant masses.
- Defending a materialist worldview.
- Protecting their comfort zone.
- Going with the flow.

Let's investigate these motives.

FIGHTING AGAINST IGNORANCE

For thousands of years, superstitious people tended to attribute anything they didn't understand, any gaps in their knowledge of the natural world, to God or the gods. That has been referred to as "God-of-the-gaps" argumentation. Over time, many of these gaps were filled with scientific explanations instead of the supernatural.

In today's culture, many scientists believe that their work must be based strictly upon materialism—meaning they think that any threat to materialism directly threatens science. Since ID points to evidence challenging materialism, critics conclude that it is an "anti-science" resurgence of ignorance and superstition based on God-of-the-gaps thinking.

Philosopher and political scientist Marshall Berman expresses those fears:

> The current Intelligent Design movement poses a threat to all of science and perhaps to secular democracy itself.... Replacing sound science and engineering with pseudo-science, polemics, blind faith, and wishful thinking won't save you when the curtain of "Dark Ages II" begins to fall![2]

Does ID really threaten to bring on a new Dark Ages? Of course not.

ID challenges a reigning scientific paradigm. But as sociologist Steve Fuller says, ID is not *anti-science*, but rather *anti-establishment*.[3] ID theorists want more scientific investigation, not less. They simply want the freedom to follow the evidence without harassment or philosophical restrictions.

As elaborated in Appendix A, an ID-based view of science promises to open new avenues of scientific investigation. Without materialist paradigms governing science, perhaps more scientists would have sought function for structures like "junk" DNA and vestigial organs, rather than assuming they were non-functional evolutionary relics.

Now that we've addressed the "Dark Ages" issue, let's consider a more sensible question: *Does ID present a God-of-the-gaps argument?* Again, it does not. First of all, ID refers to an intelligent cause and does not identify the designer as "God."

Moreover, ID theory is not based on what we *don't* know (gaps), but on what we *do* know (evidence). For example, we know from experience that high levels of complex and specified information come from the action of an intelligent agent. When we find objects in nature with high CSI, we have positive evidence for design. (We'll explore the positive case for design further in Chapter 20.)

There will, of course, always be gaps in scientific knowledge. But by insisting that all gaps must be filled with materialistic explanations, ID critics are engaging in "materialism-of-the-gaps" thinking. While ID critics may think they are protecting science, they are actually hindering it by restricting scientific inquiry.

In contrast, ID rejects **gaps-based reasoning** of all kinds, and suggests we should follow the evidence where it leads—free from philosophical blinders.

DEFENDING A MATERIALIST WORLDVIEW

Some critics oppose ID not simply because they think they're saving science, but also to validate their personal worldview. This may not be true for all ID critics, but the advocacy of evolution has a long history that is intertwined with promoting materialism. Richard Dawkins showed this when he said "Darwin made it possible to become an intellectually fulfilled atheist."[4]

Likewise, Harvard paleontologist and author Richard Lewontin states that materialism must be protected at all costs:

> [W]e have a prior commitment, a commitment to materialism. It is not that the methods and institutions of science somehow compel us to accept a material explanation of the phenomenal world, but, on the contrary, that we are forced by our *a priori* adherence to material causes to… produce material explanations… [T]hat materialism is absolute, for we cannot allow a Divine Foot in the door.[5]

Lewontin is not alone in this view. Some scientists, educators and journalists have become so entrenched in seeing the world through a materialist prism, that they are no longer open to contrary evidence. As Darwinian philosopher Michael Ruse suggests, "for many evolutionists, evolution has functioned as… a secular religion."[6]

We're not suggesting that Darwinian evolution *is* a religion—we believe it is a science that is refuted by much evidence. Nonetheless, many materialists *treat* Darwinism like a religion to be defended at all costs.

COMFORT ZONE

But not all scientists who reject ID are attempting to save science or to protect their philosophy. There also seems to be something in human nature that prevents us from considering ideas that challenge reigning paradigms. The famous historian of science, Thomas Kuhn, explained this phenomenon:

> No part of the aim of normal science is to call forth new sorts of phenomena; indeed those that will not fit the box are often not seen at all. Nor do scientists normally aim to invent new theories, and they are often intolerant of those invented by others.[7]

As a new theory that challenges conventional wisdom, ID faces this very type of intolerance. For example, in 2002, the American Association for the Advancement of Science (AAAS) issued an official statement that "the ID movement has failed to offer credible scientific evidence."[8]

But when some of the AAAS board members were surveyed, it was found that they voted to declare "intelligent design as unscientific without actually reading for themselves the academic books and articles by scientists proposing the theory."[9]

Calm, objective scientific groups don't normally issue official declarations against a theory, much less without having studied it. Politics, rather than science, is driving the behavior of these ID critics.

Evolutionary biologist Stephen Jay Gould explained why such political tactics are dangerous to the progress of science:

> Automatically rejecting dissenting views that challenge the conventional wisdom is a dangerous fallacy, for almost every generally accepted view was once deemed eccentric or heretical.[10]

Some mistakenly believe that scientists are always open-minded and objective. Again, Gould dispelled this idea by observing that "[t]he stereotype of a fully rational and objective 'scientific method,' with individual scientists as logical (and interchangeable) robots, is self-serving mythology."[11]

According to the German philosopher Arthur Schopenhauer, "All truth passes through three stages":

- First, it is ridiculed.
- Second, it is violently opposed.
- Third, it is accepted as being self-evident.[12]

As philosopher Jay Richards has observed, intelligent design is currently somewhere between the first and second stage. The vehement opposition to ID is not necessarily a sign that the theory lacks merit. Instead, it might be, as Gould suggested, that many critics automatically reject views that challenge the conventional wisdom.

Now we'll turn our attention from those who tenaciously fight against ID because they think they are saving science, protecting their philosophy, or maintaining their comfort zone—and we'll consider scientists and educators who are just following the majority.

GOING WITH THE FLOW

Because materialism has become so infused in the culture of the scientific community, monetary grants tend go to those who promise to advance materialist ideas, or, more likely, simply leave those ideas unquestioned. This creates a powerful motivation to be a true believer, or at least simply go along with the majority viewpoint.

Biologist Lynn Margulis recounted a lecture in which Richard Lewontin admitted that the mathematical predictions of Darwinian evolution were not being confirmed by field and laboratory studies:

> So I said, "Richard Lewontin, you are a great lecturer to have the courage to say it's gotten you nowhere. But then why do you continue to do this work?" And he looked around and said, "It's the only thing I know how to do, and if I don't do it I won't get grant money."[13]

There is another reason why scientists follow the majority. Most scientists are highly specialized and often have limited scientific knowledge outside their area of focus. It's not unusual for a scientist to realize that materialistic mechanisms are weak within his specialty, but then to assume that they must be valid for other areas. If everyone assumes "someone else has figured it all out," then a paradigm can persist through groupthink despite being full of fractures.

Yet another reason for going with the flow is that many (probably most) scientists have no reason to concern themselves with questions about origins. National Academy of Sciences (NAS) member Philip Skell surveyed scientists from many fields on whether they needed Darwinian evolution to conduct their research, and reported:

> I found that Darwin's theory had provided no discernible guidance, but was brought in, after the breakthroughs, as an interesting narrative gloss.[14]

It's much easier to endorse the leading materialist paradigms than to stick your neck out by bucking the trend.

We've examined various reasons for scientific opposition to ID, from a desire to save science from a new Dark Ages to just going with the flow. This leads to the second question raised at the beginning of the chapter: *How has materialism become so entrenched in our culture?* The answer can be summed up in one word: intimidation. As we'll see in the next section, many scientists who dissent from the mainstream Darwinian view have faced harsh consequences for doing so.

WE HAVE WAYS TO MAKE YOU SHUT UP

The 2008 documentary *Expelled: No Intelligence Allowed* observes that scientific ideas begin with scholars in the **academy**, but if the ideas are to get into the minds of the public, they must pass through a series of checkpoints. Members of the **Darwin lobby** use these checkpoints to stop ID-friendly arguments in their tracks (Figure 19-1).

> **Darwin Lobby:**
>
> ***A coalition of scientific, educational, legal, and political activist groups that work to censor non-evolutionary viewpoints from the public.***

In the film, the first checkpoint is the academy itself, which controls peer-reviewed journals and research grants, and decides who gets promoted.

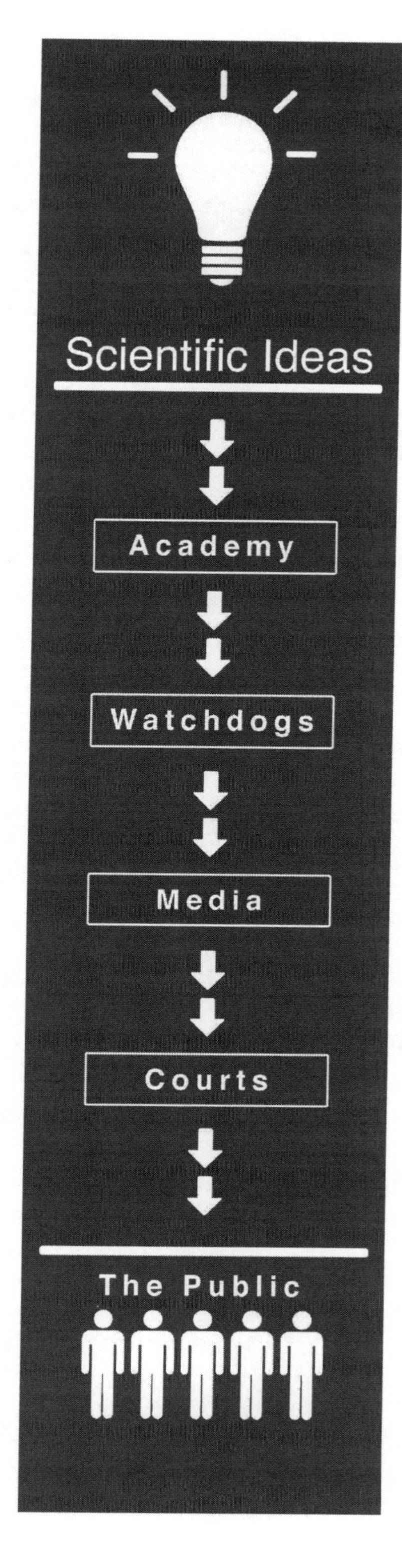

Figure 19-1: How the Darwin Lobby Censors New Scientific Ideas.

In their book *What Darwin Got Wrong*, materialist academics Jerry Fodor and Massimo Piattelli-Palmarini explain the intimidation tactics used to silence critics of Darwinian evolution:

> [W]e've been told by more than one of our colleagues that, even if Darwin was substantially wrong to claim that natural selection is the mechanism of evolution, nonetheless we shouldn't say so. Not, anyhow, in public. To do that is, however inadvertently, to align oneself with the Forces of Darkness, whose goal is to bring Science into disrepute.[15]

The second checkpoint is composed of watchdog groups that organize and kindle opposition against skeptics of materialism. Some of the most prominent of these activist groups are the National Center for Science Education (NCSE), the American Civil Liberties Union (ACLU), and Americans United for Separation of Church and State.

The next checkpoint is the media, which carefully selects the sources of information it will broadcast to the public on this issue. When all those checkpoints fail, the final checkpoint is the courts.

Today, materialists hold an immense amount of cultural power, which rests largely upon the facade that, as Eugenie Scott of the NCSE put it in a news article, "[t]here are no weaknesses in the theory of evolution."[16] Likewise, the National Academy of Sciences boldly expressed this view when they declared that ID "is not supported by scientific evidence" and "[t]here is no scientific controversy about the basic facts of evolution," because neo-Darwinian evolution is "so well established that no new evidence is likely to alter" it.[17]

Such rhetoric is the basis for many misunderstandings about ID. No one, including ID theorists, argues against all types of evolution—

because that would include microevolution. But the message in the NAS's statement is that neo-Darwinian theory should be accepted entirely, without question.

When scientists no longer have the freedom to challenge prevailing ideas, the progress of science is threatened. The simple fact is that ID arguments pose a strong challenge to the conventional power structure, and many materialists are fighting to preserve it. Let's examine a few of the tactics they use:

- Fallacies of reasoning.
- *Inherit the Wind* stereotype.
- Academic and career concerns.

FALLACIES OF REASONING

In a fair debate, both sides present arguments based on reason and evidence. But when the facts are not on your side, it can be tempting to resort to **fallacies of reasoning.** We described one fallacy, circular reasoning, in Chapter 2. Here are a few others.[18]

- ***Ad Hominem* Attack**: A personal attack intended to discredit a person rather than an argument. This is an extremely common tactic used by ID critics to intimidate those who support ID.

  > *Intelligent design proponents are either basically ignorant of science, seriously deluded, or fundamentally dishonest.*[19]

- **Genetic Fallacy:** Attacking the origin of an argument rather than the argument itself.

  > *Armed with another new idea from the Discovery Institute, that bastion of ignorance, right-wing political ideology, and pseudo-scientific claptrap, the creationist movement has mounted yet another assault on science.*[20]

A related fallacy is **Questioning Motives,** in which one attacks another person's reason for making an argument rather than the actual argument.

- **Straw Man**: Misrepresenting an opponent's argument, and then attacking the false version.

  > *ID says that if evolution is incorrect, we can invoke a cause outside of nature, a supernatural creator. ID is thus religion and not science.*[21]

- **Guilt by Association:** A logical fallacy that claims, in effect, "A has some similarity or association with B, therefore A is the same as B."

  > *Creationist "A" once wrote about Topic Y.* Discovering Intelligent Design *also discusses Topic Y. Therefore* Discovering Intelligent Design *is promoting creationism.*[22]

- **False Choice:** Portraying two options as mutually exclusive when they really are not.

  > *Either a person has scientific integrity, or he accepts ID. He can't do both.*[23]

- **Unwarranted Conclusion**: Over-extrapolating from the evidence to make an unrelated or unjustified argument.

  > *If microevolution is supported by evidence, then macroevolution must also be true.*[24]

- **Professional Intimidation:** Citing one's educational training or professional accomplishments to silence another's viewpoint.

  I am a trained biologist capable of speaking with authority about evolution. You are not in a position to evaluate the central theory of my field.[25]

- **Bullying Tactic:** Using outlandish rhetoric to play on emotions and force someone into agreement.

  Intelligent design theory and its companions are nasty, cramping, soul-destroying reversions to the more unfortunate aspects of 19th century America. Intelligent design is a moral evil.[26]

- **Argument from Popularity:** Urging people to believe an idea because it is popular. A variation of this is the **Appeal to Authority.**

  Virtually all educated people—including almost all scientists—believe that life appeared gradually in an evolutionary process, therefore so should you.[27]

ID critics commonly make appeals to authority. The goal is to discourage you from thinking for yourself, from looking closely at the scientific evidence. Instead, they would have you accept something simply because someone else does. However, scientific debates should be determined by evidence, not a popularity contest. British atheist Bertrand Russell put it this way: "The fact that an opinion has been widely held is no evidence whatever that it is not utterly absurd."[28]

There are many other fallacies of reasoning, but we've pointed out some that are used most often by ID critics. Now, let's look at another materialist tactic.

Figure 19-2: John Scopes.

INHERIT THE WIND STEREOTYPE

In 1925, the Tennessee State Legislature passed a law making it illegal to teach human evolution in public schools. That same year, a substitute biology teacher named John Scopes (Figure 19-2) used a biology textbook that promoted both evolution and eugenics.[29] Scopes soon found himself at the center of the nation's attention after being recruited to challenge the Tennessee law by the ACLU.

Obviously, criminalizing the teaching of human evolution was a terrible idea. Students should have full freedom to learn about scientific theories, whether we agree with them or not. The jury in the Scopes trial found the teacher guilty, but his conviction was overturned on a technicality. The result was a cultural earthquake.

In 1955, a heavily fictionalized account of the trial was made into a play, and five years later, a movie—both of which were titled *Inherit the Wind.*[30] A new character was created, the Reverend Jeremiah Brown—a vindictive religious zealot who leads the charge to ban Darwin's theory from schools.

The play is often performed, including a 2007 Broadway production. The film version of *Inherit the Wind* is also commonly shown in high school and college classrooms across the nation, spreading the message that opponents of Darwinian evolution are ignorant bigots who want to shut down scientific thought. The evolutionists, on the other hand, are portrayed as intelligent, articulate, and open-minded. As Phillip Johnson writes:

> *Inherit the Wind* is a masterpiece of propaganda, promoting a stereotype… The rationalists have all the good lines and all the virtues. [The opponents of Darwinism]... are a combination of folly, pride, and malice and their followers are so many mindless puppets.[31]

The audience is led to believe that the only way to be rational and fair-minded is to join the followers of Darwin and mock the backwards religious fundamentalists who oppose them. After all, no one wants to be associated with the uneducated and close-minded Jeremiah Brown.

A lot has changed in the more than eighty years since the Scopes trial. There are still many angry and intolerant people who want to shut down scientific debate. But the cast of the modern drama has changed—and it isn't fiction. Writing about those who dissent from the Darwinian consensus, anti-ID biologist P. Z. Myers declared:

> The only appropriate responses should involve some sort of righteous fury, much butt-kicking, and the public firing and humiliation of some teachers, many school board members, and vast numbers of sleazy, far-right politicians.[32]

Likewise, University of Chicago evolutionary biologist Jerry Coyne unapologetically states: "adherence to ID… should be absolute grounds for not hiring a science professor."[33]

In the twenty-first century, it is the materialists who are trying to silence and intimidate legitimate scientists who wish to consider the possibility of intelligent design in their scientific studies.

ACADEMIC AND CAREER CONCERNS

In recent years, there has been a widespread pattern of attacks on academic freedom for proponents of intelligent design. These attacks seek to censor any scientific evidence that challenges neo-Darwinian evolution, while targeting academics who support intelligent design.

There are many examples of such academic resistance to ID, but here are four instances where would-be censors sought to prevent students from hearing about the science behind intelligent design:

- In 2005, the NCSE, ACLU, and other members of the Darwin lobby convinced a federal judge to ban intelligent design from science classrooms in Dover, Pennsylvania. (We'll say more about this lawsuit in the next chapter.)

- That same year, the president of the University of Idaho instituted a campus-wide classroom speech code, in which "evolution" was "the only curriculum that is appropriate" for professors to cover in science classes.[34]

- In 2006, a professor of biochemistry and leading biochemistry textbook author at the University of Toronto stated that a major public research university "should never have admitted" students who support ID, and should "just flunk the lot of them and make room for smart students."[35]

- In 2011 students at the University of Waikato were told that if they "use examples such as the bacterial flagellum to advance an ID view then they should expect to be marked down."[36]

As shown in the documentary *Expelled*, many scientists, educators, and students who supported ID have sustained great damage to their careers as a result.[37] For example:

- In 2004, after Smithsonian research biologist Richard Sternberg allowed a peer-reviewed, pro-ID paper to be published in a scientific journal, his superiors investigated his religious and political affiliations in violation of the First Amendment. According to an investigation by U.S. Congressional Subcommittee Staff, Smithsonian officials "explicitly acknowledged in emails their intent to pressure Sternberg to resign because of his role in the publication of [the pro-ID] paper and his views on evolution."[38]
- Pro-ID adjunct biology professor Caroline Crocker lost her job at George Mason University in 2005 after teaching students the evidence both for and against Darwinism in the classroom, and *mentioning* ID as a possible alternative.[39]
- At Iowa State University (ISU), over 120 faculty members signed a petition denouncing ID and calling on "all faculty members to… reject efforts to portray Intelligent Design as science."[40] The campaign focused on astronomer Guillermo Gonzalez, co-author of the book *The Privileged Planet*, who was up for tenure. As one ISU scientist said, the school "is not a friendly place for him to develop further his IDeas."[41] In 2007, Gonzalez was denied tenure.

ID critics in some arenas have become so intolerant that in 2007, the Council of Europe, the leading European "human rights" organization, adopted a resolution calling the teaching of ID a potential "threat to human rights"![42]

People who are confident that the evidence is on their side don't usually struggle so hard to muzzle opposing views.

CONCLUSION

Throughout this book we have explained why some scientists believe the evidence supports ID. While some opposition to ID is based upon the science, much of it is not. When critics claim ID is wrong, but then refuse to discuss the scientific evidence, it's safe to suspect that other factors are at work.

Rather than winning the debate through reasoned scientific arguments, materialists commonly rely on logical fallacies to attack

Group	Teach Evidence Supporting Darwin?	Teach Evidence Against Darwin?	Require Intelligent Design?	Expose Students To Different Scientific Views (Including Those You Disagree With)?	% Of Public That Agrees With Educational Policy
Darwin Lobby	Yes	No	No—ban it	No	14%
ID Movement	Yes	Yes	No—but don't ban it either	Yes	78%

Figure 19-3: Educational Philosophies of the Darwin Lobby vs. the ID Movement.

ID. They use bully tactics, make *ad hominem* attacks, offer straw-man definitions that distort ID, base many of their claims upon unwarranted conclusions, and increasingly resort to censorship.

In 1925, critics of Darwin sought to ban the teaching of evolution in public schools. But today, the vast majority of Americans—including leading proponents of intelligent design—don't want to ban anything and simply want a robust debate over evolution. As a 2009 Zogby poll revealed, 78% of American voters agree that "[b]iology teachers should teach Darwin's theory of evolution, but also the scientific evidence against it."[43]

Unfortunately, the Darwin lobby doesn't seem to want real scientific debate. First, they seek to marginalize proponents of intelligent design in the scientific community by restricting academic freedom. Second, in the context of school curricula, they try to censor any scientific critiques of the evolutionary viewpoint.

Figure 19-3 summarizes the positions of the ID movement compared to the Darwin lobby on how public schools should handle this debate.

Modern defenders of Darwin would reject the advice of John Scopes who warned:

> If you limit a teacher to only one side of anything, the whole country will eventually have only one thought.... I believe in teaching every aspect of every problem or theory.[44]

According to Supreme Court Justice Antonin Scalia, today we have "*Scopes*-in reverse,"[45] where viewpoints that do not support Darwinian evolution are excluded from classrooms through a climate of fear and intimidation.

DISCUSSION QUESTIONS

1. Which of the fallacies of reasoning have you used or had used against you? Explain what happened in these cases. Do you think any of these fallacies are effective if used against intelligent design? Explain.

2. If you randomly polled scientists, a majority would probably say they disagree with intelligent design. Does this mean the majority viewpoint is correct? Why or why not?

3. Have you ever seen *Inherit the Wind*, or heard references to it? Is it a fair portrayal of people who oppose Darwinian evolution?

4. The "God-of-the-gaps" charge has been used for years against arguments that challenge Darwinian theory. Do you agree with this argument against ID? Explain why or why not.

5. What is the basic difference between how proponents of intelligent design and materialists treat views they disagree with?

6. Does disagreeing with a viewpoint imply you should censor it from academic debate? What are the consequences of censorship?

CHAPTER 20

ANSWERING THE CRITICS

Question:

If someone said that ID is just religion, and not science, how would you reply?

ANSWERING THE CRITICS

"If a book be false in its facts, disprove them; if false in its reasoning, refute it. But for God's sake, let us freely hear both sides if we choose."

—Thomas Jefferson[1]

Imagine that a long-time, highly respected running champion, Matt, and a newcomer, Indy, are competing in a renowned marathon. When Indy unexpectedly wins, Matt accuses him of using performance-enhancing drugs and demands that he be disqualified.

Indy insists he ran a fair race, and provides test results showing he wasn't using drugs. But Matt—citing irrelevant facts and no pertinent evidence—insists the test is invalid. Due to his prestige and influence, the media sides with Matt. Feeling pressured, the race officials disqualify Indy and declare Matt the winner.

Perhaps Matt won on paper, but did the best runner really get the award?

As this book has shown, intelligent design competes well in a head-to-head scientific contest with materialistic explanations of origins. Consequently, one of the main strategies of materialists is to define the rules of the debate so that ID is disqualified. In essence, they want to win the debate without actually having one.

There are three common objections that materialists use when trying to get ID disqualified from a scientific debate:

- ID isn't science.
- ID is religion.
- ID is just politics.

In this chapter we'll assess these claims, one at a time.

IS INTELLIGENT DESIGN SCIENCE?

Perhaps the most common objection to intelligent design is the claim that it is not science. The answer to this argument is simple: ID is science because it uses the **scientific method** to make its claims. Its proponents have done a considerable amount of testing and experimentation, and have published peer-reviewed articles backing ID.

The scientific method is usually described as a four-step process consisting of:

1. Observation
2. Hypothesis
3. Experimentation
4. Conclusion

In addition, scientists often hold that theories should make falsifiable predictions, and undergo peer review. As we'll see, ID follows this precise methodology.

THE POSITIVE CASE FOR DESIGN

When arguing that ID is not science, critics often contend that ID is merely a negative argument against evolution. Providing an example of this argument, biologist Kenneth Miller says ID "is always negative, and it basically says, if evolution is incorrect, the answer must be design."[2]

However, as this book has repeatedly shown, ID is based upon the positive argument that nature contains information which, in our experience, comes from the action of intelligence. The rest of this section will show how ID uses the scientific method to make positive cases for design in biochemistry, paleontology, systematics, and genetics.

Example 1: Biochemistry

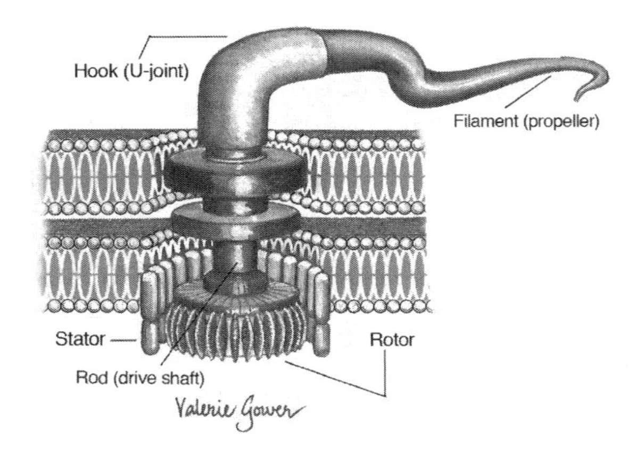

Observation: Intelligent agents solve complex problems by acting with a goal in mind, producing high levels of CSI. As Stephen Meyer explains, in our experience, systems with large amounts of specified complexity—such as codes and languages—invariably originate from an intelligent source.[3] Likewise, in our experience, intelligence is the only known cause of irreducibly complex machines.[4]

Hypothesis (Prediction): Natural structures will be found that contain many parts arranged in intricate patterns (including irreducible complexity) that perform a specific function—indicating high levels of CSI.

Experiment: Experimental investigations of DNA indicate that it is full of a CSI-rich, language-based code. Biologists have performed mutational sensitivity tests on proteins and determined that their amino acid sequences are highly specified. Additionally, genetic knockout experiments and other studies have shown that some molecular machines, like the flagellum, are irreducibly complex.[5]

Conclusion: The high levels of CSI—including irreducible complexity—in biochemical systems are best explained by the action of an intelligent agent.

Example 2: Paleontology

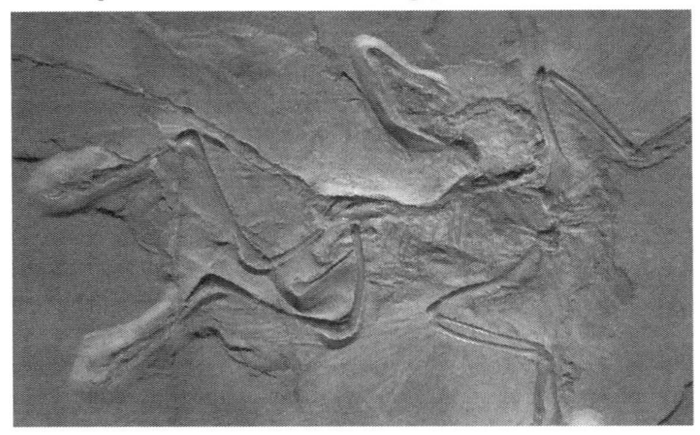

Observation: Intelligent agents rapidly infuse large amounts of information into systems. As four ID theorists write: "intelligent design provides a sufficient causal explanation for the origin of large amounts of information… the intelligent design of a blueprint often precedes the assembly of parts in accord with a blueprint or preconceived design plan."[6]

Hypothesis (Prediction): Forms containing large amounts of novel information will appear in the fossil record suddenly and without similar precursors.

Experiment: Studies of the fossil record show that species typically appear abruptly without similar precursors.[7] The Cambrian explosion is a prime example, although there are other examples of explosions in life's history. Large amounts of complex and specified information had to arise rapidly to explain the abrupt appearance of these forms.[8]

Conclusion: The abrupt appearance of new fully formed body plans in the fossil record is best explained by intelligent design.

Example 3: Systematics

Observation: Intelligent agents often reuse functional components in different designs. As Paul Nelson and Jonathan Wells explain: "An intelligent cause may reuse or redeploy the same module in different systems… [and] generate identical patterns independently."[9]

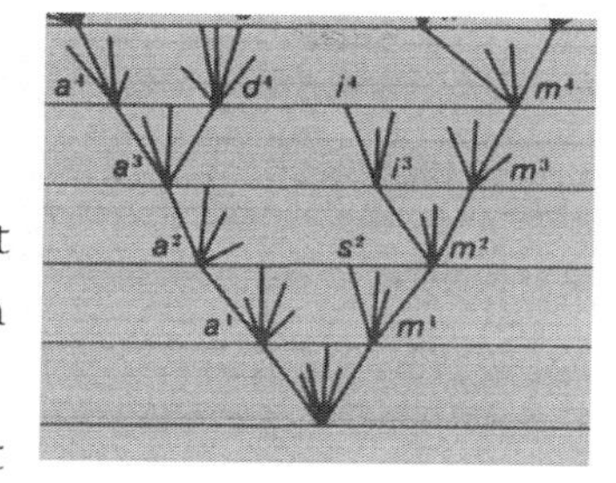

Hypothesis (Prediction): Genes and other functional parts will be commonly reused in different organisms.[10]

Experiment: Studies of comparative anatomy and genetics have uncovered similar parts commonly existing in widely different organisms. Examples of extreme convergent evolution show reusage of functional genes and structures in a manner not predicted by common ancestry.[11]

Conclusion: The reusage of highly similar and complex parts in widely different organisms in non-treelike patterns is best explained by the action of an intelligent agent.

Example 4: Genetics

Observation: Intelligent agents construct structures with purpose and function. As William Dembski argues: "Consider the term 'junk DNA.'… [O]n an evolutionary view we expect a lot of useless DNA. If, on the other hand, organisms are designed, we expect DNA, as much as possible, to exhibit function."[12]

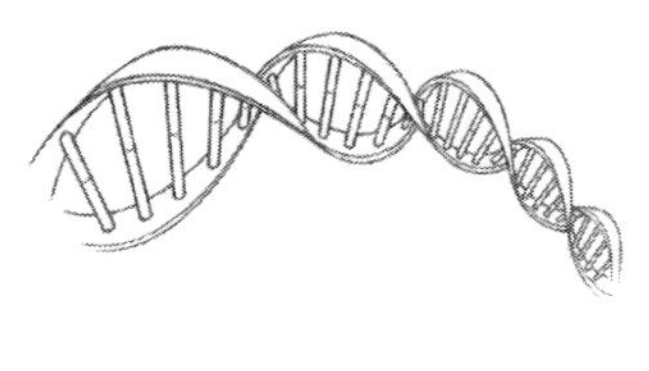

Hypothesis (Prediction): Much so-called "junk DNA" will turn out to perform valuable functions.

Experiment: Numerous studies have discovered functions for "junk DNA." Examples include functions for pseudogenes, introns, and repetitive DNA.[13]

Conclusion: The discovery of function for numerous types of "junk DNA" was successfully predicted by intelligent design.

While the evidence points strongly to design in these four areas, like all scientific conclusions, the conclusion of design is subject to future scientific discoveries.

Intelligent design uses the scientific method to make its claims and is based upon a positive argument grounded in testable and falsifiable predictions. But there is an additional requirement often applied to the scientific method.

PEER REVIEW

Peer review is a process by which scientists double-check the work of their peers to make sure it has been done correctly. It's not a perfect process,[14] but it can help ensure that published scientific papers meet high standards. However, in the case of ID, many scientific journals are controlled by critics who claim ID isn't science. Consequently, article manuscripts drawing ID-friendly conclusions are often dismissed, sometimes with outright hostility.

Many materialists demand that, to be considered scientific, the case for design *must* be published in peer-reviewed scientific literature. This leads to a circular argument: ID isn't science because it isn't published in peer-reviewed journals, and it can't be published in peer-reviewed journals because it (allegedly) isn't science.

For example, when Michael Behe submitted to a mainstream biology journal a paper critical of Darwinian evolution, he was told that it could not be published because "your unorthodox theory would have to displace something" defending the "current paradigm."[15] Likewise, as discussed in the previous chapter, after Richard Sternberg allowed a pro-ID paper in a Smithsonian biology journal, his supervisors harassed him in the hope he would resign.

Because peer review is an imperfect system, the U.S. Supreme Court has ruled that "in some instances well-grounded but innovative theories will not have been published."[16] Likewise, Stephen Jay Gould recognized:

> The quality of a scientific approach or opinion depends on the strength of its factual premises and on the depth and consistency of its reasoning, not on its appearance in a particular journal or on its popularity among other scientists.[17]

However, even though the requirement for peer-reviewed publication has been abused to block pro-ID views, pro-ID scientists *have* nonetheless published multiple peer-reviewed articles in scientific journals.[18] As of the publication of this book, the total number is over 50 peer-reviewed scientific papers.[19]

Some of these peer-reviewed, pro-ID articles have been cited and discussed throughout this book. A more complete list can be found on the Discovery Institute's website (see the box below). This indicates that scientists in various fields have examined ID arguments and found them to be scientifically defensible.

For additional information on ID research projects, visit the following websites:

- List of Peer-Reviewed Pro-ID Papers: **www.discovery.org/a/2640**
- Biologic Institute: **www.biologicinstitute.org**
- *BIO-Complexity* Journal: **www.bio-complexity.org**
- Evolutionary Informatics Lab: **www.evoinfo.org**

These pro-ID papers could not have been published without a growing body of ID-friendly scientists conducting ID-related research. For example, Biologic Institute is a research lab that is "developing and testing the scientific case for intelligent design in biology."[20] Its research program includes:

- Building and testing computer models that study the ability of unguided mechanisms compared to intelligent causes to produce new information.

- Examining the cosmological, physical, and biological fine-tuning required for life.

- Investigating the ways that humans go about designing complex structures, so that scientists can recognize the hallmarks of design.[21]

Some significant, peer-reviewed ID research studies include:

- Mutational sensitivity tests performed by Douglas Axe and published in *Journal of Molecular Biology* show that functional protein folds in enzymes contain high CSI and could not arise by unguided evolution.[22]

- William Dembski and Robert Marks's Evolutionary Informatics Lab has published papers in various science and engineering journals analyzing computer simulations of evolution, and showing that intelligence, not natural selection, creates new information.[23]

- Work published by Michael Behe in *Quarterly Review of Biology* and *Protein Science* shows that Darwinian evolution is unlikely to produce new complex biochemical features.[24]

A common theme running through all of these papers is that unguided evolutionary mechanisms cannot generate complex biological features. Some intelligent cause is required to generate new information.

Clearly ID has a bona fide, peer-reviewed scientific research program.

MOVING THE GOALPOSTS

The scientific method was developed to allow an unhindered empirical search for truth. Nothing in its procedures prohibits experimental results from pointing toward any testable conclusion—including intelligent design. As we have seen, ID uses the scientific method to make testable, falsifiable predictions that are confirmed by experiments. ID arguments and research have been published in peer-reviewed scientific journals.

By any reasonable definition, ID should qualify as science. Many materialists can't tolerate this, so they respond by attempting to change the definition of science to specifically exclude ID. In their view, *only materialistic explanations may be considered scientifically valid*.

We saw such thinking in Chapter 1 where one biologist published a letter in *Nature* stating, "Even if all the data point to an intelligent designer, such an hypothesis is excluded from science because it is not naturalistic."[25]

This attempt by materialists has created an ongoing debate over the proper definition of science. The ID position contends that restricting experimental results in advance for philosophical reasons is not science. In fact, it hinders science.

This exemplifies how materialists seek to define the terms of the debate so they can win it without ever actually debating.

IS INTELLIGENT DESIGN RELIGION?

Critics often level the charge that ID is religion disguised as science—usually equating ID with creationism. We addressed this argument in Chapter 1, but it's worth considering in more detail.

Some critics try to link ID to creationism simply by using the label "intelligent design creationism" or by talking about "creationists" while attacking ID. Another argument holds that if a creationist and an ID proponent ever talk about the same topic, then ID must be creationism.[26] This is like saying that if dogs and cats both eat meat, then dogs are cats.

However, there are at least two major differences between ID and creationism:

- Creationism starts with a religious text, such as the Bible, and tries to fit the findings of science to it. In contrast, the theory of ID does not rely on any religious text. ID starts with the empirical data and relies only on the scientific evidence to detect design.

- Creationism holds that a supernatural or divine power created life.[27] ID merely refers to an intelligent cause, and does not attempt to address religious questions about the identity of the designer.

Because ID respects the limits of scientific inquiry and does not inject theology into scientific investigations, it lacks the key characteristics that cause creationism—and religion in general—to be unscientific and unconstitutional to teach in public schools.

The assertion that ID is religion received support in 2005 when a federal district court judge in Pennsylvania ruled that ID is a religious concept, and banned ID from science classrooms in the Dover School District.[28] His ruling gained instant popularity among those seeking to dismiss ID. But were the judge's claims correct, and what is the impact of this ruling?

IT'S NOT OVER BECAUSE OF DOVER

In 2004, members of the Dover Area School Board adopted a policy requiring that a four-paragraph statement mentioning ID be read to students in biology class. The board members approached the subject from a Biblical creationist perspective and knew little about intelligent design.[29]

Leading ID groups oppose pushing ID into public schools, and they strongly urged the Dover school board to repeal its ill-advised policy.[30] The board refused, and attorneys working with the ACLU and the NCSE seized the opportunity to take the Dover School District to court.

Federal district court Judge John E. Jones III decided to use the lawsuit as an occasion to rule on the scientific status of ID, concluding that "ID is an interesting theological argument, but... not science."[31] He felt it was his place to define science, declaring that it must be "based upon natural explanations."[32]

Since that trial, materialists have been proclaiming that the ID debate is over because a federal judge decided that it's a religious concept, not a scientific one.[33] But judges are not inerrant. In fact, the *Kitzmiller v. Dover* ruling contains many factual and legal mistakes. Judge Jones's errors include:

- Adopting a false definition of ID by claiming that ID requires supernatural (i.e., divine) causation, that it is nothing more than creationism relabeled, and that it is a strictly negative argument against evolution.

- Denying the existence of pro-ID, peer-reviewed, scientific publications and research that were testified about in his own courtroom.

- Adopting an unfair double standard of legal analysis where religious implications, beliefs, and motives count against ID but never against materialist theories of origins.

- Presuming it is permissible for a federal judge to try to define science, settle controversial scientific questions, and explain the proper relationship between evolution and religion.

- Attempting to turn science into a voting contest by claiming that popularity is required for an idea to be scientific.

The fact that one federal judge disagrees with ID does not settle this debate. In America's three-tiered system of federal courts, the *Kitzmiller v. Dover* ruling was issued from the lowest level. No other case to date has squarely dealt with teaching ID in public schools. While Judge Jones did successfully ban ID in Dover, no other school district in the U.S. is subject to the ruling, and there is no national law that prohibits teaching ID.

A court ruling cannot negate the scientific evidence pointing to intelligent design. Rather than showing that ID is not science, the *Kitzmiller v. Dover* decision shows that scientific and academic freedom in this country are at risk.

Commenting on Judge Jones's activism, anti-ID legal scholar Jay Wexler stated that "the part of *Kitzmiller* that finds ID not to be science is unnecessary, unconvincing, not particularly suited to the judicial role, and even perhaps dangerous both to science and to freedom of religion."[34]

DO RELIGIOUS BELIEFS MATTER?

Judge Jones stated in his ruling that ID is a religious concept, in part because "Professors Behe and Minnich admitted their personal view is that the designer is God."[35] Is that a valid objection?

Of course not. Everyone has personal beliefs. But personal beliefs don't change the evidence.

In science, only one thing ultimately matters: the evidence. The claim that people of faith have no business commenting on scientific matters is both hypocritical and logically absurd.

Many proponents of intelligent design are religious and believe in God, just like the vast majority of people around the world. However, there are also noteworthy non-religious scholars who defend ID as a legitimate intellectual position.[36] What all ID proponents share is an awareness that the universe shows scientifically detectable evidence for intelligent design.

If religious motives or beliefs are sufficient reasons to dismiss scientific theories, then many of the foundational figures of modern science would have to be erased from the record.

For example, Johannes Kepler and Isaac Newton are among many scientists who were inspired to their scientific work by religious conviction that God would create an orderly, comprehensible universe. Their scientific theories turned out to be correct[37]—not because of their religious beliefs but because the scientific evidence validated their hypotheses.

Furthermore, when materialists base their objections upon the alleged religious motives of ID proponents, they also make a hypocritical argument. After all, many leading evolutionists have passionately expressed *anti-religious* beliefs and motives.

For example, Richard Dawkins holds the dual honor of being the world's most famous evolutionist and the world's most famous atheist. His personal stated goal is "to kill religion,"[38] and he asserts that "faith is one of the world's great evils, comparable to the smallpox virus but harder to eradicate."[39]

Another materialist and ardent evolutionist, Nobel Laureate Steven Weinberg, has stated:

> [T]he teaching of modern science is corrosive of religious belief, and I'm all for that! One of the things that in fact has driven me in my life, is the feeling that this is one of the great social functions of science—to free people from superstition.[40]

The NCSE's executive director, Eugenie Scott, was called by *Nature* "the nation's most high-profile Darwinist,"[41] and she is a signer of the Third Humanist Manifesto.[42] The manifesto aims to promote a worldview "without supernaturalism," where humans are "the result of unguided evolutionary change," and the universe is "self-existing."[43]

Many evolutionary biologists agree with Dr. Scott's outlook. A poll published in *Nature* found that only 5.5% of National Academy of Sciences biologists believe in God.[44] Likewise, a survey of evolutionary biologists published in *The Scientist* found that only two out of 149 polled described themselves as "full theists."[45]

ID and neo-Darwinian evolution are both scientific theories that can have larger philosophical or religious implications. But the fact that they may have implications for or against theism does not disqualify either of them from being scientific.

Science should be an empirically based search for truth. If ID critics want to harp upon the religious beliefs and motives of ID proponents, then they should be forewarned that the argument cuts both ways.

Scientists, whatever their personal beliefs, should have every right to express their views. Their personal religious or anti-religious views are irrelevant as long as they are making sound scientific claims.

IS INTELLIGENT DESIGN POLITICS?

Having failed to prove either that ID is religion or that it lacks scientific merit, some critics fall back to a weaker argument that ID is simply a political movement.

If ID is nothing more than politics, then critics must explain why leading ID organizations, such as the Discovery Institute, oppose pushing ID into public schools. This is precisely because ID proponents want the debate over ID to be scientific, not political.

Some critics argue that the ID movement's occasional involvement in lawsuits or public education debates shows that it is political. But Darwin lobbyists regularly become involved in such battles.[46] If ID's involvement in such activities makes it political, then the same charge must also apply to neo-Darwinism.

Additionally, there is a fundamental difference between the educational and court battles initiated by the ID movement,[47] and those coming from the evolution advocates.

- Darwin lobby: Seeks to *restrict* free speech and intellectual inquiry, stopping people from discussing non-evolutionary views.

- ID movement: Seeks to *expand* intellectual inquiry and free-speech rights, defending academic freedom to debate evolution.[48]

Critics of ID should have every right to express their views, but they often try to suppress the pro-ID viewpoint. When the free speech rights of ID proponents are violated, sometimes lawsuits and educational initiatives are necessary to protect objective scientific inquiry. There's nothing inappropriately political about this.

Finally, the vibrant ID research program shows that the movement has invested its resources in scientifically investigating the truth about origins. The charge that ID is nothing more than politics ignores this growing body of research.

THE NEXT STEPS

> "The outcome of this scientific revolution will be decided by young people who have the courage to question dogmatism and follow the evidence wherever it leads."
> —Biologist Jonathan Wells[49]

Today there is much talk about "science literacy." Most everyone wants to be scientifically literate, but Darwin defenders have attempted to redefine the concept to mean "accepting neo-Darwinism."[50] The true meaning of scientific literacy is being an independent thinker who understands science and forms his or her own informed opinions.

We hope that after completing this book, you are scientifically literate in the debate over materialism (including neo-Darwinism) and intelligent design. For many readers, the information presented so far will be sufficient. Others, however, may want to dig deeper.

There are three practical steps you can take to stay informed about the debate in the future.

Step #1: Think critically and question assumptions.

Materialism depends on your letting down your guard and allowing its assumptions to slip into your thought processes. It's vital that you think for yourself and question assumptions. As this book has shown, critical thinking reveals that neo-Darwinian evolution and other materialistic theories are full of questionable assumptions, not compelling conclusions.

Step #2: Continue to learn.

New scientific discoveries are always being made, and the key to understanding this debate is to keep learning.

It's easy to stay informed about the pro-Darwinian viewpoint—we're bombarded with it every day. However, the Darwinian educational establishment and the media make it difficult to learn the facts about intelligent design. If you want to base your views on a full understanding of the scientific evidence, you will need to research and investigate pro-ID arguments.

With a little proactive self-education, critical thinking, and patience, you can keep yourself informed. Many of the websites and resources listed after this chapter in Recommended Resources contain helpful information about Darwinian evolution and intelligent design.

Step #3: Get involved.

As long as materialists are allowed to claim sole ownership of scientific knowledge, they will continue to win court battles and control education and the media. Staying informed about this debate is only part of the solution. There are many ways you can get involved, and there is much you can do.

Action Items for Everyone:

- Share what you've learned with family and friends. An easy way to do this is by acquiring the DVDs listed in the Recommended Resources and sharing them.

- Join and support established ID groups like Discovery Institute.

- Start an Intelligent Design & Evolution Awareness (IDEA) Club in your community.

- Become a voice for objective evolution education in your local school, school district, university, or state. Contact Discovery Institute for important information on how to positively impact public education.

- Put on an extracurricular ID event in your local school, church, or other community organization.

- Write letters to the editor, making concise, clear points based on scientific evidence.

- Stay informed by regularly visiting Evolution News & Views (**www.evolutionnews.org**) and subscribing to the ID the Future Podcast (**www.idthefuture.com**). Also, sign up for free email newsletters offered by Discovery Institute (**www.discovery.org/csc/notabene**).

- Work to elect leaders and judges who share your views.

Action Items for Students:

- Share pro-ID books, videos, and other information with your friends.

- Start an Intelligent Design & Evolution Awareness (IDEA) Club on your high school or college campus. For details see: **www.ideacenter.org/clubs**

- Attend Discovery Institute's Summer Seminars on Intelligent Design (for junior-class level college students and above). For details, see: **www.discovery.org/sem**

- Consider pursuing a career in the sciences, helping to advance the scientific case for intelligent design.

Action Items for Parents:

- Stay informed about the debate so you can teach your kids about the evidence for ID that they may be missing at school. Help them develop critical thinking skills and understand the issues.

- Consider attitudes towards materialist philosophy, evolution, and intelligent design as major factors when choosing a university for yourself or your kids.

- Join your local PTA, attend school board meetings, and make your views known.

CONCLUSION

Materialists claim that ID is a purely negative argument, but they ignore the positive case for design. They charge that it is not science but instead is religion or politics. The evidence cited in this chapter demonstrates that these criticisms are incorrect. These common objections are designed to avoid discussing scientific challenges and to allow materialism to win the debate over ID without actually having one.

Leading ID theorist William Dembski offers good advice on handling such criticism:

> Our critics have, in effect, adopted a *zero-concession* policy toward intelligent design. According to this policy, absolutely nothing is to be conceded to intelligent design and its proponents. This is especially difficult for novices to accept.... The point is not to induce a cognitive shift in our critics, but instead to clarify our arguments, to address weaknesses in our own position, to identify areas requiring further work and study, and, perhaps most significantly, to appeal to the undecided middle that is watching this debate and trying to sort through the issues.[51]

There is nothing scientific about excluding a possible explanation for philosophical reasons. If the empirical evidence, probability calculations, and logical arguments based upon the scientific method all point toward design, then it should be accepted as a reasonable answer until new data indicate otherwise.

Discovering intelligent design can be a lifelong journey. The more you study this topic, the more you will understand the powerful case for ID in nature. Today it's more important than ever to stand up for academic freedom in science and help the "undecided middle" understand the case for design, free of false caricatures and fallacious objections.

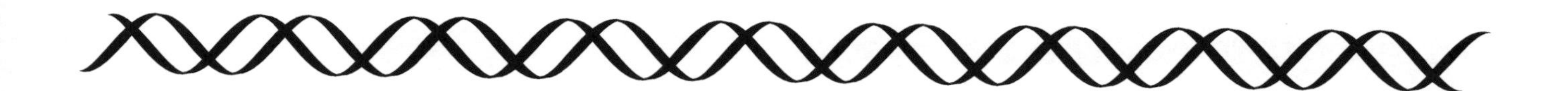

DISCUSSION QUESTIONS

1. Imagine you are a pro-ID scientist. What kinds of experiments would you suggest to test the theory of intelligent design?

2. Of the choices given—science, religion, or politics—which do you think most strongly applies to the theory of intelligent design? Explain your answer.

3. Is ID a positive argument, a negative argument, or both? Discuss your answer.

4. In this chapter, it was stated that "only the evidence matters." Do you agree? Why or why not?

5. The court ruling by Judge Jones in the *Kitzmiller v. Dover* case equated intelligent design with creationism and prohibited it from being taught in public schools in the district. Do you agree with the ruling? Explain your answer.

6. Imagine you are debating intelligent design and your opponent accuses you of having religious (or anti-religious) motives. How would you respond?

7. The scientific method allows scientists to explore the world around them with observation, hypothesis, experimentation, and conclusions. Do you think this is a good methodology? Are you in favor of the requirement that "only material explanations may be considered scientifically valid"? Explain your reasons.

SECTION VII

APPENDICES AND ENDNOTES

RECOMMENDED RESOURCES

The websites and resources listed below contain helpful information and resources for learning more about Darwinian evolution and intelligent design.

WEBSITES:

- Intelligent Design Gateway Portal: **www.intelligentdesign.org**
- Evolution News & Views: **www.evolutionnews.org**
- ID the Future Podcast: **www.idthefuture.com**
- Discovery Institute: **www.discovery.org**
- IDEA Student Clubs: **www.ideacenter.org**
- Access Research Network: **www.arn.org**
- Uncommon Descent (Blog): **www.uncommondescent.com**
- The College Student's Back to School Guide to Intelligent Design: **www.evolutionnews.org/BacktoSchoolGuide.pdf**
- A Parents' Guide to Intelligent Design **www.evolutionnews.org/parentsguide.pdf**

DVDs:

- *Unlocking the Mystery of Life* (Illustra Media, 2002). **www.unlockingthemysteryoflife.com**
- *Icons of Evolution* (Coldwater Media, 2002). **www.discovery.org/a/2125**
- *The Privileged Planet* (Illustra Media, 2004). **www.theprivilegedplanet.com**
- *Darwin's Dilemma* (Illustra Media, 2009). **www.darwinsdilemma.org**
- *Programming of Life* (LaBarge Media, 2011). **www.programmingoflife.com**
- *Metamorphosis* (Illustra Media, 2011). **www.metamorphosisthefilm.com**
- *Expelled: No Intelligence Allowed* (Premise Media, 2008). **www.ncseexposed.org**

BOOKS:

More information about many of the books below, including links for ordering, can be found at **www.discovery.org/csc/books.**

Introductory and Intermediate Science Books:

- *Darwin on Trial*, by Phillip E. Johnson (InterVarsity Press, 1993).

- *Darwin's Black Box: The Biochemical Challenge to Evolution*, by Michael J. Behe (Free Press, 1996).

- *The Design Revolution: Answering the Toughest Questions About Intelligent Design*, by William A. Dembski (InterVarsity, 2004).

- *The Edge of Evolution: The Search for the Limits of Darwinism*, by Michael J. Behe (The Free Press, 2007).

- *Evolution: A Theory in Crisis*, by Michael Denton (Adler & Adler, 1986).

- *Explore Evolution: The Arguments For and Against Neo-Darwinism*, by Stephen C. Meyer, Scott Minnich, Jonathan Moneymaker, Paul Nelson, and Ralph Seelke (Hill House, 2007).

- *Icons Of Evolution: Why Much of What We Teach about Evolution is Wrong*, by Jonathan Wells (Regnery Publishing, Inc., 2000).

- *Intelligent Design 101: Leading Experts Explain the Key Issues*, edited by H. Wayne House (Kregel Publications, 2008).

- *Intelligent Design Uncensored: An Easy-to-Understand Guide to the Controversy*, by William A Dembski and Jonathan Witt (InterVarsity, 2010).

- *The Myth of Junk DNA*, by Jonathan Wells (Discovery institute Press, 2011).

- *The Politically Incorrect Guide to Darwinism and Intelligent Design*, by Jonathan Wells (Regnery Publishing Inc., 2006).

- *The Privileged Planet: How Our Place in the Cosmos is Designed for Discovery*, by Guillermo Gonzales and Jay W. Richards (Regnery Publishing, Inc., 2004).

- *Probability's Nature and Nature's Probability: A Call to Scientific Integrity*, by Donald E. Johnson (Booksurge Publishing, 2009).

- *Science and Human Origins*, by Ann Gauger, Douglas Axe, and Casey Luskin (Discovery Institute Press, 2012).

- *Signature in the Cell: DNA and the Evidence for Intelligent Design*, by Stephen C. Meyer (HarperOne, 2009).

- *Uncommon Dissent: Intellectuals Who Find Darwinism Unconvincing*, edited by William Dembski (ISI Books, 2004).

- *What Darwin Didn't Know*, by Geoffrey Simmons (Harvest House Publishers, 2004).

- *What's Darwin Got to Do with It? A Friendly Conversation About Evolution*, by Robert C. Newman and John L. Wiester, with Janet and Jonathan Moneymaker (InterVarsity Press, 2000).

Advanced Science Books:

- *Darwinism, Design, and Public Education*, edited by John Angus Campbell and Stephen C. Meyer (Michigan State University Press, 2003).

- *Darwin's Doubt: The Explosive Origin of Animal Life and the Case for Intelligent Design* by Stephen C. Meyer (HarperOne, 2013).

- *Debating Design: From Darwin to DNA*, edited by William Dembski and Michael Ruse (Cambridge University Press, 2004).

- *The Design Inference: Eliminating Chance Through Small Probabilities*, by William Dembski (Cambridge University Press, 1998).

- *The Design of Life: Discovering Signs of Intelligence in Biological Systems*, by William A. Dembski and Jonathan Wells (The Foundation for Thought and Ethics, 2008).

- *The Mystery of Life's Origin: Reassessing Current Theories*, by Charles B. Thaxton, Walter L. Bradley, and Roger L. Olsen (Philosophical Library, 1984).

- *No Free Lunch: Why Specified Complexity Cannot Be Purchased Without Intelligence*, by William Dembski (Rowman & Littlefield, 2002).

Arts, Culture, Law, and Education:

- *A Meaningful World: How Arts and Science Reveal the Genius of Nature*, by Benjamin Wiker and Jonathan Witt (IVP Academic, 2006).

- *Darwin Day in America: How Our Politics and Culture Have Been Dehumanized in the Name of Science*, by John G. West (ISI Books, 2007).

- *Defeating Darwinism by Opening Minds*, by Phillip E. Johnson (InterVarsity Press, 1997).

- *Reason in the Balance: The Case Against Naturalism in Science, Law and Education*, by Phillip E. Johnson (InterVarsity Press, 1995).

- *Traipsing Into Evolution: Intelligent Design and the Kitzmiller v. Dover Decision*, by David K. DeWolf, John G. West, Casey Luskin, and Jonathan Witt (Discovery Institute Press, 2006).

APPENDIX A

THE SCIENTIFIC VALUE OF INTELLIGENT DESIGN

Intelligent design promises to lead to new avenues of research, new scientific knowledge, and scientific discoveries in a variety of fields.

In the context of historical sciences such as neo-Darwinian evolution or intelligent design, new knowledge can take the form of practical insights into the workings of biology in the present day (potentially leading to insights into fighting disease), as well as new knowledge about biological history and the origin of natural structures.

Below are ten examples of areas in which ID is helping science to generate new knowledge. Each example includes citations to technical scientific articles and publications by ID proponents that elaborate on ID's contributions to these fields:

- **Biochemistry:** ID encourages scientists to recognize and understand the origin of complex and specified information in biology in the form of finely tuned DNA, RNA, protein sequences, and biochemical pathways. This has practical implications not just for explaining biological origins but also for engineering enzymes and anticipating or fighting the future evolution of diseases.[1]

- **Genetics and Physiology:** ID's predictions have guided scientists to do research seeking function for non-coding "junk DNA" or allegedly "vestigial" structures, allowing us to understand development, cellular biology, and the function of many biological systems.[2]

- **Systematics:** ID encourages scientists to understand whether similarities between living species, including examples of extreme genetic "convergent evolution," are best explained by ID or by Darwinian evolution. ID has also encouraged scientists to see life as being front-loaded with information such that it is designed to evolve within limits. This would explain previously unanticipated "out of place" genes in various taxa.[3]

- **Cell biology:** ID directs scientists to view the cell as having been built from "designed structures rather than accidental by-products of neo-Darwinian evolution." This has led researchers to reverse-engineer irreducibly complex molecular machines like the bacterial flagellum, to understand their function as machines. In turn, scientists are casting light on how the machine-like properties of life allow biological systems to function, thus allowing researchers to better understand molecular machines and propose testable hypotheses about the causes of cancer and other diseases.[4]

- **Microbiology:** ID has produced research into the limited power of Darwinian mechanisms to evolve complex traits that require multiple mutations to function. This research has practical implications for responding to public health challenges such as antibiotic resistance.[5]

- **Systems biology:** ID leads biologists to look at biological systems as integrated components of larger systems that are designed to work together in a top-down, coordinated fashion.[6]

- **Bioinformatics:** ID has led scientists to investigate the computer-like properties of DNA and the genome in the hopes of better understanding genetics and the origin of biological systems. This includes seeking new layers of information and functional language embedded in the genetic codes, as well as other codes and sources of information within living organisms.[7]

- **Information theory:** ID has helped researchers to understand intelligence as a mechanism which generates certain types of information and complexity. This mechanism can be studied scientifically, leading to theoretical research into the information-generative powers of intelligent causes as compared to Darwinian searches, leading in turn to the finding that the search abilities of Darwinian processes are limited. This has practical implications for the viability of using genetic algorithms to solve problems. ID has also motivated scientists to study proper measures of biological information, leading to concepts like complex and specified information or functional sequence complexity. This allows us to better quantify complexity and understand what features are, or are not, within the reach of Darwinian evolution.[8]

- **Paleontology:** ID helps scientists to understand how the irreducibly complex nature of biological systems can predict patterns of gain and loss of biodiversity, due to "explosions" of diversity, mass extinction, or stasis throughout the history of life.[9]

- **Physics and Cosmology:** ID directs scientists to investigate instances of cosmic fine-tuning which are required for the existence of advanced forms of life (discovering, e.g., the Galactic Habitable Zone). This has significant implications for proper cosmological models of the universe, hints at proper avenues for successful "theories of everything" which must accommodate fine-tuning, and has other implications for theoretical physics.[10]

COSMOLOGICAL CALCULATIONS

The Cosmological Constant

The probability for the cosmological constant to have a value in a life-compatible range can be calculated using quantum field theory, the mathematical formalism that describes the rates for processes in which "particles" can interact, be created, or be destroyed.

In the mathematical description that quantum field theory gives us, random fluctuations of fields manifest themselves as particles of different kinds.

When the probability for a life-compatible value of the cosmological constant is calculated, it turns out to be one part in 10^{120}, or:

1
―――――――――――――――――
1,000,000,000,000,000,000,000,000,000,000,000,
000,000,000,000,000,000, 000,000,000,000,000,
000,000,000,000,000,000,000,000,000,000,000,
000,000,000,000,000,000,000

To give you an idea of how improbable this is, consider the fact that the number of baryons (protons and neutrons) in the observable universe is only about 10^{80}. To get 10^{120} of these particles, you would need 10^{40} universes the size of our own.

The probability of getting a life-compatible cosmological constant is the same as reaching into a grab-bag of 10^{40} universes, each containing 10^{80} particles, and pulling out one specific particle that has been specified in advance.

The Initial Entropy of the Universe

The calculation of the initial entropy of the universe involves considering it as emerging from a singularity, which can be understood as a black hole, the entropy of which is calculated using the Bekenstein-Hawking formula. Black hole entropy measures the amount of "disorder" that must be assigned to a black hole *internally or on its boundary* in order for it to comply with the laws of thermodynamics (conservation of energy, increasing entropy) as they are interpreted by observers *external* to the black hole.

The fine-tuning of the initial entropy of the universe comes about by comparing the massive entropy of the initial black hole singularity from which our universe began in a "Big Bang" with the observed statistical entropy of our present-day universe, realizing that our universe exhibits *far more order than it should* given the way it began. The ratio of the observed statistical entropy to the initial entropy is what gives Roger Penrose his startling figure for entropic fine-tuning:

$$\frac{1}{10^{10^{123}}}$$

INFLATION AND STRING THEORIES

Inflation and string theories involve varying degrees of uncertainty and controversy. However, one or both of them may ultimately be useful in answering questions about the Big Bang event.

Meanwhile, some materialists attempt to use the theories to explain how the universe could have been created by random events. Many lay people, unfamiliar with the basic concepts, may be intimidated or confused by materialists' claims in this area. With that in mind, here's a very brief introduction to the theories.

Inflation Theory

This theory proposes that during a tiny fraction of a second after the Big Bang, the universe expanded at an extraordinarily fast rate and then slowed to a more "normal" rate. The decay of this initial "inflation field" is postulated to produce innumerable "pocket" or "bubble" universes as it continues to expand and decay locally.

The cosmic inflation proposal: (1) has very slender contact with observable consequences (which have other possible explanations); (2) has severe fine-tuning problems of its own that exceed those it was originally invoked to explain away; (3) does not avoid the necessity of an absolute beginning to the multiverse; and (4) leads to the destruction of scientific reasoning altogether by multiplying probabilistic resources without bound.

At best this theory resolves only one (of many) finely tuned parameters, and critics argue that it increases the need for fine-tuning rather than reducing it.

An extended critique of inflationary cosmology can be found in Bruce Gordon's essay "Balloons on a String: A Critique of Multiverse Cosmology" in Bruce L. Gordon and William A. Dembski, eds., *The Nature of Nature: Examining the Role of Naturalism in Science* (Wilmington, DE: ISI Books, 2010), 561-605.

String Theory

Scientists have tried to understand how, right after the Big Bang, the force of gravity was united with other forces in nature described by quantum theory.

Solving this problem was difficult when scientists took the traditional approach and treated fundamental particles as points. The problem disappears, however, when the particles are treated as very short, vibrating, one-dimensional strings that are either open-ended or closed in the form of loops.

In order for the mathematics of string theory to work, the universe has to have nine space dimensions (forget trying to visualize this, you can't), plus one time dimension. According to string theory, right after the Big Bang, only the three space dimensions familiar to us (plus time, creating four-dimensional space-time) expanded to give us the universe in which we live.

The other six space dimensions still exist, but they are extremely small, on the order of the size of the strings themselves.

If string theory is correct—and many physicists have a healthy skepticism about this—we live in ten-dimensional space-time, with the extra six spatial dimensions curled up on an unimaginably small scale.

String theory is highly speculative and controversial among many physicists. Moreover, it is woefully inadequate to explain the degree of fine-tuning observed in the universe.

THE "RNA WORLD" HYPOTHESIS

In Chapter 8, we presented four reasons why the RNA world hypothesis is inadequate. One of those reasons—the inability of the RNA world to explain the origin of the genetic code—deserves amplification.

Before the RNA world evolved the ability to produce proteins, it would have to perform all necessary functions using RNA alone. But the RNA world hypothesis provides no explanation of the most crucial step in protein production: matching amino acids to their corresponding commands in the genetic code.

The biochemical language of the genetic code uses short strings of three nucleotides (called codons) to symbolize commands—including start commands, stop commands, and codons that signify each of the 20 amino acids used in life. After the information in DNA is transcribed into mRNA, a series of codons in the mRNA molecule instructs the ribosome which amino acids are to be strung in which order to build a protein.

Translation works by using another type of RNA molecule called transfer RNA (tRNA). During translation, tRNA molecules ferry needed amino acids to the ribosome so the protein chain can be assembled.

Each tRNA molecule is linked to a single amino acid on one end, and at the other end exposes three nucleotides (called an anti-codon). At the ribosome, small free-floating pieces of tRNA bind to the mRNA. When the anti-codon on a tRNA molecule binds to matching codons on the mRNA molecule at the ribosome, the amino acids are broken off the tRNA and linked up to build a protein.

For the genetic code to be translated properly, each tRNA molecule must be attached to the proper amino acid that corresponds to its anti-codon as specified by the genetic code. If this critical step does not occur, then the language of the genetic code breaks down, and there is no way to convert the information in DNA into properly ordered proteins.[1] So how do tRNA molecules become attached to the right amino acid?

Cells use special proteins called **aminoacyl tRNA synthetase (aars)** enzymes to attach tRNA molecules to the "proper" amino acid under the language of the genetic code. Most cells use 20 different aaRS enzymes, one for each amino acid used in life.[2] These aaRS enzymes are key to ensuring that the genetic code is correctly interpreted in the cell.

Yet these aaRS enzymes themselves are encoded by the genes in the DNA. This forms the essence of a "chicken-egg problem": aaRS enzymes themselves are necessary to perform the very task that constructs them.

How could such an integrated, language-based system arise in a step-by-step fashion? If any component is missing, the genetic information cannot be converted into proteins, and the message is lost. The RNA world is unsatisfactory because it provides no explanation for how the key step of the genetic code—linking amino acids to the correct tRNA—could have arisen.[3]

Furthermore, for the language of DNA to be accurately transcribed and translated into proteins, a whole suite of enzymes and molecular machines must be present and functioning properly. As two theorists observed in a 2004 article in *Cell Biology International:*

> The letters of any alphabet used in words have no prescriptive function unless the destination reading those words first knows the language convention.[4]

In other words, without the hardware to convert the information in DNA into protein, the message is lost. Yet this hardware itself is generated by the very processes and functions it fulfills in the cell.

The article further explains the scope of the complexity of this system:

> The nucleotide sequence is also meaningless without a conceptual translative scheme and physical "hardware" capabilities. Ribosomes, tRNAs, aminoacyl tRNA synthetases, and amino acids are all hardware components of the Shannon message "receiver." But the instructions for this machinery is itself coded in DNA and executed by protein "workers" produced by that machinery. Without the machinery and protein workers, the message cannot be received and understood. And without genetic instruction, the machinery cannot be assembled.[5]

This integrated system is a highly specified irreducibly complex circuit, suggesting intelligent design.

APPENDIX E

THE PRECAMBRIAN FOSSIL RECORD

When trying to explain the Cambrian explosion, materialists will sometimes mention a small number of Precambrian fossils that they claim demonstrate the existence of ancestors to the complex animals that appear in the Cambrian. A small number of Precambrian animal fossils do exist; this has been known for many decades, from fossil sites around the world. But no ID proponent has ever argued that there are no Precambrian fossils. Rather, ID proponents observe that there are no clear evolutionary precursors to the Cambrian fauna, where nearly all of the major living animal phyla appear in an abrupt fashion without any evolutionary antecedents.

That the precursors to the Cambrian groups are indeed missing from the fossil record is widely accepted among paleontologists. This is not the controversial aspect of the ID position. About the missing precursors at the base of the tree of the animal phyla, the leading paleontologist James Valentine notes:

> … many of the branches, large as well as small, are cryptogenetic (cannot be traced into ancestors). Some of these gaps are surely caused by the incompleteness of the fossil record, but that cannot be the sole explanation for the cryptogenetic nature of some families, many invertebrate orders, all invertebrate classes, and all metazoan phyla.[1]

Paleontologist Charles Marshall concurs:

> While the fossil record of the well-skeletonized animal phyla is pretty good, we have virtually no fossils that are unambiguously assignable to the most basal stem groups [putative ancestors] of these phyla, those first branches that lie between the last common ancestor of all bilaterians and the last common ancestor of the living representatives of each of the phyla…. [T]heir absence is striking. Where are they?[2]

Likewise, a 2009 paper in *BioEssays* states[:]

> [A]s explained on an intelligent-design t-shirt:
>
> > Fact: Forty phyla of complex animals suddenly appear in the fossil record, no forerunners, no transitional forms leading to them; "a major mystery," a "challenge." The *Theory* of Evolution—exploded again (idofcourse.com).
>
> Although we would dispute the numbers, and aside from the last line, there is not much here that we would disagree with. Indeed, many of Darwin's contemporaries shared these sentiments, and we assume—if Victorian fashion dictated—that they would have worn this same t-shirt with pride.[3]

To be clear: Valentine, Marshall, and the authors of the *BioEssays* paper all oppose ID theory. By contrast, some ID theorists argue that the most likely reason we have no evidence of these ancestral fossils is that they never existed. The best explanation of this abrupt bioinformational explosion of new body plans, they argue, is intelligent design. This is the controversial aspect of the ID position on the Cambrian explosion.

Some materialists may cite papers like J. William Schopf's 2000 article in *Proceedings of the National Academy of Sciences,* "Solution to Darwin's dilemma: Discovery of the missing Precambrian record of life,"[4] claiming it shows the missing Precambrian ancestors. Anyone who does claim this, however, probably did not read past the paper's title, or has not been paying close attention to this debate in recent years. Contrary to what Schopf's title suggests, the fossils cited in this paper are bacterial and other unicellular fossils, or even non-fossils, that do not provide the "solution to Darwin's dilemma." For example: fossils cited by Schopf include:

- *Eozoon canadense*, which was found to be not a fossil but a rock produced by metamorphism. Early evolutionary biologists wrongly presumed it was a fossil, thinking that it would solve "Darwin's dilemma."
- *Cryptozoon*, which is thought to be a stromatolite—a bacterial mat. At best, these stromatolites show the existence of bacteria and are not true multicellular fossils that would have been directly ancestral to the Cambrian fauna. Like the prior example, this turned out to be a case where, according to Schopf, "Mineralic, purely inorganic objects had been misinterpreted as fossil." Thus faulty evolutionary assumptions led to faulty conclusions about these fossils.
- *Chuaria*, which is a single-celled algae, originally wrongly thought to be a shelly invertebrate due to more misguided attempts to solve "Darwin's dilemma."
- Barghoorn Gunflint microfossils, which are bacterial stromatolites that do not serve as precursors to the Cambrian fauna.
- Bitter Springs Chert, which are microbe fossils, not clear evolutionary precursors to the Cambrian fauna.
- Other organisms at Ediacara, also called the Ediacaran fauna, which are enigmatic fossils generally not thought to be ancestral to the Cambrian fauna.

The last example just listed, the Ediacaran fauna, are often cited by those who discuss Precambrian fossils. But these fossils do not help materialists, because they are not thought to be ancestral to the modern phyla that appear explosively in the Cambrian. As evolutionary paleontologist Peter Ward writes:

> [L]ater study cast doubt on the affinity between these ancient remains preserved in sandstones and living creatures of today; the great German paleontologist A. Seilacher, of Tübingen University, has even gone so far as to suggest that the Ediacaran fauna has no relationship whatsoever with any currently living creatures. In this view, the Ediacaran fauna was completely annihilated before the start of the Cambrian fauna.[5]

Seilacher's skepticism is shared by a number of modern evolutionary scientists today, as paleontologist Richard Fortey explains:

> The beginning of the Cambrian period, some 545 million years ago, saw the sudden appearance in the fossil record of almost all the main types of animals (phyla) that still dominate the biota today. To be sure, there are fossils in older strata, but they are either very small (such as bacteria and algae), or their relationships to the living fauna are highly contentious, as is the case with the famous soft-bodied fossils from the late Precambrian Pound Quartzite, Ediacara, South Australia.[6]

Likewise, Andrew Knoll and Sean Carroll observe that "It is genuinely difficult to map the characters of Ediacaran fossils onto the body plans of living invertebrates" and thus evidence of these fossils being precursors to Cambrian fauna "remains equivocal."[7] A Blackwell Scientific invertebrate biology textbook concurs that the Ediacaran fauna do not solve Darwin's dilemma:

> Whether [the Ediacaran fossils] were, in fact, early members of any phyla still living today and possible ancestral forms, or were members of phyla long since extinct, is a question of considerable current debate. At any rate, they shed little light on the question of which phyla were ancestral to other phyla, or if indeed, animals have a common ancestry.[8]

Finally, the prominent paleontologists Valentine, Erwin, and Jablonski are hesitant to claim that the Ediacaran fossils bear any ancestral relation to

Cambrian fauna, stating that "the relations of any of these fossils to Cambrian bilaterians remains uncertain and awaits further collecting and critical analysis."[9] Most evolutionary scientists believe that the Ediacaran fossils neither solve Darwin's dilemma nor document the evolution of the Cambrian fauna.

In fact, the lack of Precambrian fossil precursors to the Cambrian fauna is all but admitted in Schopf's aforementioned *PNAS* paper, which concedes that:

> Megascopic eukaryotes, the large organisms of the Phanerozoic, are now known not to have appeared until shortly before the beginning of the Cambrian—except in immediately sub-Cambrian strata, the hunt for large body fossils in Precambrian rocks was doomed from the outset.[10]

This is a striking admission: paleontologists do not know of fossils that serve as clear evolutionary precursors to the explosion of biodiversity that appears in the Cambrian. Schopf states that "those of us who wonder about life's early history can be thankful that what was once 'inexplicable' to Darwin is no longer so to us."[11] However, this is difficult to accept, because "Darwin's dilemma" is not solved by these bacterial, uniceullular, and other ambiguous fossils yielded by Precambrian rocks.

The fact that essentially all of the Precambrian fossils cited by Schopf are unicellular organisms or other enigmatic fossils, and the fact that he admits a lack of Precambrian "Megascopic eukaryotes," shows that evolutionary paleontologists still face a disturbing enigma that is highly resistant to Darwinian solutions.

For a more detailed discussion of the Precambrian fossil record and why it does not solve Darwin's dilemma, see Stephen C. Meyer's book *Darwin's Doubt: The Explosive Origin of Animal Life and the Case for Intelligent Design* (HarperOne, 2013).

ENDNOTES

Introduction

1. See the Intelligent Design and Evolution Awareness (IDEA) Center's IDEA Club program at www.ideacenter.org.
2. "Discovery Institute's Science Education Policy," Discovery.org, accessed March 21, 2013, http://www.discovery.org/a/3164.

Chapter 1

1. Socrates in Plato's dialogue *The Republic* (394d) as paraphrased by Antony Flew and Roy Abraham Varghese, *There Is A God: How the World's Most Notorious Atheist Changed His Mind* (New York: HarperOne, 2007), 89.
2. See William A. Dembski, *No Free Lunch: Why Specified Complexity Cannot Be Purchased without Intelligence* (Plymouth, U.K.: Rowman & Littlefield, 2002).
3. Stephen C. Meyer, "The origin of biological information and the higher taxonomic categories," *Proceedings of the Biological Society of Washington*, 117 (2004): 213–239.
4. For a discussion, see Chapters 19 and 20. See also *Expelled: No Intelligence Allowed* (Premise Media, 2008); Jonathan Wells, *The Politically Incorrect Guide to Darwinism and Intelligent Design* (Washington D.C.: Regnery, 2006).
5. Scott C. Todd, "A view from Kansas on that evolution debate," *Nature*, 401 (September 30, 1999): 423.

Chapter 2

1. Charles Darwin, Introduction, *The Origin of Species* (1859), Literature.org, accessed March 14, 2012, http://www.literature.org/authors/darwin-charles/the-origin-of-species/introduction.html.
2. According to neo-Lamarckism, new discoveries in genetics and cell biology have revealed some very limited types of traits that might be influenced by the environment and passed on to offspring. Nonetheless, the vast majority of inheritance appears to be non-Lamarckian.
3. There are many definitions of the term "species." This book will adopt the most common definition, the biological species concept, which refers to a reproductively isolated population of organisms that interbreed in the wild. For a thorough discussion of this and other definitions, see William Dembski and Jonathan Wells, Chapter 4, *The Design of Life: Discovering Signs of Intelligence in Biological Systems* (Dallas, TX: Foundation For Thought and Ethics, 2008).
4. Michael Flannery, "The First Challenge to Darwin, the Preamble to Intelligent Design," Intelligent Design the Future Podcast (August 5, 2009), accessed September 14, 2012, http://www.idthefuture.com/2009/08/the_first_challenge_to_darwin.html.
5. Douglas Axe, "The Science of Denial," Evolution News & Views (October 7, 2009), accessed March 14, 2012, http://www.evolutionnews.org/2009/10/the_science_of_denial026381.html (emphasis in original).

Chapter 3

1. Stephen Hawking and Roger Penrose, *The Nature of Space and Time* (Princeton, NJ: Princeton University Press, 1996), 20.
2. Jeffrey Bennett, Megan Donahue, Nicholas Schneider, and Mark Voit, *The Essential Cosmic Perspective*, 4th ed. (San Francisco, CA: Pearson/Addison Wesley, 2008), 13.
3. Charles H. Lineweaver and Tamara M. Davis, "Misconceptions about the Big Bang," *Scientific American* (March, 2005): 36–45.
4. Charles Q. Choi, "Discovery May Triple the Number of Stars In the Universe," LiveScience.com (December 1, 2010), accessed February 18, 2013, http://www.livescience.com/9044-discovery-triple-number-stars-universe.html.
5. The central insights of the *kalam* argument trace back to the Alexandrian Christian philosopher and experimentalist John Philoponus (490–570 AD).
6. See William Lane Craig, *The Kalam Cosmological Argument* (Eugene, OR: Wipf & Stock Publishers, 2000).
7. Donald Goldsmith, *Einstein's Greatest Blunder? The Cosmological Constant and Other Fudge Factors in the Physics of the Universe* (Cambridge, MA: Harvard University Press, 1995), 7.
8. See Georges Lemaître, *Annales de la Société Scientifique de Bruxelles Tomme XLVII, Series A, Première Partie* (April, 1927); Georges Lemaître, "A Homogenous Universe of Constant Mass and Increasing Radius accounting for the Radial Velocity of Extra-galactic Nebulae," *Monthly Notices of the Royal Astronomical Society*, 91 (March 1931): 483–490, translated by Abbé G. Georges Lemaître.
9. See Chapter 9, Guillermo Gonzalez and Jay W. Richards, *The Privileged Planet: How Our Place in the Cosmos is Designed for Discovery* (Washington D.C.: Regnery, 2004).
10. Hubert P. Yockey, *Information Theory and Molecular Biology* (Cambridge: Cambridge University Press, 1992), 212.
11. Sir Arthur S. Eddington, "The End of the World: From the Standpoint of Mathematical Physics," *Nature* 127 (March 21, 1931): 447–455.
12. For a discussion, see Granville Sewell, *In The Beginning and Other Essays on Intelligent Design* (Seattle, WA: Discovery Institute Press, 2010).
13. See for example N. W. Boggess *et al.*, "The COBE Mission: Its Design and Performance Two Years After Launch," *The Astrophysical Journal*, 397 (1992): 420–429; J. C. Mather *et al.*, "A Preliminary Measurement of the Cosmic Microwave Background Spectrum by the *Cosmic Background Explorer (COBE)* Satellite," *The Astrophysical Journal*, 354 (May 10, 1990): L37–L40; D. J. Fixsen *et al.*, "The Cosmic Microwave Background Spectrum from the Full COBE FIRAS Data Set," *The Astrophysical Journal*, 473 (December 20, 1996): 576–587. See also "Echo of the Big Bang wins US pair Nobel Prize," PhysOrg (October 3, 2006), accessed March 14, 2012 http://www.physorg.com/news79074220.html.
14. Neil F. Comins, *Discovering the Essential Universe*, 4th ed. (New York: W. H. Freeman, 2009), 406.
15. Stephen Hawking, quoted in "Professor George Smoot," *Current Biography*, 55 (April, 1994), accessed March 14, 2012, http://aether.lbl.gov/www/personnel/Smoot-bio.html.
16. George Smoot quoted in Dennis Overbye, "2 Americans Whose Work Buttressed the Big Bang Theory Win Nobel," *New York Times* (October 4, 2006), accessed March 14, 2012, http://select.nytimes.com/gst/abstract.html?res=FB0B15F63F540C778CDDA90994DE404482.
17. Antony Flew and Roy Abraham Varghese, *There Is A God: How the World's Most Notorious Atheist Changed His Mind* (New York: HarperOne, 2007), 136, 145.

Chapter 4

1. Charles Townes quoted in Bonnie Azab Powell, "'Explore as much as we can': Nobel Prize winner Charles Townes on evolution, intelligent design, and the meaning of life," *UC Berkeley News Center* (June 17, 2005), accessed March 14, 2012, http://www.berkeley.edu/news/media/releases/2005/06/17_townes.shtml.
2. Steven Weinberg, *The First Three Minutes: A Modern View of the Origin of the Universe* (New York: Basic Books, 1993), 5.
3. See Chapter 9, Guillermo Gonzalez and Jay W. Richards, *The Privileged Planet: How Our Place in the Cosmos Is Designed for Discovery* (Washington D.C.: Regnery, 2004).

4. Ibid. See also Guillermo Gonzalez, Donald Brownlee, and Peter D. Ward, "Refuges for Life in a Hostile Universe," *Scientific American* (October, 2001): 62–67.

5. See Robert G. Strom *et al.*, "The Origin of Planetary Impactors in the Inner Solar System," *Science*, 309 (September 16, 2005): 1847–1850; Norman H. Sleep *et al.*, "Annihilation of ecosystems by large asteroid impacts on the early Earth," *Nature*, 342 (November 9, 1989): 139–142. See also William A. Dembski and Jonathan Wells, *The Design of Life: Discovering Signs of Intelligence in Biological Systems* (Dallas, TX: The Foundation for Thought and Ethics, 2008), 253.

6. For a discussion of this idea, see Chapter 10, Gonzalez and Richards, *The Privileged Planet: How Our Place in the Cosmos is Designed for Discovery*.

7. Paul Davies, "The Unreasonable Effectiveness of Science," in *Evidence of Purpose: Scientists Discover the Creator*, ed. John Marks Templeton (New York: The Continuum Publishing Company, 1994), 49.

8. Much of the information in this list is drawn from Gonzalez and Richards, *The Privileged Planet: How Our Place in the Cosmos Is Designed for Discovery*; Peter D. Ward and Donald Brownlee, *Rare Earth: Why Complex Life Is Uncommon in the Universe* (New York: Copernicus Books, 2000); Hugh Ross, *Why the Universe is the Way It Is* (Grand Rapids, MI: Baker Books, 2008); Granville Sewell, *In the Beginning and Other Essays on Intelligent Design* (Seattle, WA: Discovery Institute Press, 2010); John D. Barrow and Frank J. Tipler, *The Anthropic Cosmological Principle* (Oxford: Oxford University Press, 1986); Paul Davies, *The Accidental Universe* (Cambridge, MA: Cambridge University Press, 1982).

9. Gonzalez and Richards, *The Privileged Planet: How Our Place in the Cosmos Is Designed for Discovery*, 186.

10. Robin Collins, "God, Design, and Fine-Tuning," in *God Matters: Readings in the Philosophy of Religion*, eds. Raymond Martin and Christopher Bernard (New York: Longman Press, 2002), 122.

11. See Roger Penrose and Martin Gardner, *The Emperor's New Mind: Concerning Computers, Minds, and the Laws of Physics* (Oxford: Oxford University Press, 2002), 444–445.

12. Ibid., 445–446.

Chapter 5

1. Joslyn Pine (Editor), *Wit and Wisdom of the American Presidents: A Book of Quotations* (Mineola, NY: Dover Thrift Editions, 2009), 55.

2. Stephen Hawking and Leonard Mlodinow, *The Grand Design* (New York: Bantam, 2010), 180.

3. John Lennox, "As a scientist I'm certain Stephen Hawking is wrong. You can't explain the universe without God," *Daily Mail Online* (September 3, 2010), accessed March 14, 2012, http://www.dailymail.co.uk/debate/article-1308599/Stephen-Hawking-wrong-You-explain-universe-God.html.

4. Peter Atkins, *Creation Revisited: The Origin of Space, Time and the Universe* (Oxford: W. H. Freeman, 1992), 149.

5. For a discussion of quantum theory and its significance in cosmology, see Bruce L. Gordon, "A Quantum-Theoretic Argument Against Naturalism," in The *Nature of Nature: Examining the Role of Naturalism in Science*, eds. Bruce L. Gordon and William A. Dembski (Wilmington, DE: ISI Books, 2011).

6. William Lane Craig and James D. Sinclair, "The *Kalam* Cosmological Argument," in *The Blackwell Companion to Natural Theology*, eds. William Lane Craig and J. P. Moreland (Sussex: Wiley-Blackwell, 2009), 183.

7. William Lane Craig quoted in Lee Strobel, *The Case for a Creator* (Grand Rapids, MI: Zondervan, 2004), 101.

8. Bruce L. Gordon, "A Quantum-Theoretic Argument Against Naturalism," in *The Nature of Nature: Examining the Role of Naturalism in Science*, eds. Bruce L. Gordon and William A. Dembski (Wilmington, DE: ISI Books, 2011), 179.

9. Craig and Sinclair, "The *Kalam* Cosmological Argument," 158, 170.

10. "Evidence mounts that the expansion of the Universe is accelerating," PhysicsWorld (November 6, 1998), accessed March 14, 2012, http://physicsworld.com/cws/article/news/3137.

11. Alan H. Guth and Marc Sher, "The Impossibility of a Bouncing Universe," *Nature*, 302 (April 7, 1983): 505–506; Sidney A. Bludman, "Thermodynamics and the End of a Closed Universe," *Nature*, 308 (March 22, 1984): 319–322.

12. For a more extensive discussion of the multiverse, see Guillermo Gonzales and Jay W. Richards, *The Privileged Planet: How Our Place in the Cosmos Is Designed for Discovery* (Washington D.C.: Regnery Publishing, Inc., 2004), 268–271; Robin Collins, "The teleological argument: an exploration of the fine-tuning of the universe," in *The Blackwell Companion to Natural Theology*, eds. William Lane Craig and J. P. Moreland eds. (Sussex: Wiley-Blackwell Publishing, 2009); Stephen C. Meyer, *Signature in the Cell: DNA and the Evidence for Intelligent Design* (New York: HarperOne, 2009), 499–508; Michael Behe, *The Edge of Evolution: The Search for the Limits of Darwinism* (New York: Free Press, 2007), 221–227.

13. See comments from Nobel Prize winning theorist David Gross and astronomer Bernard Carr, quoted in Geoff Brumfiel, "Our Universe: Outrageous fortune," *Nature*, 439 (January 5, 2006): 10–12.

14. George F. R. Ellis, "Does the Multiverse Really Exist?," *Scientific American* (August, 2011).

Chapter 6

1. "Home, Sweet Home," song lyrics, by John Howard Payne.

2. See Carl Sagan and Iosef S. Shklovskii, *Intelligent Life in the Universe* (San Francisco, CA: Holden-Day, 1966), 345, 413.

3. Carl Sagan, *Cosmos* (New York: Ballantine, 1985), 159.

4. The SETI project searches the cosmos for signs of information generated by intelligence. Some might argue about the project's goals, but few would say it is unscientific. Yet that is what critics say when ID theorists use similar methods to search for signs of intelligence in nature.

5. See Keith Cooper, "SETI: The State of the Art," *Astronomy Now Magazine* (April 7, 2010), accessed March 14, 2012, http://www.astronomynow.com/news/n1004/07seti2/.

6. Many of these items are discussed in Guillermo Gonzalez and Jay W. Richards, *The Privileged Planet: How Our Place in the Cosmos Is Designed for Discovery* (Washington D.C.: Regnery, 2004), 33–35.

7. See Sagan and Shklovskii, *Intelligent Life in the Universe*, 343–350.

8. See Peter Douglas Ward and Donald Brownlee, *Rare Earth: Why Complex Life Is Uncommon in the Universe* (New York: Copernicus Books, 2000), 221–234.

9. George W. Wetherill, "Possible consequences of absence of 'Jupiters' in planetary systems," *Astrophysics and Space Science*, 212 (1994): 23–32. See also Joe Holley, "George Wetherill; Leader in Study of Planets," *The Washington Post* (July 25, 2006), accessed March 14, 2012, http://www.washingtonpost.com/wp-dyn/content/article/2006/07/21/AR2006072101558.html.

10. See Ward and Brownlee, *Rare Earth: Why Complex Life Is Uncommon in the Universe*, 240.

11. Gonzalez and Richards, *The Privileged Planet: How Our Place in the Cosmos Is Designed for Discovery*, x.

12. Ibid. (the subtitle of the book is "*How Our Place in the Cosmos Is Designed for Discovery*").

13. Stephen Hawking in the Discovery Channel documentary, *Alien Life*. For a discussion, see Casey Luskin "Stephen Hawking's Materialist Logic: 'We Don't Understand How Life Formed,' but It 'Must Have Spontaneously Generated Itself,'" Evolution News & Views (March 4, 2011), accessed March 14, 2012, http://www.evolutionnews.org/2011/03/stephen_hawkings_materialist_l044481.html.

14. Ibid.

Chapter 7

1. Massimo Pigliucci, "Where Do We Come From? A Humbling Look at the Biology of Life's Origin," in *Darwin Design and Public Education*, eds. John Angus Campbell and Stephen C. Meyer (East Lansing, MI: Michigan State University Press, 2003), 196.
2. Yogi Berra, *The Yogi Book: "I Really Didn't Say Everything I Said"* (New York: Workman Publishing Company, Inc., 1998), 83.
3. George Wald, "The Origin of Life," *Scientific American*, 191 (August, 1954): 45–53.
4. Stephen Jay Gould, "An Early Start," *Natural History* (February, 1978): 10.
5. Cyril Ponnamperuma quoted in Sir Fred Hoyle and Chandra Wickramasinghe, *Evolution from Space* (New York: Simon & Schuster, 1981), 76.
6. "Pre-History of Life: Elegantly Simple Organizing Principles Seen in Ribosomes," *ScienceDaily* (April 12, 2010), accessed March 14, 2012, http://www.sciencedaily.com/releases/2010/04/100412151823.htm.
7. Stephen C. Meyer, Scott Minnich, Jonathan Moneymaker, Paul A. Nelson and Ralph Seelke, *Explore Evolution: The Arguments For and Against Neo-Darwinism* (Malvern, Australia: Hillhouse, 2007), 52.
8. See Marco Piccolino, "Biological machines: from mills to molecules," *Nature Reviews Molecular Cell Biology*, 1 (November, 2000): 149–153.
9. Bruce Alberts, "The Cell as a Collection of Protein Machines: Preparing the Next Generation of Molecular Biologists," *Cell*, 92 (February 6, 1998): 291–294.
10. Those instructions are sent to the ribosome in the form of mRNA. This is discussed in detail in the next chapter.
11. See "Life: What A Concept!," ed. John Brockman (New York, NY: The Edge Foundation, 2008), accessed March 14, 2012, http://www.edge.org/documents/life/Life.pdf.
12. Ada Yonath, "Supervisor's Foreword," in Chen Davidovich, *Targeting Functional Centers of the Ribosome* (Heidelberg: Springer-Verlag, 2011), vii.
13. Ibid.
14. David Goodsell, "The ATP Synthase," Molecule of the Month at Protein Data Bank (December, 2005), accessed March 14, 2012, http://www.rcsb.org/pdb/education_discussion/molecule_of_the_month/download/ATPSynthase.pdf.
15. See Juli Pereto, Jeffrey L. Bada and Antonio Lazcano, "Charles Darwin and the Origin of Life," *Origins of Life and Evolution of Biospheres*, 39 (October, 2009): 395–406.
16. See Chapter 6, *Unlocking the Mystery of Life* (Illustra Media, 2003).
17. See A. Nissenbaum, "Scavenging of Soluble Organic Matter from the Prebiotic Oceans," *Origins of Life and Evolution of Biospheres*, 7 (1976): 413–416.
18. For a detailed explanation of Oparin's ideas, see Stephen C. Meyer, *Signature in the Cell: DNA and the Evidence for Intelligent Design* (New York: HarperOne, 2009), 44–57.
19. See Stanley L. Miller, "A Production of Amino Acids under Possible Primitive Earth Conditions," *Science*, 117 (May 15, 1953): 528–529.
20. See Jonathan Wells, *Icons of Evolution: Why Much of What We Teach About Evolution Is Wrong* (Washington D.C.: Regnery, 2000), 12–22. See also David W. Deamer, "The First Living Systems: a Bioenergetic Perspective," *Microbiology & Molecular Biology Reviews*, 61 (June, 1997): 239–261.
21. See Wells, *Icons of Evolution: Why Much of What We Teach About Evolution Is Wrong* 12–18.
22. Jon Cohen, "Novel Center Seeks to Add Spark to Origins of Life," *Science*, 270 (December 22, 1995): 1925–1926.
23. Antonio C. Lasaga, H. D. Holland, and Michael J. Dwyer, "Primordial Oil Slick," *Science*, 174 (October 1, 1971): 53–55.
24. See Gordon Schlesinger and Stanley L. Miller, "Prebiotic Synthesis in Atmospheres Containing CH_4, CO, and CO_2," *Journal of Molecular Evolution*, 19 (1983): 383–390; Stanley L. Miller and Leslie E. Orgel, *The Origins of Life on the Earth* (Englewood Cliffs, NJ: Prentice Hall, 1974), 33. See also Wells, *Icons of Evolution: Why Much of What We Teach About Evolution Is Wrong*, 18–19.
25. For example, a 2011 article in *Skeptic Magazine* states: "The pioneering Miller-Urey Experiment created amino acids—the building blocks of life—in an environment that simulated atmospheric conditions on the early Earth." Paul F. Deisler Jr., "How Did Life Begin?: A Perspective on the Nature and Origin of Life," *Skeptic Magazine*, 16 (January, 2011): 34–40. Likewise, multiple textbooks continue to claim the Miller-Urey Experiment accurately simulated conditions on the early Earth. David Savada, H. Craig Heller, Gordon H. Orians, William K. Purves, David M. Hillis, *Life: The Science of Biology* (Sunderland, MA: Sinauer Associates, 8th ed., 2008), 62, states: "The Miller-Urey experiment simulated possible atmospheric conditions on primitive Earth and obtained some of the molecular building blocks of biological systems." The caption claims the experiment used "conditions similar to those that existed on primitive Earth" and concludes, "The chemical building blocks of life could have been generated in the probable atmosphere of early Earth." Likewise, Alton Biggs, Whitney Crispen Hagins, Chris Kapicka, Linda Lundgren, Peter Rillero, Kathleen G. Tallman, and Dinah Zike, *Biology: The Dynamics of Life*, Florida Edition (Columbus, OH: Glencoe, 2006), 382, states: "Miller and Urey's experiments showed that under the proposed conditions on early Earth, small organic molecules, such as amino acids, could form."
26. Committee on the Limits of Organic Life in Planetary Systems, Committee on the Origins and Evolution of Life, National Research Council, *The Limits of Organic Life in Planetary Systems* (Washington D.C.: National Academy Press, 2007), 60.
27. For more information, see William Dembski and Jonathan Wells, *The Design of Life: Discovering Signs of Intelligence in Biological Systems* (Dallas, TX: The Foundation for Thought and Ethics, 2007), 237–238.
28. Ibid., 232.
29. Statements made by Stanley Miller at a lecture he gave at a UCSD Origins of Life seminar class on January 19, 1999 attended by Casey Luskin. See also Alonso Ricardo and Jack W. Szostak, "Origin of Life on Earth," *Scientific American* (September 2009): 54–61 ("under the right conditions some building blocks of proteins, the amino acids, form easily from simpler chemicals, as Stanley L. Miller and Harold C. Urey of the University of Chicago discovered in pioneering experiments in the 1950s. But going from there to proteins and enzymes is a different matter."); Peter Radetsky, "How Did Life Start?," *Discover Magazine* (November 1, 1992), accessed March 14, 2012, http://discovermagazine.com/1992/nov/howdidlifestart153/ ("Even Miller throws up his hands at certain aspects of it. The first step, making the monomers, that's easy. We understand it pretty well. But then you have to make the first self-replicating polymers. That's very easy, he says, the sarcasm fairly dripping. Just like it's easy to make money in the stock market—all you have to do is buy low and sell high. He laughs. Nobody knows how it's done.").
30. J. Craig Venter and Daniel Gibson, "How We Created the First Synthetic Cell," *The Wall Street Journal* (May 26, 2010).
31. Ibid.
32. Michael Denton, *Evolution: A Theory in Crisis* (Chevy Chase, MD: Adler & Adler, 1986), 250.
33. See Meyer, *Signature in the Cell: DNA and the Evidence for Intelligent Design*.
34. George M. Whitesides, "Revolutions In Chemistry: Priestley Medalist George M. Whitesides' Address," *Chemical and Engineering News*, 85 (March 26, 2007): 12–17.

Chapter 8

1. Bill Gates, Nathan Myhrvold, and Peter Rinearson, *The Road Ahead: Completely Revised and Up-To-Date* (New York: Penguin Books, 1996), 228.
2. Carl Sagan, "Life," in *Encyclopedia Britannica: Macropaedia Vol. 10* (Encyclopedia Britannica, Inc., 1984), 894.

3. Bernd-Olaf Küppers, *Information and the Origin of Life* (Cambridge, MA: MIT Press, 1990), 170.

4. Nearly all cells contain DNA. But mature red blood cells, for example, do not.

5. When the nucleotide bases are stored in DNA or RNA, they are combined with a sugar molecule (ribose) and a phosphate group to form a slightly larger molecule called a nucleotide. However, when discussing the information in DNA and RNA, the terms "nucleotide base" and "nucleotide" are often used interchangeably.

6. See Leroy Hood and David Galas, "The digital code of DNA," *Nature*, 421 (January 23, 2003): 444–448.

7. Richard Dawkins, *River Out of Eden: A Darwinian View of Life* (New York: Basic Books, 1995), 17.

8. See Michael Polanyi, "Life's Irreducible Structure," *Science*, 160 (June 21, 1968): 1308–1312.

9. Even though this organism is classified as "free-living," it is actually a parasite, meaning that it lives off the energy provided by another organism. See Claire M. Fraser *et al.*, "The Minimal Gene Complement of *Mycoplasma genitalium*," *Science*, 270 (October 20, 1995): 397–403.

10. Alonso Ricardo and Jack W. Szostak, "Origin of Life on Earth," *Scientific American* (September, 2009): 54–61. See also Robert Shapiro, "A Simpler Origin for Life," *Scientific American* (June, 2007): 46–53.

11. Ibid.

12. Editors' note, Ricardo and Szostak, "Origin of Life on Earth," 54–61.

13. Richard Van Noorden, "RNA world easier to make," *Nature News* (May 13, 2009), accessed March 14, 2012, http://www.nature.com/news/2009/090513/full/news.2009.471.html.

14. See Stephen C. Meyer, *Signature in the Cell: DNA and the Evidence for Intelligent Design* (New York: HarperOne, 2009), 304.

15. Jack W. Szostak, David P. Bartel, and P. Luigi Luisi, "Synthesizing Life," *Nature*, 409 (January 18, 2001): 387–390.

16. See, for example, Shapiro, "A Simpler Origin for Life," 46–53 (stating, "Some chance event or circumstance may have led to the connection of nucleotides to form RNA" and "some of these compounds join together in a chain, by chance forming a molecule—perhaps some kind of RNA—capable of reproducing itself").

17. The problem is actually much worse for materialism. Our calculation ignores the fact that in the RNA world, the synthesis of nucleotides is blind and while there are many nucleotide bases, only four are used in RNA. Likewise, there are many sugars that might be produced, but RNA uses only ribose. And even if only ribose were available, it has many places where the nucleotide bases could attach, and where the phosphate group can bind, but only certain positions work for building RNA. These severe chemical obstacles are overlooked in mathematical calculations like ours. In sum, there is not just the problem of getting the sequence right, but also the problem of building the RNA chain properly. Without intelligent guidance, you wind up with a non-functional knotted mess.

18. Shapiro, "A Simpler Origin for Life," 46–53.

19. For detailed discussions of why the RNA world cannot explain the origin of information in life or the genetic code, see Meyer, *Signature in the Cell: DNA and the Evidence for Intelligent Design*, 296–323; Stephen C. Meyer and Paul Nelson, "Can the Origin of the Genetic Code Be Explained by Direct RNA Templating?," *BIO-Complexity* 2011 (2): 1–10.

Chapter 9

1. Franklin M. Harold, *The Way of the Cell: Molecules, Organisms and the Order of Life* (Oxford: Oxford University Press, 2001), 205.

2. Michael J. Behe, *Darwin's Black Box: The Biochemical Challenge to Evolution* (New York: Free Press, 1996), 9.

3. Charles Darwin, Chapter 6, *The Origin of Species* (1859), Literature.org, accessed March 14, 2012, http://www.literature.org/authors/darwin-charles/the-origin-of-species/chapter-06.html.

4. This assumes, of course, that the mutation had its effect prior to reproduction. A mutation that causes death or sterility only after an organism is able to reproduce could still be passed on.

5. Jerry A. Coyne, "The Great Mutator," *The New Republic* (June 18, 2007): 38–44. Coyne goes on to assert that he knows of no example where this is the case.

6. Lynn Margulis quoted in Darry Madden, "UMass Scientist to Lead Debate on Evolutionary Theory," *Brattleboro (Vt.) Reformer* (February 3, 2006).

7. David S. Goodsell, *The Machinery of Life*, 2nd ed. (New York: Springer, 2009), 17.

8. Michael J. Behe, *The Edge of Evolution: The Search for the Limits of Darwinism* (New York, Free Press, 2007), 22.

9. Ibid.

10. See Douglas D. Axe, "Estimating the Prevalence of Protein Sequences Adopting Functional Enzyme Folds," *Journal of Molecular Biology*, 341 (2004): 1295–1315; Douglas D. Axe, "Extreme Functional Sensitivity to Conservative Amino Acid Changes on Enzyme Exteriors," *Journal of Molecular Biology*, 301 (2000): 585–595.

11. Douglas Axe, quoted in Chapter 5, "Biological Information," *Darwin's Dilemma: The Mystery of the Cambrian Fossil Record* (Illustra Media, 2009).

12. Douglas D. Axe, "The Limits of Complex Adaptation: An Analysis Based on a Simple Model of Structured Bacterial Populations," *BIO-Complexity*, 2010 (4): 1–10.

13. Ann K. Gauger and Douglas D. Axe, "The Evolutionary Accessibility of New Enzyme Functions: A Case Study from the Biotin Pathway," *BIO-Complexity*, 2011 (1): 1–17.

14. Michael J. Behe and David W. Snoke, "Simulating Evolution by Gene Duplication of Protein Features that Require Multiple Amino Acid Residues," *Protein Science*, 13 (2004): 2651–2664.

15. Rick Durrett and Deena Schmidt, "Waiting for Two Mutations: With Applications to Regulatory Sequence Evolution and the Limits of Darwinian Evolution," *Genetics*, 180 (November, 2008): 1501–1509.

16. Michael Behe quoted in "Is There an 'Edge' to Evolution?," FaithandEvolution.org, accessed March 14, 2012, http://www.faithandevolution.org/debates/is-there-an-edge-to-evolution.php.

17. Behe, *Darwin's Black Box: The Biochemical Challenge to Evolution*, 39 (emphasis in original).

18. David J. DeRosier, "The Turn of the Screw: The Bacterial Flagellar Motor," *Cell*, 93 (April 3, 1998): 17–20. Note: DeRosier is not pro-intelligent design.

19. Ibid.

20. Transcript of testimony of Scott Minnich, *Kitzmiller et al. v. Dover Area School Board* (M.D. Pa., PM Testimony, November 3, 2005), 103–112. Other experimental studies have identified over 30 proteins necessary to form flagella. See Table 1 in Robert M. Macnab, "Flagella," in *Escherichia Coli and Salmonella Typhimurium: Cellular and Molecular Biology Vol. 1*, eds. Frederick C. Neidhardt, John L. Ingraham, K. Brooks Low, Boris Magasanik, Moselio Schaechter, and H. Edwin Umbarger (Washington D.C.: American Society for Microbiology, 1987), 73–74.

21. Ibid. See also William A. Dembski, *No Free Lunch: Why Specified Complexity Cannot Be Purchased without Intelligence* (Lanham, MD: Rowman & Littlefield, 2002), 239–310.

22. William Dembski and Jonathan Witt, *Intelligent Design Uncensored: An Easy-to-Understand Guide to the Controversy* (Downers Grove, IL: InterVarsity Press, 2010), 54.

23. Harold, *The Way of the Cell: Molecules, Organisms and the Order of Life*, 205 (internal citations omitted).

24. Behe, *Darwin's Black Box: The Biochemical Challenge to Evolution*, 15.

Chapter 10

1. Richard Dawkins, *The Blind Watchmaker* (New York: W. W. Norton, 1986), 1.

2. Michael Behe, *The Edge of Evolution: The Search for the Limits of Darwinism* (New York: Free Press, 2007), 85

3. Richard O. Prum and Alan H. Brush, "Which Came First, the Feather or the Bird?," *Scientific American* (March, 2003): 84–93.

4. While most food webs require organisms that use photosynthesis, there are a few deep water organisms—particularly near deep sea thermal vents—that don't use photosynthesis and don't directly depend on organisms that do use the process.

5. Jin Xiong and Carl E. Bauer, "Complex Evolution of Photosynthesis," *Annual Review of Plant Biology*, 53 (2002): 503–521.

6. Brian Capon, *Botany for Gardeners*, 3rd ed. (Portland, OR: Timber Press, 2010), 188.

7. Ibid., 60.

8. See Tony Tilford, *The Complete Book of Hummingbirds* (San Diego, CA: Thunder Bay Press, 2008), 15–17.

9. David Berlinski, "Keeping an Eye on Evolution: Richard Dawkins, a relentless Darwinian spear carrier, trips over Mount Improbable, Review of *Climbing Mount Improbable* by Richard Dawkins (W. H. Norton & Company, Inc. 1996)," *The Globe & Mail* (November 2, 1996).

10. Karen McGhee and George McKay, *Encyclopedia of Animals* (Washington D.C.: National Geographic Society, 2007), 49; Fred Cooke and Jenni Bruce, *The Encyclopedia of Animals: A Complete Visual Guide* (Berkeley, CA: University of California Press, 2004), 164.

11. David Goodsell, "Calcium Pump," Molecule of the Month at Protein Data Bank (March, 2004), accessed March 14, 2012, http://www.rcsb.org/pdb/101/motm.do?momID=51. See also C. Mavroidis, A. Dubey, and M. L. Yarmush, "Molecular Machines," *Annual Review of Biomedical Engineering*, 6 (2004): 363–395; Ronald D. Vale, "The Molecular Motor Toolbox for Intracellular Transport," *Cell*, 112 (February 21, 2003): 467–480.

12. John F. Morrissey and James L. Sumich, *Introduction to the Biology of Marine Life*, 9th ed. (Sudbury, MA: Jones and Bartlett, 2009), 205.

13. David H. Evans, Peter M. Piermarini, and W. T. W. Potts, "Ionic Transport in the Fish Gill Epithelium," *Journal of Experimental Zoology*, 283 (1999): 641–652.

14. "With the exception of the sodium pump and the apical chloride channel, the sequence and physical structure of transporters involved in ion transport in fish have not been characterized. More direct evidence is needed to establish the roles and location of these transporters in teleosts." Stephen D. McCormick, "Endocrine Control of Osmoregulation in Teleost Fish," *American Zoologist*, 41 (2001): 781–794 (internal citations omitted).

15. Joyce Poole, *Elephants* (Stillwater, MN: WorldLife Library, 1997), 11–12.

16. Believe it or not, the analogy between organisms and machines is inadequate because organisms are *much more* than complex than machines. Nonetheless, they contain many systems which operate like machines.

17. Christopher Taylor, "The Really Abominable Mystery," *Catalogue of Organisms* (January 21, 2009), accessed March 14, 2012, http://coo.fieldofscience.com/2009/01/really-abominable-mystery.html.

18. For a more expansive discussion of this topic, see *Metamorphosis: The Case for Intelligent Design in a Chrysalis*, ed. David Klinghoffer (Seattle, WA: Discovery Institute Press, 2011).

19. Paul Nelson and Jonathan Wells, "Homology in Biology," in *Darwinism, Design, and Public Education*, eds. John A. Campbell and Stephen C. Meyer (East Lansing, MI: Michigan State University Press, 2003), 316.

Chapter 11

1. Isaac Asimov, "In the game of energy and thermodynamics you can't even break even," *Smithsonian*, 1 (August, 1970): 4–11.

2. See Arthur S. Seiderman and Steven E. Marcus, *20/20 Is Not Enough: The New World of Vision* (New York: Knopf, 1989), 6.

3. Michael J. Behe, *Darwin's Black Box: The Biochemical Challenge to Evolution* (New York: Free Press, 1996), 18–22.

4. Charles Darwin, Chapter 6, *The Origin of Species* (1859), Literature.org, accessed March 14, 2012, http://www.literature.org/authors/darwin-charles/the-origin-of-species/chapter-06.html.

5. Charles Darwin, Chapter 6, *The Origin of Species* (1872), Literature.org, 6th edition, accessed March 14, 2012, http://www.literature.org/authors/darwin-charles/the-origin-of-species-6th-edition/chapter-06.html.

6. Sean B. Carroll, *The Making of the Fittest: DNA and the Ultimate Forensic Record of Evolution* (New York: W. W. Norton, 2006), 197.

7. John Whitfield, "Biological theory: Postmodern evolution?," *Nature*, 455 (September 17, 2008): 281–284.

8. David Berlinski, "Keeping an Eye on Evolution: Richard Dawkins, a relentless Darwinian spear carrier, trips over Mount Improbable, Review of *Climbing Mount Improbable* by Richard Dawkins (W. H. Norton & Company, Inc. 1996)," *The Globe & Mail* (November 2, 1996).

9. Behe, *Darwin's Black Box: The Biochemical Challenge to Evolution*, 22.

10. Henry Gee, *In Search of Deep Time: Beyond the Fossil Record to a New History of Life* (New York: Free Press, 1999), 109.

11. Mark Ridley, *The Cooperative Gene: How Mendel's Demon Explains the Evolution of Complex Beings* (New York: Free Press, 2001), 111.

12. Mark Twain quoted in Henry O. Dormann, *The Speaker's Book of Quotations* (New York: Ballantine Books, 2000), 55.

Chapter 12

1. Charles Darwin, "Chapter 21—General Summary and Conclusion," *The Descent of Man* (1871), accessed March 14, 2012, http://www.literature.org/authors/darwin-charles/the-descent-of-man/chapter-21.html.

2. Kenneth R. Miller, "Life's Grand Design," PBS Evolution, accessed March 14, 2012, http://www.pbs.org/wgbh/evolution/change/grand/, adapted from Kenneth R. Miller, "Life's Grand Design," *Technology Review*, 97 (February / March, 1994): 24–32.

3. Ibid.

4. Richard Dawkins, *The Blind Watchmaker: Why the Evidence of Evolution Reveals a Universe Without Design* (New York: W. W. Norton, 1996), 93.

5. Richard Dawkins, *The Greatest Show on Earth: The Evidence for Evolution* (New York: Free Press, 2009), 354.

6. Dawkins, *The Blind Watchmaker: Why the Evidence of Evolution Reveals a Universe Without Design*, 93.

7. Ibid.

8. George Ayoub, "On the Design of the Vertebrate Retina," *Origins & Design*, 17 (Winter, 1996): 1, accessed March 12, 2012, http://www.arn.org/docs/odesign/od171/retina171.htm.

9. Kate McAlpine, "Evolution gave flawed eye better vision," *New Scientist* (May 6, 2010), accessed March 14, 2012, http://www.newscientist.com/article/mg20627594.000-evolution-gave-flawed-eye-better-vision.html.

10. Miller, "Life's Grand Design," 24–32.

11. A. M. Labin and E. N. Ribak, "Retinal Glial Cells Enhance Human Vision Acuity," *Physical Review Letters*, 104 (2010).

12. William Dembski and Sean McDowell, *Understanding Intelligent Design: Everything You Need to Know in Plain Language* (Eugene, OR: Harvest House, 2008), 54.

13. Stephen Jay Gould, *The Panda's Thumb: More Reflections in Natural History* (New York: W. W. Norton, 1980), 20–21.

14. Miller, "Life's Grand Design," 24–32.

15. Stephen Jay Gould, "The Panda's Peculiar Thumb," *Natural History*, 87 (November, 1978): 30.

16. Miller, "Life's Grand Design," 24–32.

17. Hideki Endo, Daishiro Yamagiwa, Yoshihiro Hayashi, Hiroshi Koie, Yoshiki Yamaya, Junpei Kimura, "Role of the giant panda's 'pseudo-thumb,'" *Nature*, 397 (January 28, 1999): 309–310.

18. Jerry Coyne, *Why Evolution is True* (New York: Viking, 2009), 82.

19. Richard Dawkins, quoted in "Richard Dawkins demonstrates laryngeal nerve of the giraffe," accessed February 8, 2011, http://www.youtube.com/watch?v=cO1a1Ek-HD0.

20. See Wolf-Ekkehard Lönnig, "The Laryngeal Nerve of the Giraffe: Does it Prove Evolution?" (September 6, 2010), accessed March 14, 2012, http://www.weloennig.de/LaryngealNerve.pdf.

21. "As the recurrent nerve hooks around the subclavian artery or aorta, it gives off several cardiac filaments to the deep part of the cardiac plexus. As it ascends in the neck it gives off branches, more numerous on the left than on the right side, to the mucous membrane and muscular coat of the esophagus; branches to the mucous membrane and muscular fibers of the trachea; and some pharyngeal filaments to the Constrictor pharyngis inferior." Henry Gray, "5j. The Vagus Nerve," *Anatomy of the Human Body*, accessed March 14, 2012, http://www.bartleby.com/107/205.html.

22. For further documentation and discussion, see Casey Luskin, "The Recurrent Laryngeal Nerve Does Not Refute Intelligent Design," IDEA Center, accessed March 14, 2012, http://www.ideacenter.org/contentmgr/showdetails.php/id/1507.

23. Miller, "Life's Grand Design," *Technology Review*, 24–32.

24. Examples taken from Bharat Bhushan, "Biomimetics: lessons from nature—an overview," *Philosophical Transactions of the Royal Society* A, 367 (2009): 1445–1486; "Whales And Dolphins Influence New Wind Turbine Design," *ScienceDaily* (July 8, 2008), accessed March 14, 2012, http://www.sciencedaily.com/releases/2008/07/080707222315.htm; Tudor Vieru, "Hippo Sweat to Be Turned into Sunscreen," *Softpedia* (March 17, 2009), accessed March 14, 2012, http://news.softpedia.com/news/Hippo-Sweat-To-Be-Turned-into-Sunscreen-106946.shtml.

25. "Flower power," *The Economist* (January 21, 2012), accessed July 16, 2012, http://www.economist.com/node/21543123.

26. Julie Steenhuysen, "Eye spy: U.S. scientists develop eye-shaped camera," Reuters (August 6, 2008), accessed March 14, 2012, http://www.reuters.com/article/2008/08/06/us-camera-eye-idUSN0647922920080806.

27. Richard Dawkins, quoted in Christopher Hitchens, "Losing Sight of Progress: How blind salamanders make nonsense of creationists' claims," *Slate.com* (July 21, 2008), accessed March 14, 2012, http://www.slate.com/id/2195683/.

28. David G. Addiss, Nathan Shaffer, Barbara S. Fowler, and Robert V. Tauxe, "The Epidemiology of Appendicitis and Appendectomy in the United States," *American Journal of Epidemiology*, 132 (1990): 910–925.

29. Douglas Theobald, "29+ Evidences for Macroevolution," TalkOrigins.org, accessed March 14, 2012, http://www.talkorigins.org/faqs/comdesc/section2.html.

30. Charles Q. Choi, "The Appendix: Useful and in Fact Promising," *LiveScience* (August 24, 2009).

31. See Loren G. Martin, "What is the function of the human appendix? Did it once have a purpose that has since been lost?," *Scientific American* (October, 21, 1999), accessed March 14, 2012, http://www.scientificamerican.com/article.cfm?id=what-is-the-function-of-t.

32. See Johan Styrud, Staffan Eriksson, Ingemar Nilsson, Gunnar Ahlberg, Staffan Haapaniemi, Gunnar Neovius, Lars Rex, Ibrahim Badume, Lars Granström, "Appendectomy versus Antibiotic Treatment in Acute Appendicitis. A Prospective Multicenter Randomized Controlled Trial," *World Journal of Surgery*, 30 (April 27, 2006): 1033–1037.

33. William Parker quoted in Choi, "The Appendix: Useful and in Fact Promising."

34. Horatio Hackett Newman, quoted in *The World's Most Famous Court Trial: Tennessee Evolution Case*, 2nd ed. (Dayton, TN: Bryan College, 1990), 268. See also Robert Wiedersheim, *The Structure of Man: An Index to His Past History* (London: MacMillan and Co, 1895; reprinted by Kessinger, 2007).

35. Laura Spinney, "Vestigial organs: Remnants of evolution," *New Scientist*, 2656 (May 14, 2008), accessed March 14, 2012, http://www.newscientist.com/article/mg19826562.100-vestigial-organs-remnants-of-evolution.html.

36. Sylvia S. Mader, *Inquiry into Life*, 10th ed. (McGraw Hill, 2003), 293.

37. Laura Spinney, "The Five things humans no longer need," *NewScientist* (May 19, 2008), accessed March 14, 2012, http://www.newscientist.com/article/dn13927-five-things-humans-no-longer-need.html.

38. Spinney, "Vestigial organs: Remnants of evolution."

39. Miller, "Life's Grand Design," 24–32. Miller cites "orphaned genes" but these are not normally understood to be functionless genes. Rather, orphan genes are functional genes that have no known homology to any other gene. Such orphan genes provide evidence *for* intelligent design because there is no plausible material source for their information.

40. Francis Collins, *The Language of God: A Scientist Presents Evidence for Belief*, (New York: Free Press, 2006), 136.

41. Ibid., 137.

42. Jonathan Wells, "Using Intelligent Design Theory to Guide Scientific Research," *Progress in Complexity, Information, and Design*, 3.1.2 (November, 2004).

43. This includes repetitive DNA, LINE and SINE sequences, *Alu* sequences, endogenous retroviruses (ERVs), introns, and pseudogenes. Many dozens of papers document such functions, including: Richard Sternberg, "On the Roles of Repetitive DNA Elements in the Context of a Unified Genomic-Epigenetic System," *Annals of the NY Academy of Science*, 981 (2002): 154–188; James A. Shapiro, and Richard Sternberg, "Why repetitive DNA is essential to genome function," *Biological Reviews of the Cambridge Philosophical Society*, 80 (2005): 227–250; Tammy A. Morrish *et al.*, "DNA repair mediated by endonuclease-independent LINE-1 retrotransposition," *Nature Genetics*, 31 (June, 2002): 159–165; Galit Lev-Maor, Rotem Sorek, Noam Shomron, and Gil Ast, "The birth of an alternatively spliced exon: 3' splice-site selection in Alu exons," *Science*, 300 (May 23, 2003): 1288–1291; Wojciech Makalowski, "Not junk after all," *Science*, 300 (May 23, 2003): 1246–1247; Andrew B. Conley, Jittima Piriyapongsa, and I. King Jordan, "Retroviral promoters in the human genome," *Bioinformatics*, 24 (2008): 1563–1567; Geoffrey J. Faulkner *et al.* "The regulated retrotransposon transcriptome of mammalian cells," *Nature Genetics*, 41 (April 19, 2009): 563–571; W. Wayt Gibbs, "The Unseen Genome, Gems Among the Junk," *Scientific American* (November, 2003): 46–53.

44. Morrish *et al.*, "DNA repair mediated by endonuclease-independent LINE-1 retrotransposition," 159–165; Annie Tremblay, Maria Jasin, and Pierre Chartrand, "A Double-Strand Break in a Chromosomal LINE Element Can Be Repaired by Gene Conversion with Various Endogenous LINE Elements in Mouse Cells," *Molecular and Cellular Biology*, 20 (January, 2000): 54–60; Ulf Grawunder, Matthias Wilm, Xiantuo Wu, Peter Kulesza, Thomas E. Wilson, Matthias Mann, and Michael R. Lieber, "Activity of DNA ligase IV stimulated by complex formation with XRCC4 protein in mammalian cells," *Nature*, 388 (July 31, 1997): 492–495; Thomas E. Wilson, Ulf Grawunder, and Michael R. Lieber, "Yeast DNA ligase IV mediates non-homologous DNA end joining," *Nature*, 388 (July 31, 1997): 495–498.

45. Richard Sternberg and James A. Shapiro, "How repeated retroelements format genome function," *Cytogenetic and Genome Research*, 110 (2005): 108–116.

46. Jeffrey S. Han, Suzanne T. Szak, and Jef D. Boeke, "Transcriptional disruption by the L1 retrotransposon and implications for mammalian transcriptomes," *Nature*, 429 (May 20, 2004): 268–274; Bethany A. Janowski, Kenneth E. Huffman, Jacob C. Schwartz, Rosalyn Ram, Daniel Hardy, David S. Shames, John D. Minna, and David R. Corey, "Inhibiting gene expression at transcription start sites in chromosomal DNA with antigene RNAs," *Nature Chemical Biology*, 1 (September, 2005): 216–222; J. A. Goodrich, and J. F. Kugel, "Non-coding-RNA regulators of RNA polymerase II transcription," *Nature Reviews Molecular and Cell Biology*, 7 (August, 2006): 612–616; L. C. Li, S. T. Okino, H. Zhao, H., D. Pookot, R. F. Place, S. Urakami, H. Enokida, and R. Dahiya, "Small dsRNAs induce transcriptional activation in human cells," *Proceedings of the National Academy of Sciences USA*, 103 (November 14, 2006): 17337–17342; A. Pagano, M. Castelnuovo, F. Tortelli, R. Ferrari, G. Dieci, and R. Cancedda, "New small nuclear RNA

gene-like transcriptional units as sources of regulatory transcripts," *PLoS Genetics*, 3 (February, 2007): e1; L. N. van de Lagemaat, J. R. Landry, D. L. Mager, and P. Medstrand, "Transposable elements in mammals promote regulatory variation and diversification of genes with specialized functions," *Trends in Genetics*, 19 (October, 2003): 530–536; S. R. Donnelly, T. E. Hawkins, and S. E. Moss, "A Conserved nuclear element with a role in mammalian gene regulation," *Human Molecular Genetics*, 8 (1999): 1723–1728; C. A. Dunn, P. Medstrand, and D. L. Mager, "An endogenous retroviral long terminal repeat is the dominant promoter for human B1,3- galactosyltransferase 5 in the colon," *Proceedings of the National Academy of Sciences USA*, 100 (October 28, 2003): 12841–12846; B. Burgess-Beusse, C. Farrell, M. Gaszner, M. Litt, V. Mutskov, F. Recillas-Targa, M. Simpson, A. West, and G. Felsenfeld, "The insulation of genes from external enhancers and silencing chromatin," *Proceedings of the National Academy of Sciences USA*, 99 (December 10, 2002): 16433–16437; P. Medstrand, Josette- Renée Landry, and D. L. Mager, "Long Terminal Repeats Are Used as Alternative Promoters for the Endothelin B Receptor and Apolipoprotein C-I Genes in Humans," *Journal of Biological Chemistry*, 276 (January 19, 2001): 1896–1903; L. Mariño-Ramírez, K. C. Lewis, D. Landsman, and I. K. Jordan, "Transposable elements donate lineage-specific regulatory sequences to host genomes," *Cytogenetic and Genome Research*, 110 (2005): 333–341.

47. C. Bell, A. G. West, and G. Felsenfeld, "Insulators and Boundaries: Versatile Regulatory Elements in the Eukaryotic Genome," *Science*, 291 (January 19, 2001): 447–450; M.-L. Pardue and P. G. DeBaryshe, "*Drosophila* telomeres: two transposable elements with important roles in chromosomes," *Genetica*, 107 (1999): 189–196; S. Henikoff, "Heterochromatin function in complex genomes," *Biochimica et Biophysica Acta*, 1470 (February, 2000): O1–O8; L. M. Figueiredo, L. H. Freitas-Junior, E. Bottius, Jean-Christophe Olivo-Marin, and A. Scherf, "A central role for *Plasmodium falciparum* subtelomeric regions in spatial positioning and telomere length regulation," *The EMBO Journal*, 21 (2002): 815–824; Mary G. Schueler, Anne W. Higgins, M. Katharine Rudd, Karen Gustashaw, and Huntington F. Willard, "Genomic and Genetic Definition of a Functional Human Centromere," *Science*, 294 (October 5, 2001): 109–115.

48. Ling-Ling Chen, Joshua N DeCerbo, and Gordon G Carmichael, "*Alu* element-mediated gene silencing," *The EMBO Journal*, 27 (2008): 1694–1705; Jerzy Jurka, "Evolutionary impact of human Alu repetitive elements," *Current Opinion in Genetics & Development*, 14 (2004): 603–608; Lev-Maor, Sorek, Shomron, and Ast, "The birth of an alternatively spliced exon: 3' splice-site selection in Alu exons," 1288–1291; E. Kondo-Iida, K. Kobayashi, M. Watanabe, J. Sasaki, T. Kumagai, H. Koide, K. Saito, M. Osawa, Y. Nakamura, and T. Toda, "Novel mutations and genotype-phenotype relationships in 107 families with Fukuyama-type congenital muscular dystrophy (FCMD)," *Human Molecular Genetics*, 8 (1999): 2303–2309; John S. Mattick and Igor V. Makunin, "Non-coding RNA," *Human Molecular Genetics*, 15 (2006): R17–R29.

49. M. Mura, P. Murcia, M. Caporale, T. E. Spencer, K. Nagashima, A. Rein, and M. Palmarini, "Late viral interference induced by transdominant Gag of an endogenous retrovirus," *Proceedings of the National Academy of Sciences USA*, 101 (July 27, 2004): 11117–11122; M. Kandouz, A. Bier, G. D Carystinos, M. A Alaoui-Jamali, and G. Batist, "Connexin43 pseudogene is expressed in tumor cells and inhibits growth," *Oncogene*, 23 (2004): 4763–4770.

50. K. A. Dunlap, M. Palmarini, M. Varela, R. C. Burghardt, K. Hayashi, J. L. Farmer, and T. E. Spencer, "Endogenous retroviruses regulate periimplantation placental growth and differentiation," *Proceedings of the National Academy of Sciences USA*, 103 (September 26, 2006): 14390–14395; L. Hyslop, M. Stojkovic, L. Armstrong, T. Walter, P. Stojkovic, S. Przyborski, M. Herbert, A. Murdoch, T. Strachan, and M. Lako, "Downregulation of NANOG Induces Differentiation of Human Embryonic Stem Cells to Extraembryonic Lineages," *Stem Cells*, 23 (2005): 1035–1043; E. Peaston, A. V. Evsikov, J. H. Graber, W. N. de Vries, A. E. Holbrook, D. Solter, and B. B. Knowles, "Retrotransposons Regulate Host Genes in Mouse Oocytes and Preimplantation Embryos," *Developmental Cell*, 7 (October, 2004): 597–606.

51. The ENCODE Project Consortium, "An integrated encyclopedia of DNA elements in the human genome," *Nature*, 489 (September 6, 2012): 57–74.

52. Ewan Birney and Tom Gingeras quoted in Ed Yong, "ENCODE: the rough guide to the human genome," *Discover Magazine* (September 5, 2012), accessed September 10, 2012, http://blogs.discovermagazine.com/notrocketscience/2012/09/05/encode-the-rough-guide-to-the-human-genome/.

53. Joseph R. Ecker, "Serving up a genome feast," *Nature*, 489 (September 6, 2012): 52–53.

54. Gina Kolata, "Bits of Mystery DNA, Far From 'Junk,' Play Crucial Role," *The New York Times* (September 5, 2012), accessed September 11, 2012, http://www.nytimes.com/2012/09/06/science/far-from-junk-dna-dark-matter-proves-crucial-to-health.html.

55. Alok Jha, "Breakthrough study overturns theory of 'junk DNA' in genome," *The Guardian* (September 5, 2012), accessed September 10, 2012, http://www.guardian.co.uk/science/2012/sep/05/genes-genome-junk-dna-encode.

56. Richard Dawkins, "The Information Challenge," *The Skeptic*, 18 (December, 1998). For a full response to this article, see Casey Luskin, "A Response to Dr. Dawkins' 'The Information Challenge'," Evolution News & Views (October 4, 2007), accessed March 14, 2012, http://www.discovery.org/a/4278.

57. Ibid.

58. For examples of some papers that have found function or pseudogenes, see D. Zheng and M. B. Gerstein, "The ambiguous boundary between genes and pseudogenes: the dead rise up, or do they?," *Trends in Genetics*, 23 (May, 2007): 219–224; S. Hirotsune *et al.*, "An expressed pseudogene regulates the messenger-RNA stability of its homologous coding gene," *Nature*, 423 (May 1, 2003): 91–96; O. H. Tam *et al.*, "Pseudogene-derived small interfering RNAs regulate gene expression in mouse oocytes," *Nature*, 453 (May 22, 2008): 534–538; D. Pain *et al.*, "Multiple Retropseudogenes from Pluripotent Cell-specific Gene Expression Indicates a Potential Signature for Novel Gene Identification," *The Journal of Biological Chemistry*, 280 (February 25, 2005): 6265–6268; J. Zhang *et al.*, "NANOGP8 is a retrogene expressed in cancers," *FEBS Journal*, 273 (2006): 1723–1730.

59. Evgeniy S. Balakirev and Francisco J. Ayala, "Pseudogenes, Are They 'Junk' or Functional DNA?," *Annual Review of Genetics*, 37 (2003): 123–151.

60. Ryan Charles Pink, Kate Wicks, Daniel Paul Caley, Emma Kathleen Punch, Laura Jacobs, and David Paul Francisco Carter, "Pseudogenes: Pseudo-functional or key regulators in health and disease?," *RNA*, 17 (2011): 792–798.

61. Yan-Zi Wen, Ling-Ling Zheng, Liang-Hu Qu, Francisco J. Ayala and Zhao-Rong Lun, "Pseudogenes are not pseudo any more," *RNA Biology*, 9 (January, 2012): 27–32.

62. Makalowski, "Not Junk After All," 1246–1247.

63. Ibid.

64. Gibbs, "The Unseen Genome: Gems among the Junk," 46–53 (quoting John Mattick, internal quotations omitted).

65. Jonathan Wells, *The Myth of Junk DNA* (Seattle, WA: Discovery Institute Press, 2011), 107.

66. Philip Ball, "What a shoddy piece of work is man," *Nature* (May 3, 2010), accessed March 14, 2012, http://www.nature.com/news/2010/030510/full/news.2010.215.html.

Chapter 13

1. Partly quoting Eric Bapteste, in Graham Lawton, "Why Darwin was wrong about the tree of life," *New Scientist*, 2692 (January 21, 2009), accessed February 18, 2013, http://www.newscientist.com/article/mg20126921.600-why-darwin-was-wrong-about-the-tr (internal quotations omitted).

2. The drawing portrayed here shows part of the "tree of life" from Darwin's *Origin of Species*. His full drawing, however, was not necessarily a single tree of life. Darwin suggested there might have been a single grand tree of life, but he wasn't completely sure. In his time, scientists knew little about microorganisms, such as many types of bacteria. What Darwin was clear about was his view that most major known groups of

animals (e.g., humans, frogs, fish, flies, and worms) were related. This view remains a core part of the "tree of life" today.

3. Charles Darwin, Chapter 14, *The Origin of Species* (1859), Literature.org, accessed February 25, 2013, http://www.literature.org/authors/darwin-charles/the-origin-of-species/chapter-14.html.

4. While the vast majority of materialists accept universal common ancestry, it should be noted that a growing minority is beginning to doubt that there is a single, grand "tree of life."

5. Perry Mann, "The Dinky Insect That Helps Demonstrate Darwin's Theory," *Huntington News* (April 27, 2009), accessed February 18, 2013, http://archives.huntingtonnews.net/columns/090427-mann-columnsmanntalk.html.

6. Karl Giberson, *Saving Darwin: How to Be a Christian and Believe in Evolution* (New York: HarperOne, 2008), 53.

7. For example, one article states: "molecular systematics is (largely) based on the assumption, first clearly articulated by Zuckerkandl and Pauling (1962), that degree of overall similarity reflects degree of relatedness." Jeffrey H. Schwartz, Bruno Maresca, "Do Molecular Clocks Run at All? A Critique of Molecular Systematics," *Biological Theory*, 1 (2006): 357–371. One textbook states: "The key assumption made when constructing a phylogenetic tree from a set of sequences is that they are all derived from a single ancestral sequence, i.e., they are homologous." Marketa Zvelebil and Jeremy O. Baum, *Understanding Bioinformatics* (New York: Garland Science, 2008), 239. Likewise, another treatise states: "Cladistics can run into difficulties in its application because not all character states are necessarily homologous. Certain resemblances are convergent—that is, the result of independent evolution. We cannot always detect these convergences immediately, and their presence may contradict other similarities, 'true homologies' yet to be recognized. Thus, we are obliged to assume at first that, for each character, similar states are homologous, despite knowing that there may be convergence among them." Guillaume Lecointre and Hervé Le Guyader, *The Tree of Life: A Phylogenetic Classification* (Cambridge, MA: Harvard University Press, 2006), 16.

8. Richard Dawkins quoted in "Richard Dawkins answers reddit question about evolution," accessed February 26, 2011, http://www.youtube.com/watch?v=5PlqNoCAIgA (emphasis added).

9. For example:
   - A 2009 paper in *Trends in Ecology and Evolution* notes that: "A major challenge for incorporating such large amounts of data into inference of species trees is that conflicting genealogical histories often exist in different genes throughout the genome." James H. Degnan and Noah A. Rosenberg, "Gene tree discordance, phylogenetic inference and the multispecies coalescent," *Trends in Ecology and Evolution*, 24 (2009): 332–340.
   - Similarly, a paper in the journal *Genome Research* studied the DNA sequences in various animal groups and found that "different proteins generate different phylogenetic tree[s]." Arcady R. Mushegian, James R. Garey, Jason Martin and Leo X. Liu, "Large-Scale Taxonomic Profiling of Eukaryotic Model Organisms: A Comparison of Orthologous Proteins Encoded by the Human, Fly, Nematode, and Yeast Genomes," *Genome Research*, 8 (1998): 590–598.
   - A study published in *Science* in 2005 tried to construct a phylogeny of animal relationships but concluded that "[d]espite the amount of data and breadth of taxa analyzed, relationships among most [animal] phyla remained unresolved." Again, the problem lies in the fact that trees based upon one gene or protein often conflict with trees based upon other genes. Their study tried to avoid this problem by using a many-gene technique, yet still found that "[a] 50-gene data matrix does not resolve relationships among most metazoan phyla." Antonis Rokas, Dirk Krüger, Sean B. Carroll, "Animal Evolution and the Molecular Signature of Radiations Compressed in Time," *Science*, 310 (December 23, 2005): 1933–1938.
   - Another study published in *Science* found that the molecular data implied that six-legged arthropods, or hexapods (i.e., insects) are not monophyletic, a conclusion that differed strikingly from virtually all previous wisdom. The article concluded "Although this tree shows many interesting outcomes, it also contains some evidently untenable relationships, which nevertheless have strong statistical support." Francesco Nardi, Giacomo Spinsanti, Jeffrey L. Boore, Antonio Carapelli, Romano Dallai, Francesco Frati,, "Hexapod Origins: Monophyletic or Paraphyletic?," *Science*, 299 (March 21, 2003): 1887–1889.
   - A paper in the *Journal of Molecular Evolution* found that molecule-based phylogenies conflicted sharply with previously established phylogenies of major mammal groups, concluding that this anomalous tree "is not due to a stochastic error, but is due to convergent or parallel evolution." Ying Cao, Axel Janke, Peter J. Waddell, Michael Westerman, Osamu Takenaka, Shigenori Murata, Norihiro Okada, Svante Pääbo, Masami Hasegawa, "Conflict Among Individual Mitochondrial Proteins in Resolving the Phylogeny of Eutherian Orders," *Journal of Molecular Evolution*, 47 (1998): 307–322.
   - A study in *Proceedings of the National Academy of Sciences USA* explains that when biologists tried to construct a phylogenetic tree for the major groups of birds using mitochondrial DNA (mtDNA), their results conflicted sharply with traditional notions of bird relationships. Strikingly, they even find "convergent" similarity between some bird mtDNA and the mtDNA of distant species such as snakes and lizards. The article suggests bird mtDNA underwent "multiple independent originations," with their study making a "finding of multiple independent origins for a particular mtDNA gene order among diverse birds." David P. Mindell, Michael D. Sorenson, and Derek E. Dimcheff, "Multiple independent origins of mitochondrial gene order in birds," *Proceedings of the National Academy of Sciences USA*, 95 (September, 1998): 10693–10697.

10. Partly quoting Eric Bapteste, in Lawton, "Why Darwin was wrong about the tree of life," (internal quotations omitted).

11. Partly quoting Michael Syvanen, in Lawton, "Why Darwin was wrong about the tree of life," (internal quotations omitted).

12. Michael Syvanen, quoted in Lawton, "Why Darwin was wrong about the tree of life."

13. Elie Dolgin, "Rewriting Evolution," *Nature*, 486 (June 28, 2012): 460–462.

14. Antonis Rokas and Sean B. Carroll, "Bushes in the Tree of Life," *PLoS Biology*, 4 (November, 2006): 1899–1904.

15. Stephen C. Meyer, Scott Minnich, Jonathan Moneymaker, Paul A. Nelson, and Ralph Seelke, *Explore Evolution: The Arguments For and Against Neo-Darwinism* (Melbourne: Hill House, 2007), 45.

16. Peter Atkins, *Galileo's Finger: The Ten Great Ideas of Science* (Oxford: Oxford University Press, 2003), 16.

17. Trisha Gura, "Bones, Molecules or Both?," *Nature*, 406 (July 20, 2000): 230–233.

18. For example, see *BSCS Biology: A Molecular Approach* (Glencoe/McGraw Hill, 2006), 227; Sylvia S. Mader, Jeffrey A. Isaacson, Kimberly G. Lyle-Ippolito, Andrew T. Storfer, *Inquiry Into Life*, 13th ed. (McGraw Hill, 2011), 550.

19. Michael S. Y. Lee, "Molecular phylogenies become functional," *Trends in Ecology and Evolution*, 14 (May 5, 1999): 177–178 .

20. Figure 13-3 used with permission from American Association for the Advancement of Science, Figure 3, W. Ford Doolittle, "Phylogenetic Classification and the Universal Tree," *Science*, 284 (June 25, 1999): 2124–2128. Copyright 1999.

21. Doolittle, "Phylogenetic Classification and the Universal Tree," 2124–2128.

22. It's important to note that horizontal gene transfer does not create new genes. It simply involves the transfer of *pre-existing* genetic information from one microorganism to another.

23. For a discussion, see Casey Luskin, "Inconvenient Fungus Genetic Data Leads to Epicycles in the Tree of Life," Evolution News & Views (March 29, 2011), accessed March 15, 2012, http://www.evolutionnews.org/2011/03/inconvenient_fungus_genetic_da045301.html.

24. Lawton, "Why Darwin was wrong about the tree of life" ("More fundamentally, recent research suggests that the evolution of animals and plants isn't exactly tree-like either. 'There are problems even in that

little corner,' says Dupré. Having uprooted the tree of unicellular life, biologists are now taking their axes to the remaining branches.").

25. *See* Mark A. Ragan and Robert G. Beiko, "Lateral genetic transfer: open issues," *Philosophical Transactions of the Royal Society B*, 364 (2009): 2241–2251 ("In the phylogenetic approach, each instance of topological discordance between a gene tree and a trusted reference tree is taken as a prima facie instance of LGT. Discordance can be found throughout the entire range of nodal depths within these trees, from recent (genera, species) to older, presumably reflecting a commerce in genetic material that has been ongoing since pre-genomic times (Woese 2000). Viewed in this way, every genome has LGT in its ancestry"). Note: Horizontal gene transfer is often also called "lateral gene transfer," or "LGT," as seen in the quote here.

26. See Paul Nelson and Jonathan Wells, "Homology in Biology," in *Darwinism, Design, and Public Education*, eds. John Angus Campbell and Stephen C. Meyer (East Lansing: Michigan State University Press, 2003), 303–322; Rebecca Quiring, Uwe Walldorf, Urs Kloter, and Walter J. Gehring, "Homology of the eyeless Gene of *Drosophila* to the *Small eye* Gene in Mice and *Aniridia* in humans," *Science*, 265 (August 5, 1994): 785–789; David B. Wake, Marvalee H. Wake, and Chelsea D. Specht, "Homoplasy: From Detecting Pattern to Determining Process and Mechanism of Evolution," *Science*, 331 (February 25, 2011): 1032–1035. See also Ernst Mayr, *What Evolution Is* (New York: Basic Books, 2001), 113 ("It had been shown that by morphological-phylogenetic research that photoreceptor organs (eyes) had developed at least 40 times independently during the evolution of animal diversity. A developmental geneticist, however, showed that all animals with eyes have the same regulator gene, *Pax* 6, which organizes the construction of the eye. It was therefore concluded at first concluded that all eyes were derived from a single ancestral eye with the *Pax* 6 gene. But then the geneticist also found *Pax* 6 in species without eyes, and proposed that they must have descended from ancestors with eyes. However, this scenario turned out to be quite improbable and the wide distribution of *Pax* 6 required a different explanation. It is now believed that *Pax* 6, even before the origin of eyes, had an unknown function in eyeless organisms, and was subsequently recruited for its role as an eye organizer.").

27. See "In Bats and Whales, Convergence in Echolocation Ability Runs Deep" *ScienceDaily* (January 25, 2010), accessed March 14, 2012, http://www.sciencedaily.com/releases/2010/01/100125123219.htm ("The discovery represents an unprecedented example of adaptive sequence convergence between two highly divergent groups and suggests that such convergence at the sequence level might be more common than scientists had suspected.... 'We were surprised by the strength of support for convergence between these two groups of mammals and, related to this, by the sheer number of convergent changes in the coding DNA that we found.'"). See also Ying Li, Zhen Liu, Peng Shi, and Jianzhi Zhang, "The hearing gene Prestin unites echolocating bats and whales," *Current Biology*, 20 (January, 2010): R55–R56 ("Only microbats and toothed whales have acquired sophisticated echolocation, indispensable for their orientation and foraging. Although the bat and whale biosonars originated independently and differ substantially in many aspects, we here report the surprising finding that the bottlenose dolphin, a toothed whale, is clustered with microbats in the gene tree constructed using protein sequences encoded by the hearing gene Prestin.") (internal citations omitted).

28. Pascal-Antoine Christin, Daniel M. Weinreich, and Guillaume Besnard, "Causes and evolutionary significance of genetic convergence," *Trends in Genetics*, 26 (2010): 400–405; Li, Liu, Shi, and Zhang, "The hearing gene Prestin unites echolocating bats and whales," R55–R56.

29. Richard Dawkins, *The Blind Watchmaker: Why the Evidence of Evolution Reveals a Universe Without Design* (New York: W. W. Norton, 1996), 94.

30. Ibid.

31. Ibid.

32. "Cladistics can run into difficulties in its application because not all character states are necessarily homologous. Certain resemblances are convergent—that is, the result of independent evolution. We cannot always detect these convergences immediately, and their presence may contradict other similarities, 'true homologies' yet to be recognized. Thus, we are obliged to assume at first that, for each character, similar states are homologous, despite knowing that there may be convergence among them." Lecointre and Guyader, *The Tree of Life: A Phylogenetic Classification*, 16.

33. Meyer, Minnich, Moneymaker, Nelson, and Seelke, *Explore Evolution: The Arguments For and Against Neo-Darwinism*, 48.

34. Nelson and Wells, "Homology in Biology," 316.

35. While functional biological similarity is easily explained by common design, *non-functional* similarity can still point to common descent. However, as we saw in Chapter 12, many supposed cases of functionless "junk" DNA or "vestigial" organs have turned out to perform important biological functions.

36. Tim Berra, *Evolution and the Myth of Creationism: A Basic Guide to the Facts of the Evolution Debate* (Stanford, CA: Stanford University Press, 1990), 117.

37. While horizontal gene transfer might be a viable explanation within microorganisms, in higher organisms such as animals it is not a well-demonstrated mechanism of change.

Chapter 14

1. Charles Darwin, Chapter 21—General Summary and Conclusion, *The Descent of Man* (1871), accessed March 15, 2012, http://www.literature.org/authors/darwin-charles/the-descent-of-man/chapter-21.html.

2. See Scott F. Gilbert, "Ernst Haeckel and the Biogenetic Law," *DevBio Companion to Developmental Biology*, 9th ed. (Sunderland, MA: Sinauer Associates, 2010), accessed March 15, 2012, http://8e.devbio.com/article.php?id=219.

3. Elizabeth Pennisi, "Haeckel's Embryos: Fraud Rediscovered," *Science*, 277 (September 5, 1997): 1435.

4. Michael K. Richardson quoted in Pennisi, "Haeckel's Embryos: Fraud Rediscovered," 1435.

5. For example, Kenneth Miller's 1994 textbook *Biology: Discovering Life* contains Haeckel's original embryo drawings and promotes recapitulation theory, stating without qualification that "the embryological development of an individual repeats its species's evolutionary history." Joseph S. Levine and Kenneth R. Miller, *Biology: Discovering Life*, 2nd ed. (Lexington, MA: D.C. Heath, 1994), 162.

6. "10 Questions, and Answers, about Evolution," *New York Times* (August 23, 2008), accessed March 15, 2012, http://www.nytimes.com/2008/08/24/us/WEB-tenquestions.html.

7. See Casey Luskin, "What do Modern Textbooks Really Say about Haeckel's Embryos?," Discovery Institute (March 27, 2007), accessed March 15, 2012, http://www.discovery.org/a/3935; Casey Luskin, "Current Textbooks Misuse Embryology to Argue for Evolution," Evolution News & Views (June 18, 2010), accessed March 15, 2012, http://www.evolutionnews.org/2010/06/current_textbooks_misuse_embry035751.html. Those pages discuss some of the following examples: Peter H. Raven and George B. Johnson, *Biology*, 5th ed. (Boston, MA: McGraw Hill, 1999), 1181; Peter H. Raven and George B. Johnson, *Biology*, 6th ed. (Boston: McGraw Hill, 2002); Douglas J. Futuyma, *Evolutionary Biology*, 3rd ed. (Sunderland, MA: Sinauer, 1998), 1229; Cecie Starr and Ralph Taggart, *Biology: The Unity and Diversity of Life*, 8th ed. (Belmont, CA: Wadsworth, 1998), 317; Joseph Raver, *Biology: Patterns and Processes of Life*, Draft version presented to the Texas State Board of Education for approval in 2003 (Dallas, TX: J. M. Lebel, 2004), 100; Cecie Starr and Ralph Taggart, *Biology: The Unity and Diversity of Life*, Draft version presented to the Texas State Board of Education in 2003 (Belmont CA: Brooks Cole / Wadsworth, 2004), 315; William D. Schraer and Herbert J. Stoltze, *Biology: The Study of Life*, 7th ed. (Upper Saddle River, NJ: Prentice Hall, 1999), 583; Michael Padilla *et al.*, *Focus on Life Science: California Edition* (Needham, MA: Prentice Hall, 2001), 372; Kenneth R Miller and Joseph Levine, *Biology: The Living Science* (Upper Saddle River, NJ: Prentice Hall, 1998), 223; Kenneth R Miller and Joseph Levine, *Biology*, 4th ed. (Upper Saddle River, NJ: Prentice Hall, 1998), 283; Sylvia S. Mader, *Biology*, 10th ed. (New York: McGraw Hill, 2010), 278; Sylvia S. Mader, Jeffrey A. Isaacson, Kimberly G. Lyle-Ippolito, Andrew T. Storfer, *Inquiry into Life*, 13th ed. (New York: McGraw Hill, 2011), 549.

8. Michael Richardson quoted in Stephen Jay Gould, "Abscheulich! (Atrocious!)," *Natural History* (March, 2000): 42–49.

9. Michael K. Richardson, James Hanken, Mayoni L. Gooneratne, Claude Pieau, Albert Raynaud, Lynne Selwood, and Glenda M. Wright, "There is No Highly Conserved Embryonic Stage in the Vertebrates: Implications for Current Theories of Evolution and Development," *Anatomy and Embryology*, 196 (1997): 91–106 (internal citations omitted).

10. Gould, "Abscheulich! (Atrocious!)," 42–49.

11. Jonathan Wells, "Haeckel's Embryos & Evolution: Setting the Record Straight," *The American Biology Teacher*, 61 (May, 1999): 345–349 (internal citations removed).

12. See Mader, *Biology*, 10th ed., 278; Mader *et al.*, *Inquiry into Life*, 13th ed., 549.

13. For example, one paper states "Recent workers have shown that early development can vary quite extensively, even within closely related species, such as sea urchins, amphibians, and vertebrates in general. By early development, I refer to those stages from fertilization through neurulation (gastrulation for such taxa as sea urchins, which do not undergo neurulation). Elinson (1987) has shown how such early stages as initial cleavages and gastrula can vary quite extensively across vertebrates." Andres Collazo, "Developmental Variation, Homology, and the Pharyngula Stage," *Systematic Biology*, 49 (2000): 3 (internal citations omitted). Another paper states, "According to recent models, not only is the putative conserved stage followed by divergence, but it is preceded by variation at earlier stages, including gastrulation and neurulation. This is seen for example in squamata, where variations in patterns of gastrulation and neurulation may be followed by a rather similar somite stage. Thus the relationship between evolution and development has come to be modelled as an 'evolutionary hourglass.'" Richardson *et al.*, "There is No Highly Conserved Embryonic Stage in the Vertebrates: Implications for Current Theories of Evolution and Development," 91–106 (internal citations omitted).

14. Alex T. Kalinka et al., "Gene expression divergence recapitulates the developmental hourglass model," *Nature*, 468 (December 9, 2010): 811–814 (internal citations removed).

15. Brian K. Hall, "Phylotypic stage or phantom: is there a highly conserved embryonic stage in vertebrates?," *Trends in Ecology and Evolution*, 12 (December, 1997): 461–463.

16. Richardson *et al.*, "There is No Highly Conserved Embryonic Stage in the Vertebrates: Implications for Current Theories of Evolution and Development," 91–106. See also Steven Poe and Marvalee H. Wake, "Quantitative Tests of General Models for the Evolution of Development," *The American Naturalist*, 164 (September, 2004): 415–422; Michael K. Richardson, "Heterochrony and the Phylotypic Period," *Developmental Biology*, 172 (1995): 412–421; Olaf R. P. Bininda-Emonds, Jonathan E. Jeffery, and Michael K. Richardson, "Inverting the hourglass: quantitative evidence against the phylotypic stage in vertebrate development," *Proceedings of the Royal Society of London, B*, 270 (2003): 341–346;

17. Richardson *et al.*, "There is No Highly Conserved Embryonic Stage in the Vertebrates: Implications for Current Theories of Evolution and Development," 91–106.

18. Jonathan Wells, "Second Thoughts about Peppered Moths," *The Scientist*, 13 (May 24, 1999): 13. For an update, see Jonathan Wells, "Revenge of the Peppered Moths?," Evolution News & Views (February 12, 2012), accessed March 15, 2012, http://www.evolutionnews.org/2012/02/revenge_of_the056291.html.

19. Ibid.

20. See *Unlocking the Mystery of Life*, Chapter 2: "Darwin's Theory" (Illustra Media, 2003).

21. Kenneth R. Miller and Joseph Levine, *Biology*, Teacher's Edition (Upper Saddle River, NJ: Prentice Hall, 2008), 406.

22. Jonathan Weiner, *The Beak of the Finch: A Story of Evolution in our Time* (New York: Vintage Books, 1994), 43.

23. Jeffrey Podos and Stephen Nowicki, "Beaks, Adaptation, and Vocal Evolution in Darwin's Finches," *BioScience*, 54 (June, 2004): 501–510.

24. H. Lisle Gibbs and Peter R. Grant, "Oscillating selection on Darwin's finches," *Nature*, 327 (June 11, 1987): 511–513.

25. Robert L. Carroll, *Patterns and Processes of Vertebrate Evolution* (Cambridge: Cambridge University Press, 1997), 9.

26. Phillip Johnson, "The Church of Darwin," *The Wall Street Journal* (August, 16, 1999).

27. Henry Gee, Rory Howlett, and Philip Campbell, "15 Evolutionary Gems," *Nature* (January, 2009), accessed March 14, 2012, http://www.nature.com/nature/newspdf/evolutiongems.pdf. For a rebuttal to that document, please see Casey Luskin, "Evaluating *Nature*'s 2009 '15 Evolutionary Gems' Darwin-Evangelism Kit," accessed March 14, 2012, http://www.evolutionnews.org/NaturePacketResponse_Handout_FN.pdf.

28. Jaume Baguñà and Jordi Garcia-Fernàndez, "*Evo-Devo*: The Long and Winding Road," *International Journal of Developmental Biology*, 47 (2003): 705–713 (internal citations removed).

29. Charles Darwin, Chapter 6, *The Origin of Species* (1859), Literature.org, accessed March 14, 2012, http://www.literature.org/authors/darwin-charles/the-origin-of-species/chapter-06.html.

30. Wells, *Icons of Evolution: Why Much of What We Teach about Evolution is Wrong*, 230.

Chapter 15

1. Thomas Henry Huxley, quoted in Michael Ruse, *Darwin and Design: Does Evolution Have a Purpose* (New York: Harvard University Press, 2003), 133.

2. Charles Darwin, Chapter 9, *The Origin of Species* (1859), Literature.org, accessed March 15, 2012, http://www.literature.org/authors/darwin-charles/the-origin-of-species/chapter-09.html.

3. Ibid.

4. Ibid.

5. For example, see Niles Eldredge and Ian Tattersall, *The Myths of Human Evolution* (New York: Columbia University Press, 1982), 59 ("The record jumps, and all the evidence shows that the record is real: the gaps we see reflect real events in life's history—not the artifact of a poor fossil record").

6. M. J. Benton, M. A. Wills, and R. Hitchin, "Quality of the fossil record through time," *Nature*, 403 (February 3, 2000): 534–536. See also Mike Foote, "Sampling, Taxonomic Description, and Our Evolving Knowledge of Morphological Diversity," *Paleobiology*, 23 (Spring, 1997): 181–206 (discussing our knowledge of the completeness of the fossil record and stating "in many respects our view of the history of biological diversity is mature").

7. David S. Woodruff, "Evolution: The Paleobiological View," *Science*, 208 (May 16, 1980): 716–717.

8. Stephen Jay Gould, "Is a new and general theory of evolution emerging?," *Paleobiology*, 6 (1980): 119–130.

9. R. S. K. Barnes, P. Calow, P. J. W. Olive, D. W. Golding, and J. I. Spicer, *The Invertebrates: A New Synthesis*, 3rd ed. (Malden, MA: Blackwell Scientific Publications, 2001), 9–10.

10. Arthur N. Strahler, *Science and Earth History: The Evolution/Creation Controversy* (New York: Prometheus Books, 1987), 408–409.

11. Richard M. Bateman, Peter R. Crane, William A. DiMichele, Paul R. Kenrick, Nick P. Rowe, Thomas Speck, and William E. Stein, "Early Evolution of Land Plants: Phylogeny, Physiology, and Ecology of the Primary Terrestrial Radiation," *Annual Review of Ecology and Systematics*, 29 (1998): 263–292.

12. Stephen C. Meyer, Scott Minnich, Jonathan Moneymaker, Paul A. Nelson, and Ralph Seelke, Explore Evolution: *The Arguments For and Against Neo-Darwinism* (Malvern: Hill House, 2007), 24. See also Stefanie De Bodt, Steven Maere, and Yves Van de Peer, "Genome duplication and the origin of angiosperms," *Trends in Ecology and Evolution*, 20 (2005): 591–597 ("In spite of much research and analyses of different sources of data (e.g., fossil record and phylogenetic analyses using molecular and morphological characters), the origin of the angiosperms remains unclear. Angiosperms appear rather suddenly

in the fossil record… with no obvious ancestors for a period of 80–90 million years before their appearance").

13. Niles Eldredge, *The Monkey Business: A Scientist Looks at Creationism* (New York: Washington Square Press, 1982), 65.

14. See Alan Cooper and Richard Fortey, "Evolutionary Explosions and the Phylogenetic Fuse," *Trends in Ecology and Evolution*, 13 (April, 1998): 151–156; Frank B. Gill, *Ornithology*, 3rd ed. (New York: W. H. Freeman, 2007), 42.

15. Cleveland P. Hickman, Larry S. Roberts, and Francs M. Hickman, *Integrated Principles of Zoology*, 8th ed. (St. Louis, MO: Times Mirror/ Moseby College Publishing, 1988), 866.

16. Working Group on Teaching Evolution, National Academy of Sciences, *Teaching About Evolution and the Nature of Science* (Washington D.C.: National Academy Press, 1998), 57.

17. See Niles Eldredge and Stephen Jay Gould, "Punctuated Equilibria: An Alternative to Phyletic Gradualism," in *Models in Paleobiology*, ed. Thomas J. M. Schopf (San Francisco, CA: Freeman Cooper & Company, 1972), 82–115.

18. As Darwin wrote, "For forms existing in larger numbers will always have a better chance, within any given period, of presenting further favourable variations for natural selection to seize on, than will the rarer forms which exist in lesser numbers." Charles Darwin, Chapter 6, *The Origin of Species* (1859), Literature.org, accessed March 14, 2012, http://www.literature.org/authors/darwin-charles/the-origin-of-species/chapter-06.html.

19. Brian Charlesworth, Russell Lande, and Montgomery Slatkin, "A Neo-Darwinian Commentary on Macroevolution," *Evolution*, 36 (1982): 474–498.

20. For one example, see John Turner, "Why we need evolution by jerks," *New Scientist*, 1396 (February 9, 1984): 34–35.

21. Eörs Szathmáry, "When the means do not justify the end, Book review of *Sudden Origins: Fossils, Genes, and the Emergence of Species* by Jeffrey H. Schwartz," *Nature*, 399 (June 24, 1999): 745–746.

22. Szathmáry, "When the means do not justify the end; book review of *Sudden Origins: Fossils, Genes, and the Emergence of Species* by Jeffrey H. Schwartz," 745–746.

23. See Hopi E. Hoekstra and Jerry A. Coyne, "The Locus of Evolution: Evo Devo and the Genetics of Adaptation," *Evolution*, 61-5 (2007): 995–1016.

24. See for example, Benjamin Prud'homme, Nicolas Gompel, and Sean B. Carroll, "Emerging principles of regulatory evolution," *Proceedings of the National Academy of Sciences USA*, 104 (May 15, 2007): 8605–8612.

25. Stephen C. Meyer, Scott Minnich, Jonathan Moneymaker, Paul A. Nelson, and Ralph Seelke, *Explore Evolution: The Arguments For and Against Neo-Darwinism* (Hill House, 2007), 106 (emphasis added, internal citations omitted).

26. As a result, some critics have charged punc eq is not testable. See Philip D. Gingerich, "Darwin's Gradualism and Empiricism," *Nature*, 309 (May 10, 1984): 116 (stating that punc eq is "by nature untestable" because it "postulates how speciation takes place, based not on empirical evidence but on negative evidence - gaps in the fossil record").

27. Michael Denton, *Evolution: A Theory in Crisis* (Chevy Chase, MD: Adler & Adler, 1986), 193.

28. Stephen C. Meyer, Marcus Ross, Paul Nelson, and Paul Chien, "The Cambrian Explosion: Biology's Big Bang," in *Darwinism, Design, and Public Education*, eds. John A. Campbell and Stephen C. Meyer (East Lansing, MI: Michigan State University Press, 2003), 367 ("Intelligent design provides a sufficient causal explanation for the origin of large amounts of information, since we have considerable experience of intelligent agents generating informational configurations of matter").

29. Stephen C. Meyer, "The origin of biological information and the higher taxonomic categories," *Proceedings of the Biological Society of Washington*, 117 (2004): 213–239 ("we have repeated experience of rational and conscious agents—in particular ourselves—generating or causing increases in complex specified information, both in the form of sequence-specific lines of code and in the form of hierarchically arranged systems of parts").

30. Meyer, Ross, Nelson, and Chien, "The Cambrian Explosion: Biology's Big Bang," 386 ("We know from experience that intelligent agents often conceive of plans prior to the material instantiation of the systems that conform to the plans—that is, the intelligent design of a blueprint often precedes the assembly of parts in accord with a blueprint or preconceived design plan").

Chapter 16

1. Henry Gee, *In Search of Deep Time: Beyond the Fossil Record to a New History of Life* (Ithaca, NY: Cornell Paperbacks, 1999), 116–117.

2. Samantha Strong and Rich Schapiro, "Missing link found? Scientists unveil fossil of 47 million-year-old primate, *Darwinius masillae*," New York *Daily News* (May 19, 2009), accessed March 19, 2012, http://articles.nydailynews.com/2009-05-19/news/17922526_1_human-evolution-missing-link-jorn-hurum.

3. Vince Soodin, "Fossil is Evolution's 'Missing Link,'" *The Sun* (May 19, 2009), accessed March 19, 2012, http://www.thesun.co.uk/sol/homepage/news/article2437749.ece.

4. "Who is Ida?," The Link, accessed March 19, 2012, http://www.revealingthelink.com/who-is-ida/.

5. Some might object that the term "missing link" is not used by scientists. While we are not necessarily promoting that term, such an objection is not entirely true. The term appears in textbooks and even the occasional scientific paper. For example, see: Peter H. Raven, George B. Johnson, Kenneth A. Mason, Jonathan B. Losos, and Susan R. Singer, *Biology*, 9th ed. (New York, NY: McGraw Hill, 2011), 425; Jingxia Zhao, Yunyun Zhao, Chungkun Shih, Dong Ren and Yongjie Wang, "Transitional fossil earwigs—a missing link in Dermaptera evolution," *BMC Evolutionary Biology*, 10 (2010): 344; Michael Coates and Marcello Ruta, "Nice snake, shame about the legs," *Trends in Ecology and Evolution*, 15 (December, 2000): 503–507; Matt Kaplan, "*Archaeopteryx* no longer first bird," *Nature News* (July 27, 2011); Rex Dalton, "Fossil primate challenges Ida's place," *Nature*, 461 (2009): 1040; John Whitfield, "Almost like a Whale," *Nature News* (September 20, 2001); Ann Gibbons, "Missing Link Ties Birds," *Science*, 279 (March 20, 1998): 1851–1852. What is more, the term is commonly cited in the media whenever a supposed "missing link" is found.

6. J. Madeleine Nash, "Our Cousin, the Fishapod," *Time* (April 10, 2006), accessed March 20, 2012, http://www.time.com/time/magazine/article/0,9171,1181611,00.html.

7. Jerry A. Coyne, *Why Evolution is True* (New York: Viking, 2009), 35.

8. There are a few vertebrates classified as tetrapods—such as snakes and whales—that do not have four limbs.

9. Neil Shubin, *Your Inner Fish: A Journey into the 3.5-Billion-Year History of the Human Body* (New York: Pantheon, 2008), 38.

10. See Casey Luskin, "*Tiktaalik roseae*: Where's the Wrist? (Updated)," Evolution News & Views (July 14, 2008), accessed March 21, 2012, http://www.evolutionnews.org/2008/07/tiktaalik_roseae_wheres_the_wr008921.html; Casey Luskin, "An 'Ulnare' and an 'Intermedium' a Wrist Do Not Make: A Response to Carl Zimmer," Evolution News & Views (August 1, 2008), accessed March 21, 2012, http://www.evolutionnews.org/2008/08/an_ulnare_and_an_intermedium_a009651.html.

11. Per Erik Ahlberg and Jennifer A. Clack, "A firm step from water to land," *Nature*, 440 (April 6, 2006): 747–749.

12. Neil Shubin, quoted in "Judgment Day: Intelligent Design on Trial" (PBS / NOVA, 2007).

13. Philippe Janvier and Gaël Clément, "Muddy tetrapod origins," *Nature*, 463 (January 7, 2010): 40–41; Rex Dalton, "Discovery pushes back date of first four-legged animal," *Nature News* (January 6, 2010); John Roach, "Oldest Land-Walker Tracks Found—Pushes Back Evolution," *National Geographic News* (January 6, 2010), accessed March 21, 2012, http://news.nationalgeographic.com/news/2010/01/100106-tetrapod-tracks-oldest-footprints-nature-evolution-walking-land/.

14. Henry Gee, "First Footing," *Nature iEditor Blog* (January 6, 2010), accessed September 19, 2010, http://blogs.nature.com/henrygee/2010/01/06/first-footing.

15. The dates of hypothetical fish ancestors of tetrapods, *Tiktaalik roseae*, and clear tetrapods taken from Coyne, *Why Evolution is True*, 35–38.

16. In addition to the hoatzin, other birds with claws include owls, storks, loons, and other species. See Peter R. Stettenheim, "The Integumentary Morphology of Modern Birds—An Overview," *American Zoologist*, 40 (2000): 461–477.

17. William Dembski and Jonathan Wells, *The Design of Life: Discovering Signs of Intelligence in Biological Systems* (Dallas, TX: The Foundation for Thought and Ethics, 2007), 62.

18. Robert A. Martin, *Missing Links: Evolutionary Concepts & Transitions Through Time* (Boston, MA: Jones and Bartlett Publishers, 2004), 153. *See also* Carl C. Swisher III, Yuan-qing Wang, Xiao-lin Wang, Xing Xu, and Yuan Wang, "Cretaceous age for the feathered dinosaurs of Liaoning, China," *Nature*, 400 (July 1, 1999): 58–61.

19. Martin, *Missing Links: Evolutionary Concepts & Transitions Through Time*, 154.

20. Crocodiles do have a four-chambered heart, but crocodilians are not thought to be ancestors to birds. Since the primitive condition in reptiles is thought to be a three-chambered heart, evolution requires that a transition from a three to four-chambered heart took place somewhere in the line that led to birds. For a detailed discussion of problems with this transition, see Stephen C. Meyer, Scott Minnich, Jonathan Moneymaker, Paul A. Nelson, and Ralph Seelke, *Explore Evolution: The Arguments For and Against Neo-Darwinism* (Melbourne: Hill House, 2007), 128–139.

21. Virtually all living reptiles do not care for their young although there are some reports of python snakes that care for their young. See "Serpent First: Egg-Laying Snakes Care for Young," RedOrbit.com (September 18, 2005), accessed March 21, 2012, http://www.redorbit.com/news/science/243036/serpent_first_egglaying_snakes_care_for_young/.

22. "Discovery Raises New Doubts About Dinosaur-bird Links," *ScienceDaily* (June 9, 2009), accessed March 21, 2012, http://www.sciencedaily.com/releases/2009/06/090609092055.htm. See also Devon E. Quick and John A. Ruben, "Cardio-Pulmonary Anatomy in Theropod Dinosaurs: Implications From Extant Archosaurs," *Journal of Morphology*, 270 (October, 2009): 1232–1246; Frances C. James and John A. Pourtless IV, "Cladistics and the Origins of Birds: A Review and Two New Analyses," *Ornithological Monographs*, 66 (Columbia MO: American Ornithologists Union, 2009): 1–78.

23. For example, see L. Wang, "Dinosaur fossil yields feathery structures," *Science News*, 159 (March 10, 2001): 149.

24. Alan Feduccia, *The Origin and Evolution of Birds*, 2nd ed. (New Haven, CT: Yale University Press, 1999), 396. See also Terry D. Jones, James O. Farlow, John A. Ruben, Donald M. Henderson, and Willem J. Hillenius, "Cursoriality in bipedal archosaurs," *Nature*, 406 (August 17 2000): 716–718; J. Lee Kavanau, "Secondarily flightless birds or Cretaceous non-avian theropods?," *Medical Hypotheses*, 74 (February, 2010): 275–276; Alan Feduccia, Theagarten Lingham-Soliar, and J. Richard Hinchliffe, "Do Feathered Dinosaurs Exist? Testing the Hypothesis on Neontological and Paleontological Evidence," *Journal of Morphology*, 266 (2005): 125–166; Frances C. James and John A. Pourtless IV, "Cladistics and the Origin of Birds: A Review and Two New Analyses," *The Ornithological Monographs*, 66 (April 30, 2009): 1–78.

25. Ostriches have been compared to the secondarily flightless birds found in the fossil record. See Kavanau, "Secondarily flightless birds or Cretaceous non-avian theropods?," 275–276; Jennifer Viegas, "Some 'Non-Avian Feathered Dinosaurs' May Have Been Birds," *Discovery News* (January 7, 2010), accessed March 21, 2012, http://news.discovery.com/dinosaurs/some-feathered-dinosaurs-may-have-been-birds.html.

26. Stephen Czerkas quoted in Christopher P. Sloan, "Feathers for *T. rex*?," *National Geographic*, 196 (November, 1999): 98–107.

27. Sloan, "Feathers for *T. rex*?," 98–107.

28. Xu Xing, "Feathers for *T. rex*?," *National Geographic*, 197 (March, 2000).

29. Storrs L. Olson, "OPEN LETTER TO: Dr. Peter Raven, Secretary PRaven@nas.org, Committee for Research and Exploration, National Geographic Society" (November 1, 1999), accessed March 22, 2012, http://dml.cmnh.org/1999Nov/msg00263.html.

30. Feduccia, *The Origin and Evolution of Birds*, 396.

31. Richard O. Prum and Alan H. Brush, "Which Came First, the Feather or the Bird?," *Scientific American* (March, 2003): 84–93.

32. Ibid.

33. Ibid.

34. Frank B. Gill, *Ornithology*, 3rd ed. (New York: W. H. Freeman, 2007), 39; Prum and Brush, "Which Came First, the Feather or the Bird?," 84–93.

35. Prum and Brush, "Which Came First, the Feather or the Bird?," 84–93.

36. Gill, *Ornithology*, 7.

37. Ibid., 28.

38. Prum and Brush, "Which Came First, the Feather or the Bird?," 84–93.

39. For a detailed discussion of problems with this transition, see Meyer, Minnich, Moneymaker, Nelson, and Seelke, *Explore Evolution: The Arguments For and Against Neo-Darwinism*, 128–139.

40. For example, one paper states that horses are "a classic example of evolutionary processes." Jaco Weinstock, Eske Willerslev, Andrei Sher, Wenfei Tong, Simon Y. W. Ho, Dan Rubenstein, John Storer, James Burns, Larry Martin, Claudio Bravi, Alfredo Prieto, Duane Froese, Eric Scott, Lai Xulong, Alan Cooper, "Evolution, Systematics, and Phylogeography of Pleistocene Horses in the New World: A Molecular Perspective," *PLoS Biology*, 3 (August, 2005): 1373–1379.

41. Alan Cooper, quoted in "Ancient DNA helps clarify the origins of two extinct New World horse species," *PLoS Biology* (June 27, 2005), accessed April 9, 2012, http://www.eurekalert.org/pub_releases/2005-06/plos-adh062205.php.

42. See for example Raven, Johnson, Mason, Losos, and Singer, *Biology*, 426–427; Allan J. Tobin and Jennie Dusheck, *Asking about Life*, 3rd ed. (Belmont, CA: Thomason Brooks/Cole, 2005), 317; Burton S. Guttman, *Biology* (Boston: MA, McGraw-Hill, 1999), 485.

43. For a discussion, see Jonathan Wells, *Icons of Evolution: Why Much of What We Teach About Evolution is Wrong* (Washington D.C., Regnery, 2000), 195–207.

44. "Transitional Forms," Understanding Evolution, accessed April 9, 2012, http://evolution.berkeley.edu/evolibrary/article/lines_03.

45. Niles Eldredge, quoted in Tom Bethell, "Agnostic Evolutionists: The taxonomic case against Darwin," *Harper's Magazine* (February, 1985): 49–61.

46. W. D. Matthew quoted in David Young, *The Discovery of Evolution* (Natural History Museum Publications / Cambridge University Press, 1992), 217.

47. J. G. M. Thewissen and Sunil Bajpai, "Whale Origins as a Poster Child for Macroevolution," *BioEssays*, 51 (December, 2001): 1037–1049.

48. List provided courtesy of Richard Sternberg.

49. Alan Feduccia, "'Big bang' for tertiary birds?," *Trends in Ecology and Evolution*, 18 (2003): 172–176.

50. See Walter James ReMine, *The Biotic Message: Evolution Versus Message Theory* (Saint Paul: MN, Saint Paul Science, 1983).

51. Private communication with Richard Sternberg.

52. Thewissen and Bajpai, "Whale Origins as a Poster Child for Macroevolution," 1037–1049.

53. "Who is Ida?," The Link, accessed March 19, 2012, http://www.revealingthelink.com/who-is-ida/.

54. "Recently Analyzed Fossil Was Not Human Ancestor As Claimed, Anthropologists Say," *University of Texas at Austin News* (March 2, 2010), accessed April 9, 2012, http://www.utexas.edu/news/2010/03/02/human_ancestor_fossil/?AddInterest=1284.

Chapter 17

1. Mark Davis, "Into the Fray: The Producer's Story," NOVA Online (February 2002), accessed April 10, 2012, http://www.pbs.org/wgbh/nova/neanderthals/producer.html.

2. Constance Holden, "The Politics of Paleoanthropology," *Science*, 213 (August 14, 1981): 737–740.

3. See Bernard Wood and Mark Collard, "The Human Genus," *Science*, 284 (April 2, 1999): 65–71; F. Spoor, M. G. Leakey, P. N. Gathogo, F. H. Brown, S. C. Antón, I. McDougall, C. Kiarie, F. K. Manthi, and L. N. Leakey, "Implications of new early *Homo* fossils from Ileret, east of Lake Turkana, Kenya," *Nature*, 448 (August 9, 2007): 688–691.

4. Ian Tattersall and Jeffrey H. Schwartz, "Evolution of the Genus Homo," *Annual Review of Earth and Planetary Sciences*, 37 (2009): 67–92;

5. Sigrid Hartwig-Scherer, "Apes or Ancestors?" in *Mere Creation: Science, Faith & Intelligent Design*, ed. William Dembski (Downers Grove, IL: InterVarsity Press, 1998), 220.

6. Sigrid Hartwig-Scherer and Robert D. Martin, R. D., "Was 'Lucy' more human than her 'child'? Observations on early hominid postcranial skeletons," *Journal of Human Evolution*, 21 (1991): 439–449.

7. Eric Delson, "One skull does not a species make," *Nature*, 389 (October 2, 1997): 445–446; John Hawks, Keith Hunley, Sang-Hee Lee, and Milford Wolpoff, "Population Bottlenecks and Pleistocene Human Evolution," *Journal of Molecular Biology and Evolution*, 17 (2000): 2–22.

8. Wood and Collard, "The Human Genus," 65–71.

9. B. Arensburg, A. M. Tillier, B. Vandermeersch, H. Duday, L. A. Schepartz, and Y. Rak, "A Middle Palaeolithic human hyoid bone," *Nature*, 338 (April 27, 1989): 758–760.

10. Joe Alper, "Rethinking Neanderthals," *Smithsonian magazine* (June, 2003), accessed March 5, 2012, http://www.smithsonianmag.com/science-nature/neanderthals.html; Kate Wong, "Who were the Neandertals?," *Scientific American* (August, 2003): 28–37; Erik Trinkaus and Pat Shipman, "Neandertals: Images of Ourselves," *Evolutionary Anthropology*, 1 (1993): 194–201; Philip G. Chase and April Nowell, "Taphonomy of a Suggested Middle Paleolithic Bone Flute from Slovenia," *Current Anthropology*, 39 (August/October 1998): 549–53; Tim Folger and Shanti Menon, "... Or Much Like Us?," *Discover Magazine* (January, 1997), accessed March 5, 2012, http://discovermagazine.com/1997/jan/ormuchlikeus1026; C. B. Stringer, "Evolution of early humans," in *The Cambridge Encyclopedia of Human Evolution*, eds. Steve Jones, Robert Martin, and David Pilbeam (Cambridge, MA: Cambridge University Press, 1992), 248.

11. Michael D. Lemonick, "A Bit of Neanderthal in Us All?," *Time*, (April 25, 1999), accessed March 5, 2012, http://www.time.com/time/magazine/article/0,9171,23543,00.html.

12. References for cranial capacities cited in Table 2 are as follows:

Gorilla: Stephen Molnar, *Human Variation: Races, Types, and Ethnic Groups*, 4th ed. (Upper Saddle River: Prentice Hall, 1998), 203.

Chimpanzee: Molnar, *Human Variation: Races, Types, and Ethnic Groups*, 4th ed., 203.

Australopithecus: Glenn C. Conroy, Gerhard W. Weber, Horst Seidler, Phillip V. Tobias, Alex Kane, Barry Brunsden, "Endocranial Capacity in an Early Hominid Cranium from Sterkfontein, South Africa," *Science*, 280 (June 12, 1998): 1730–1731; Wood and Collard, "The Human Genus," 65–71.

Homo habilis: Wood and Collard, "The Human Genus," 65–71.

Homo erectus: Molnar, *Human Variation: Races, Types, and Ethnic Groups*, 4th ed., 203; Wood and Collard, "The Human Genus," 65–71.

Neanderthals: Molnar, *Human Variation: Races, Types, and Ethnic Groups*, 4th ed., 203; Molnar, *Human Variation: Races, Types, and Ethnic Groups*, 5th ed., 189.

***Homo sapiens* (modern man):** Molnar, *Human Variation: Races, Types, and Ethnic Groups*, 4th ed., 203; E. I. Odokuma, P. S. Igbigbi, F. C. Akpuaka and U. B. Esigbenu, "Craniometric patterns of three Nigerian ethnic groups," *International Journal of Medicine and Medical Sciences*, 2 (February, 2010): 34–37; Molnar, *Human Variation: Races, Types, and Ethnic Groups*, 5th ed., 189.

13. Jeremy Cherfas, "Trees have made man upright," *New Scientist*, 97 (January 20, 1983): 172–177; Brian G. Richmond and David S. Strait, "Evidence that humans evolved from a knuckle-walking ancestor," *Nature*, 404 (March 23, 2000): 382–385.

14. See Richard Leakey and Roger Lewin, *Origins Reconsidered: In Search of What Makes Us Human* (New York, NY: Anchor Books, 1993), 193–196.

15. Mark Collard and Leslie C. Aiello, "From forelimbs to two legs," *Nature*, 404 (March 23, 2000): 339–340; Brian G. Richmond and David S. Strait, "Evidence that humans evolved from a knuckle-walking ancestor," *Nature*, 404 (March 23, 2000): 382–385.

16. Collard and Aiello, "From forelimbs to two legs."

17. Fred Spoor, Bernard Wood, and Frans Zonneveld, "Implications of early hominid labyrinthine morphology for evolution of human bipedal locomotion," *Nature*, 369 (June 23, 1994): 645–648.

18. François Marchal, "A New Morphometric Analysis of the Hominid Pelvic Bone," *Journal of Human Evolution*, 38 (March, 2000): 347–365.

19. M. Maurice Abitbol, "Lateral view of Australopithecus afarensis: primitive aspects of bipedal positional behavior in the earliest hominids," *Journal of Human Evolution*, 28 (March, 1995): 211–229.

20. Leslie Aiello quoted in Leakey and Lewin, *Origins Reconsidered: In Search of What Makes Us Human*, 196.

21. Spoor *et al.*, "Implications of new early *Homo* fossils from Ileret, east of Lake Turkana, Kenya," 688–691.

22. Hartwig-Scherer and Martin, "Was 'Lucy' more human than her 'child'? Observations on early hominid postcranial skeletons," 439–449.

23. Hartwig-Scherer, "Apes or Ancestors?," 226.

24. Wood and Collard, "The Human Genus," 65–71.

25. Spoor *et al.*, "Implications of new early *Homo* fossils from Ileret, east of Lake Turkana, Kenya," 688–691.

26. Wood and Collard, "The Human Genus," 65–71.

27. Terrance W. Deacon, "Problems of Ontogeny and Phylogeny in Brain-Size Evolution," *International Journal of Primatology*, 11 (1990): 237–82. See also Terrence W. Deacon, "What makes the human brain different?," *Annual Review of Anthropology*, 26 (1997): 337–57; *Stephen Molnar, Human Variation: Races, Types, and Ethnic Groups*, 5th ed. (Upper Saddle River: Prentice Hall, 2002), 189 ("The size of the brain is but one of the factors related to human intelligence").

28. Ernst Mayr, *What Makes Biology Unique?: Considerations on the Autonomy of a Scientific Discipline* (Cambridge, MA: Cambridge University Press, 2004), 198.

29. "Hawks *et al.*, "Population Bottlenecks and Pleistocene Human Evolution," 2–22."

30. Robin Dennell and Wil Roebroeks, "An Asian perspective on early human dispersal from Africa," *Nature*, 438 (December 22, 2005): 1099–1104 (internal citations removed).

31. "New study suggests big bang theory of human evolution," University of Michigan News Service (January 10, 2000), accessed April 28, 2012, http://www.umich.edu/~newsinfo/Releases/2000/Jan00/r011000b.html.

32. For more extensive discussions of the lack of fossil evidence for human evolution, see Casey Luskin, "Human Origins and Intelligent Design," *Progress in Complexity, Information, and Design*, 4.1 (July, 2005), accessed April 29, 2012, http://www.iscid.org/papers/Luskin_HumanOrigins_071505.pdf; Casey Luskin, "Human Origins and the Fossil Record," *Science and Human Origins* (Discovery Institute Press, 2012), 45–83.

33. See for example "Ardi named 'breakthrough' of 2009," CBCNews.com (December 18, 2009), accessed July 17, 2012, http://www.cbc.ca/news/technology/story/2009/12/17/tech-science-top-10.html; Dan Vergano, "Oldest remains of a child found in Ethiopia," *USA Today* (September 20, 2006), accessed February 19, 2013, http://usatoday30.usatoday.com/tech/science/discoveries/2006-09-20-human-ancestor_x.htm.

34. See Edward J. Larson, *Summer for the Gods: The Scopes Trial and America's Continuing Debate over Science and Religion* (Cambridge, MA: Harvard University Press, 1997), 13.

35. Adrienne L. Zihlman and Jerold M. Lowenstein, "False Start of the Human Parade," *Natural History*, 88 (August / September, 1979): 86–91.

36. The story of the rise and fall of *Ramapithecus* is told in Donald Johanson and Blake Edgar, *From Lucy to Language* (New York, NY: Simon & Schuster, 1996), 33–37.

37. Ann Gibbons, "Breakthrough of the Year: *Ardipithecus ramidus*," *Science*, 326 (December 18, 2009): 1598–1599; Ann Gibbons, "In Search of the First Hominids," *Science*, 295 (February 15, 2002): 1214–1219.

38. See Michael D. Lemonick and Andrea Dorfman, "Ardi Is a New Piece for the Evolution Puzzle," *Time* Magazine (October 1, 2009), accessed April 29, 2012, http://www.time.com/time/health/article/0,8599,1927200-2,00.html.

39. Bernard Wood and Terry Harrison, "The evolutionary context of the first hominins," *Nature*, 470 (February 17, 2011): 347–352; Esteban E. Sarmiento, "Comment on the Paleobiology and Classification of *Ardipithecus ramidus*," *Science*, 328 (May 28, 2010): 1105; E. E. Sarmiento and D. J. Meldrum, "Behavioral and phylogenetic implications of a narrow allometric study of Ardipithecus ramidus," *HOMO: Journal of Comparative Human Biology*, 62 (2011): 75–108.

40. Eben Harrell, "Ardi: The Human Ancestor Who Wasn't," *Time* Magazine (May 27, 2010), accessed April 29, 2012, http://www.time.com/time/health/article/0,8599,1992115,00.html.

41. See Alton Biggs, Kathleen Gregg, Whitney Crispen Hagins, Chris Kapicka, Linda Lundgren, Peter Rillero, National Geographic Society, *Biology: The Dynamics of Life* (New York: Glencoe, McGraw Hill, 2000), 438, 442–443; Esteban E. Sarmiento, Gary J. Sawyer, and Richard Milner, *The Last Human: A Guide to Twenty-two Species of Extinct Humans* (New Haven, CT: Yale University Press, 2007), 75, 83, 103, 127, 137; Johanson and Edgar, *From Lucy to Language*, 82; Richard Potts and Christopher Sloan, *What Does it Mean to be Human?* (Washington D.C.: National Geographic, 2010), 32–33, 36, 66, 92; Carl Zimmer, *Smithsonian Intimate Guide to Human Origins* (Toronto: Madison Press, 2005), 44, 50.

42. Earnest Albert Hooton, *Up From The Ape*, Revised ed. (New York: McMillan, 1946), 329.

43. For example, see "Darwin's Predictions," *Judgment Day: Intelligent Design on Trial*, accessed April 28, 2012, http://www.pbs.org/wgbh/nova/id/pred-nf.html ("Darwin wrote, 'We must… acknowledge, as it seems to me, that man with all his noble qualities… still bears in his bodily frame the indelible stamp of his lowly origin.' Today, many a schoolchild can cite the figure perhaps most often called forth in support of this view—namely, that we share almost 99 percent of our DNA with our closest living relative, the chimpanzee.").

44. Jon Cohen, "Relative Differences: The Myth of 1%," *Science*, 316 (June 29, 2007): 1836.

45. Ibid.

46. See Erika Check, "It's the junk that makes us human," *Nature*, 444 (November 9, 2006): 130–131. See also The ENCODE Project Consortium, "An integrated encyclopedia of DNA elements in the human genome," *Nature*, 489 (September 6, 2012): 57–74.

47. Jennifer F. Hughes, Helen Skaletsky, Tatyana Pyntikova, Tina A. Graves, Saskia K. M. van Daalen, Patrick J. Minx, Robert S. Fulton, Sean D. McGrath, Devin P. Locke, Cynthia Friedman, Barbara J. Trask, Elaine R. Mardis, Wesley C. Warren, Sjoerd Repping, Steve Rozen, Richard K. Wilson and David C. Page, "Chimpanzee and human Y chromosomes are remarkably divergent in structure and gene content, " *Nature*, 463 (January 28, 2010): 536–539.

48. See Richard Buggs, "Chimpanzee?," *Reformatorisch Dagblad* (October 10, 2008), accessed June 7, 2011, http://www.refdag.nl/chimpanzee_1_282611.

49. Francis Collins, *The Language of God: A Scientist Presents Evidence for Belief* (New York, NY: Free Press, 2006), 134.

50. Jonathan Marks, *What it means to be 98% Chimpanzee: Apes, People, and their Genes* (Los Angeles, CA: University of California Press, 2003), 39 ("the fusion isn't what gives us language, or bipedalism, or a big brain, or art, or sugarless bubble gum. It's just one of those neutral changes, lacking outward expression and neither good nor bad").

51. This trait would spread most efficiently if the human population was quite small.

52. Yuxin Fan, Elena Linardopoulou, Cynthia Friedman, Eleanor Williams, and Barbara J. Trask, "Genomic Structure and Evolution of the Ancestral Chromosome Fusion Site in 2q13-2q14.1 and Paralogous Regions on Other Human Chromosomes," *Genome Research*, 12 (2002): 1651–1662. Likewise, evolutionary biologist Daniel Fairbanks admits, the location only has 158 repeats, and only "44 are perfect copies" of the sequence. See Daniel Fairbanks, *Relics of Eden: The Powerful Evidence of Evolution in Human DNA* (Amherst, NY: Prometheus, 2007), 27. Richard Sternberg, "Guy Walks Into a Bar and Thinks He's a Chimpanzee: The Unbearable Lightness of Chimp-Human Genome Similarity," Evolution News & Views (May 14, 2009), accessed March 6, 2012. http://www.evolutionnews.org/2009/05/guy_walks_into_a_bar_and_think020401.html.

53. Wesley J. Smith, *A Rat is a Pig is a Dog is a Boy: The Human Cost of the Animal Rights Movement* (New York: Encounter Books, 2010), 243–244 (emphases in original).

54. Noam Chomsky, *Language and Mind*, 3rd ed. (Cambridge: Cambridge University Press, 2006), 59.

55. Elizabeth Bates, Donna Thai, and Virginia Marchman quoted in Steven Pinker, *The Language Instinct: How the Mind Creates Language* (New York, NY: Harper Perennial, 1994), 351.

56. Smith, *A Rat is a Pig is a Dog is a Boy*, 243–244.

57. Michael Ruse and Edward O. Wilson, "The Evolution of Ethics," *New Scientist*, 108 (October 17, 1985): 50–52.

58. In 2010, the most prominent founder of kin selection theory, Edward O. Wilson, became a strong critic of the concept. In essence, he argues that field data are bearing out the mathematical predictions of kin selection theory because (1) cooperation is found among many species where individuals would not, genetically speaking, be expected to cooperate, and (2) there are many instances where cooperation is not found, even though the mathematics of kin selection theory would predict it. See Martin A. Nowak, Corina E. Tarnita, and Edward O. Wilson, "The evolution of eusociality," *Nature*, 466 (August 26, 2010): 1057–1062.

59. Nicholas Wade, "An Evolutionary Theory of Right and Wrong," *The New York Times* (October 31, 2006), accessed April 28, 2012, http://www.nytimes.com/2006/10/31/health/psychology/31book.html.

60. Jeffrey P. Schloss, "Evolutionary Accounts of Altruism & the Problem of Goodness by Design," in *Mere Creation; Science, Faith & Intelligent Design*, ed. William A. Dembski (Downers Grove, IL, Intervarsity Press, 1998), 251.

61. Francis Collins quoted in Dan Cray, "God vs. Science," *Time* Magazine (November 5, 2006), accessed April 28, 2012, http://www.time.com/time/printout/0,8816,1555132,00.html.

62. Jeffrey P. Schloss, "Emerging Accounts of Altruism: 'Love Creation's Final Law'?," in *Altruism and Altruistic Love: Science, Philosophy, & Religion in Dialogue*, eds. Stephen G. Post, Lynn G. Underwood, Jeffrey P. Schloss, and William B. Hurlbut (Oxford: Oxford University Press, 2002), 221.

63. Philip S. Skell, "Why do we invoke Darwin?," *The Scientist*, 19 (August 29, 2005): 10.

Chapter 18

1. Phillip E. Johnson, *Defeating Darwinism by Opening Minds* (Downers Grove, IL: InterVarsity Press, 1997), 83.

2. Stephen C. Meyer, "The Origin of Biological Information and the Higher Taxonomic Categories," *Proceedings of the Biological Society of Washington*, 117 (2004): 213–239.

3. "Life: What a Concept!," ed. John Brockman (New York, NY: The Edge Foundation, 2008), accessed March 14, 2012, http://www.edge.org/documents/life/Life.pdf.

4. Ada Yonath, "Supervisor's Foreword," in Chen Davidovich, *Targeting Functional Centers of the Ribosome* (Heidelberg: Springer-Verlag, 2011), vii.

5. David Goodsell, "The ATP Synthase," Molecule of the Month at Protein Data Bank (December, 2005), accessed March 14, 2012, http://www.rcsb.org/pdb/education_discussion/molecule_of_the_month/download/ATPSynthase.pdf.

Chapter 19

1. William O. Douglas, quoted in Laird Wilcox and John George, *Be Reasonable: Selected Quotations for Inquiring Minds* (Buffalo, NY: Prometheus Books, 1994), 66.

2. Marshall Berman, "Intelligent Design: The New Creationism Threatens All of Science and Society," *American Physical Society News*, 14 (October, 2005), accessed April 29, 2012, http://www.aps.org/publications/apsnews/200510/backpage.cfm.

3. Steve Fuller, *Dissent over Descent: Intelligent Design's Challenge to Darwinism* (Cambridge, UK: Icon Books, 2008), 133.

4. Richard Dawkins, *The Blind Watchmaker: Why the Evidence of Evolution Reveals a Universe Without Design* (New York, NY: W. W. Norton, 1986), 6.

5. Richard Lewontin, "Billions and Billions of Demons," *The New York Review of Books*, 44 (January 9, 1997): 28.

6. Michael Ruse, "Nonliteralist Antievolution," AAAS Symposium: "The New Antievolutionism," (February 13, 1993), accessed April 29, 2012, http://www.arn.org/docs/orpages/or151/mr93tran.htm.

7. Thomas S. Kuhn, *The Structure of Scientific Revolutions*, 3rd ed. (Chicago, IL: University of Chicago Press, 1996), 24.

8. "AAAS Board Resolution on Intelligent Design Theory" (October 18, 2002), accessed April 29, 2012, http://www.aaas.org/news/releases/2002/1106id2.shtml.

9. John G. West, "Intelligent Design Could Offer Fresh Ideas on Evolution," *Seattle Post-Intelligencer* (December 6, 2002), accessed April 29, 2012, http://www.discovery.org/a/1313.

10. Brief Amici Curiae of Physicians, Scientists, and Historians of Science in Support of Petitioners, *Daubert v. Merrell Dow Pharmaceuticals, Inc.*, 509 U.S. 579 (1993).

11. Stephen Jay Gould, "In the Mind of the Beholder," *Natural History*, 103 (February, 1994): 14–23.

12. Adapted from a quotation by Arthur Schopenhauer cited in William Dembski, *The Design Revolution: Answering the Toughest Questions About Intelligent Design* (Downer's Grove, IL: InterVarsity, 2004), 20.

13. Lynn Margulis, quoted in Dick Teresi, "The Discover Interview: Lynn Margulis," *Discover Magazine* (April, 2011): 71, accessed April 29, 2012, http://discover.coverleaf.com/discovermagazine/201104?pg=68#pg68.

14. Philip S. Skell, "Why do we invoke Darwin?," *The Scientist*, 19 (August 29, 2005): 10.

15. Jerry Fodor and Massimo Piattelli-Palmarini, *What Darwin God Wrong* (London, UK: Profile Books, 2010), xxii.

16. Eugenie Scott, quoted in Terrence Stutz, "State Board of Education debates evolution curriculum," *Dallas Morning News* (January 22, 2009).

17. National Academy of Sciences, Institute of Medicine, *Science, Evolution and Creationism* (Washington D.C.: National Academy Press, 2008), 11, 40, 52.

18. Some of these types of logical fallacies are taken from Phillip E. Johnson, *Defeating Darwinism by Opening Minds* (Downers Grove, IL: InterVarsity Press, 1997), 37–47; Carl Sagan, *The Demon-Haunted World: Science as a Candle in the Dark* (New York, NY: Ballantine Books, 1996), 212–215.

19. Adapted nearly verbatim from a comment by NCSE theologian Peter Hess, "Intelligent Design and the Constitution" symposium at University of St. Thomas, November 10, 2009, accessed April 29, 2012, http://www.evolutionnews.org/ncsetheologianparrotsdawkins.mp3 (stating "A third problem with intelligent design is that its practitioners are either ignorant of science or seriously deluded or fundamentally dishonest").

20. A direct quote from Gregory A. Petsko, "It is alive," *Genome Biology*, 9 (June 23, 2008): 106.

21. Adapted from Kenneth R. Miller's testimony during the *Kitzmiller v. Dover* trial where he stated that ID is a negative argument against evolution that appeals to the supernatural:

- "It is what a philosopher might call the argument from ignorance, which is to say that, because we don't understand something, we assume we never will, and therefore we can invoke a cause outside of nature, a supernatural creator or supernatural designer." Kenneth R. Miller, *Kitzmiller v. Dover*, Day 1 PM Testimony (September 26, 2005), 36.
- "The evidence is always negative, and it basically says, if evolution is incorrect, the answer must be design." Kenneth R. Miller, *Kitzmiller v. Dover*, Day 1 PM Testimony (September 26, 2005), 37–38.
- "I believe that intelligent design is inherently religious and it is a form of creationism. It is a classic form of creationism known as special creationism.... I think the statement by the Dover Board of Education falsely undermines the scientific status of the theory of evolution, and therefore it certainly does not promote student understanding or even critical thinking, and I think it does a great disservice to science education in Dover and to the students of Dover." Kenneth R. Miller, *Kitzmiller v. Dover*, Day 1 AM Testimony (September 26, 2005), 58.

22. This example of the "guilt by association" fallacy is borrowed from Paul Nelson, "The Creationism Gambit," accessed April 29, 2012, http://www2.exploreevolution.com/exploreEvolutionFurtherDebate/2009/03/the_creationism_gambit_1.php in response to critics at the NCSE who made this type of argument against the textbook *Explore Evolution*.

23. Adapted from Kevin Padian and Nicholas Matzke, "Darwin, Dover, 'Intelligent Design' and textbooks," *Biochemical Journal*, 417 (2009): 29–42 (asserting the "case for ID" has "collapsed" and "no one with scientific or philosophical integrity is going to take [Discovery Institute or ID] seriously in future").

24. Adapted from Karl Giberson and Francis Collins, *The Language of Science and Faith: Straight Answers to Genuine Questions* (Downers Grove, IL: InterVarsity Press, 2011), 45 ("Scientists, however, make the confident claim that macroevolution is simply microevolution writ large: add up enough small changes and we get a large change").

25. Adapted from Giberson and Collins, *The Language of Science and Faith*, 32 ("Many of the scientists listed are not trained in biology and so are not in a position to evaluate the central theory of that field. Of the two authors of this book... one is a trained biologist capable of speaking with authority about evolution").

26. Adapted from Michael Ruse, "Faith & Reason," *Playboy* Magazine (April, 2006): 54–58, 134–136 ("[I]ntelligent-design theory and its companions are nasty, cramping, soul-destroying reversions to the more unfortunate aspects of 19th century America. Although I am not a Christian, I look on these ideas as putrid scabs on the body of a great religion.... But if you are going to fight moral evil—and creationism in its various forms is a moral evil—you need to understand what you are fighting and why.").

27. Adapted from Darrel Falk, *Coming to Peace with Science: Bridging the Worlds Between Faith and Biology* (Downer's Grove, IL: InterVarsity Press, 2004), 224 ("almost all scientists (including Christian ones) believe in the gradual appearance of life").

28. Bertrand Russell quoted in Anthony St. Peter, *The Greatest Quotations of All-Time* (Xlibris, 2010), 25.

29. In this context, eugenics is the effort to use Darwin's theory as a basis for breeding "fitter" human beings. For details, see John G. West, *Darwin Day In America: How Our Politics and Culture Have Been Dehumanized in the Name of Science* (Wilmington, DE: ISI Books, 2007).

30. For an accurate account of the Scopes trial and a discussion of historical inaccuracies in *Inherit the Wind*, see Edward J. Larson, *Summer for the Gods: The Scopes Trial and America's Continuing Debate over Science and Religion* (Cambridge, MA: Harvard University Press, 1997).

31. See Johnson, *Defeating Darwinism by Opening Minds*, 25, 30.

32. P. Z Myers, "A New Recruit," PandasThumb (June 13, 2005), accessed April 30, 2012, http://www.pandasthumb.org/archives/2005/06/a_new_recruit.html#comment-35130.

33. Jerry Coyne, "New antievolution bills," Why Evolution is True (March 17, 2011), accessed April 30, 2012, http://whyevolutionistrue.wordpress.com/2011/03/17/new-antievolution-bills/.

34. Timothy P. White, "Letter to the University of Idaho Faculty, Staff and Students" (October 4, 2005).

35. Laurence A. Moran, "Flunk the IDiots," Sandwalk (November 17, 2006), accessed April 30, 2012, http://sandwalk.blogspot.com/2006/11/flunk-idiots.html.

36. Alison Campbell, "'intelligent design is not creationism in any shape or form'—yeah, right!," BioBlog (January 3, 2011), accessed April 30, 2012, http://sci.waikato.ac.nz/bioblog/2011/01/intelligent-design-is-not-crea.shtml.

37. See *Expelled: No Intelligence Allowed* (Premise Media, 2008). Some critics of ID have tried to discount the cases of discrimination discussed in this documentary. For an extensive rebuttal of these criticisms, see www.ncseexposed.com.

38. See United States House of Representatives Committee on Government Reform, "Intolerance and the Politicization of Science at the Smithsonian" (December, 2006), 4, accessed April 30, 2012, http://www.discovery.org/f/1489.

39. See Caroline Crocker, *Free to Think: Why Scientific Integrity Matters* (Southworth, WA: Leafcutter Press, 2010).

40. See "Intelligent Design torpedoes tenure," *World Net Daily* (May 19, 2007), accessed April 30, 2012, http://www.worldnetdaily.com/news/article.asp?ARTICLE_ID=55774; Nina Siegal, "Riled by Intelligent Design," *The New York Times* (November 6, 2005), accessed April 30, 2012, http://www.nytimes.com/2005/11/06/education/edlife/HedIntelligent.html.

41. E-mail from Vladimir Kogan to Bruce Harmon, Sergei Budko, Joerg Schmalian, John Clem, Doug Finnemore, and Paul Canfield (November 22, 2005), as quoted in "Intelligent Design Was the Issue After All," accessed April 30, 2012, http://www.evolutionnews.org/ID_was_the_Issue_Gonzalez_Tenure.pdf.

42. Committee on Culture, Science and Education, Mr. Guy Lengagne, Council of Europe, "The dangers of creationism in education" (September 17, 2007), accessed April 30, 2012, http://assembly.coe.int/main.asp?link=/documents/adoptedtext/ta07/eres1580.htm.

43. Zogby International, "Results from Nationwide Poll" (February 3, 2009), accessed April 30, 2012, http://www.evolutionnews.org/Zogby%20International%202009%20Poll%20Report.pdf.

44. John Scopes quoted in John Angus Campbell and Stephen C. Meyer, "How Should Schools Handle Evolution?," *USA Today* (August 14, 2005), accessed April 30, 2012, http://www.usatoday.com/news/opinion/editorials/2005-08-14-evolution-debate_x.htm.

45. *Edwards v. Aguillard*, 482 U.S. 578, 634 (1987) (Scalia, J., dissenting).

Chapter 20

1. Thomas Jefferson, quoted in William G. Hyland Jr., *In Defense of Thomas Jefferson: The Sally Hemings Sex Scandal* (New York, NY: St. Martins / Thomas Dunne, 2009), 118.

2. Kenneth R. Miller, *Kitzmiller v. Dover*, Day 1 PM Testimony (September 26, 2005), 37–38.

3. Stephen C. Meyer, "The origin of biological information and the higher taxonomic categories," *Proceedings of the Biological Society of Washington*, 117 (2004): 213–239.

4. "In all irreducibly complex systems in which the cause of the system is known by experience or observation, intelligent design or engineering played a role in the origin of the system." Scott A. Minnich and Stephen C. Meyer, "Genetic analysis of coordinate flagellar and type III regulatory circuits in pathogenic bacteria," *Proceedings of the Second International Conference on Design & Nature, Rhodes Greece* (2004), 8, accessed April 30, 2012, http://www.discovery.org/f/389.

5. See Scott A. Minnich, *Kitzmiller v. Dover*, Day 20 PM Testimony (November 3, 2005), 99–108; Casey Luskin, "Response to Barbara Forrest's *Kitzmiller* Account Part VIII: Important Facts Left Out About ID Research," Evolution News & Views (October 8, 2006), accessed April 30, 2012, http://www.evolutionnews.org/2006/10/response_to_barbara_forrests_k_7002560.html; William A. Dembski, *No Free Lunch: Why Specified Complexity Cannot Be Purchased without Intelligence* (Lanham, MD: Rowman & Littlefield, 2002), 239–310; Robert M. Macnab, "Flagella," in *Escherichia Coli and Salmonella Typhimurium: Cellular and Molecular Biology Vol. 1*, eds. Frederick C. Neidhardt, John L. Ingraham, K. Brooks Low, Boris Magasanik, Moselio Schaechter, and H. Edwin Umbarger (Washington D.C.: American Society for Microbiology, 1987), 73–74.

6. Stephen C. Meyer, Marcus Ross, Paul Nelson, and Paul Chien, "The Cambrian Explosion: Biology's Big Bang," in *Darwinism, Design, and Public Education*, eds. John A. Campbell and Stephen C. Meyer (East Lansing, MI: Michigan State University Press, 2003), 367, 386.

7. See Chapter 15. See also Meyer, Ross, Nelson, and Chien, "The Cambrian Explosion: Biology's Big Bang;" Wolf-Ekkehard Lönnig, "Dynamic genomes, morphological stasis, and the origin of irreducible complexity," *Dynamical Genetics*, eds. Valerio Parisi, Valeria De Fonzo, and Filippo Aluffi-Pentini (Kerala, India, Research Signpost, 2004), 101–119; A. C. McIntosh, "Evidence of Design in Bird Feathers and Avian Respiration," *International Journal of Design & Nature and Ecodynamics*, 4 (2009): 154–169.

8. Meyer, "The origin of biological information and the higher taxonomic categories," 213–239.

9. Paul Nelson and Jonathan Wells, "Homology in Biology," in *Darwinism, Design, and Public Education*, eds. John Angus Campbell and Stephen C. Meyer (East Lansing, MI: Michigan State University Press, 2003), 316.

10. In the case of this example 3, neo-Darwinism might make some of the same predictions. Is this a problem for the positive case for design? Not at all. The fact that another theory can explain some data does not negate ID's ability to successfully predict what we should find in nature. After all, part of making a "positive case" means that the arguments for design stand on their own and do not depend on refuting other theories. Moreover, as we saw in Chapter 14, there are many cases of supposed extreme "convergent evolution" that are better explained by common design. Additionally, regarding the predictions from examples 1 (biochemistry), 2 (paleontology), and 4 (genetics), neo-Darwinism has made *different* predictions from ID. In any case, in this example ID makes a slightly different prediction in that it does not predict that re-usage of parts must necessarily occur in a nested hierarchical pattern.

11. John A. Davison, "A Prescribed Evolutionary Hypothesis," *Rivista di Biologia / Biology Forum*, 98 (2005): 155–166; Nelson and Wells, "Homology in Biology;" Lönnig, "Dynamic genomes, morphological stasis, and the origin of irreducible complexity;" Michael Sherman, "Universal Genome in the Origin of Metazoa: Thoughts About Evolution," *Cell Cycle*, 6 (August 1, 2007): 1873–1877.

12. William A. Dembski, "Science and Design," *First Things*, 86 (October, 1998).

13. See Chapter 12. See also Richard Sternberg, "On the Roles of Repetitive DNA Elements in the Context of a Unified Genomic-Epigenetic System," *Annals of the NY Academy of Science*, 981 (2002): 154–188; James A. Shapiro, and Richard Sternberg, "Why repetitive DNA is essential to genome function," *Biological Reviews of the Cambridge Philosophical Society*, 80 (2005): 227–250; A. C. McIntosh, "Information and Entropy—Top-Down or Bottom-Up Development in Living Systems?," *International Journal of Design & Nature and Ecodynamics*, 4 (2009): 351–385; Jonathan Wells, *The Myth of Junk DNA* (Seattle, WA: Discovery Institute Press, 2011); The ENCODE Project Consortium, "An integrated encyclopedia of DNA elements in the human genome," *Nature*, 489 (September 6, 2012): 57–74.

14. For a discussion of some problems with the peer-review system, see Casey Luskin, "Is Peer-Review a Requirement of Good Science?," Discovery Institute (February 15, 2012), accessed May 1, 2012, http://www.discovery.org/a/18301.

15. Michael J. Behe, "Correspondence with Science Journals: Response to Critics Concerning Peer-Review," Discovery Institute (August 2, 2000), accessed April 30, 2012, http://www.discovery.org/a/450.

16. *Daubert v. Merrell Dow Pharmaceuticals, Inc.*, 509 U.S. 579, 593 (1993).

17. Brief Amici Curiae of Physicians, Scientists, and Historians of Science in Support of Petitioners, *Daubert v. Merrell Dow Pharmaceuticals, Inc.*, 509 U.S. 579 (1993).

18. Some of the journals that have published peer-reviewed pro-ID scientific articles include *Proceedings of the Biological Society of Washington, Annual Review of Genetics, Quarterly Review of Biology, Journal of Molecular Biology, Theoretical Biology and Medical Modelling, Journal of Advanced Computational Intelligence and Intelligent Informatics, Cell Biology International, Rivista di Biologia / Biology Forum*, and *Physics of Life Reviews*.

19. For a relatively complete listing of peer-reviewed pro-ID papers, see "Peer-Reviewed & Peer-Edited Scientific Publications Supporting the Theory of Intelligent Design (Annotated)," Discovery Institute (February 1, 2012), accessed April 30, 2012, http://www.discovery.org/a/2640.

20. About Biologic Institute, accessed February 1, 2012, http://www.biologicinstitute.org/about/.

21. Biologic Institute research initiatives adapted from "Research," accessed April 30, 2012, http://www.biologicinstitute.org/research/.

22. Douglas D. Axe, "Extreme Functional Sensitivity to Conservative Amino Acid Changes on Enzyme Exteriors," *Journal of Molecular Biology*, 301 (2000): 585–595; Douglas D. Axe, "Estimating the Prevalence of Protein Sequences Adopting Functional Enzyme Folds," *Journal of Molecular Biology*, 341 (2004): 1295–1315.

23. For example, see William A. Dembski and Robert J. Marks II, "The Search for a Search: Measuring the Information Cost of Higher Level Search," *Journal of Advanced Computational Intelligence and Intelligent Informatics*, 14 (2010): 475–486; Winston Ewert, George Montañez, William A. Dembski, Robert J. Marks II, "Efficient Per Query Information Extraction from a Hamming Oracle," *Proceedings of the 42nd Meeting of the Southeastern Symposium on System Theory, IEEE, University of Texas at Tyler*, (March 7–9, 2010): 290–297; Winston Ewert, William A. Dembski, and Robert J. Marks II, "Evolutionary Synthesis of Nand Logic: Dissecting a Digital Organism," *Proceedings of the 2009 IEEE International Conference on Systems, Man, and Cybernetics San Antonio, TX, USA* (October, 2009): 3047–3053; William A. Dembski, and Robert J. Marks II, "Bernoulli's Principle of Insufficient Reason and Conservation of Information in Computer Search," *Proceedings of the 2009 IEEE International Conference on Systems, Man, and Cybernetics San Antonio, TX, USA* (October, 2009): 2647–2652; William A. Dembski and Robert J. Marks II, "Conservation of Information in Search: Measuring the Cost of Success," *IEEE Transactions on Systems, Man and Cybernetics A, Systems & Humans*, 39 (September, 2009): 1051–1061.

24. Michael J. Behe, "Experimental Evolution, Loss-of-Function Mutations and 'The First Rule of Adaptive Evolution'," *Quarterly Review of Biology*, 85 (December, 2010): 419–445; Michael J. Behe and David W. Snoke, "Simulating Evolution by Gene Duplication of Protein Features That Require Multiple Amino Acid Residues," *Protein Science*, 13 (2004): 2651–2664.

25. Scott C. Todd, "A view from Kansas on that evolution debate," *Nature*, 401 (September 30, 1999): 423.

26. See Paul Nelson, "The Creationism Gambit," accessed April 29, 2012, http://www2.exploreevolution.com/exploreEvolutionFurtherDebate/2009/03/the_creationism_gambit_1.php

27. See National Academy of Sciences, *Science and Creationism: A View from the National Academy of Sciences*, 2nd ed. (Washington D.C.: National Academy Press, 1999), 7; Phillip E. Johnson, Darwin on Trial, 2nd ed. (Downers Grove, IL: InterVarsity Press, 1993), 4; Eugenie Scott, "Antievolutionism and Creationism in the United States," *Annual Review of Anthropology* 26 (1997): 263–289; Robert Pennock, *Intelligent Design Creationism and Its Critics: Philosophical, Theological and Scientific Perspectives*, ed. Robert Pennock (Cambridge, MA: MIT Press, 2001), 646; William A. Dembski, *The Design Revolution: Answering the Toughest Questions about Intelligent Design* (Downers Grove, IL: InterVarsity Press, 2004), 41; Barbara Forrest and Paul R. Gross, *Creationism's Trojan Horse* (New York, NY: Oxford University Press, 2004), 283.

28. *Kitzmiller v, Dover*, 400 F.Supp.2d 707 (M.D. Pa. 2005). For critical responses to the ruling, see David K. DeWolf, John G. West, and Casey Luskin, "Intelligent Design Will Survive *Kitzmiller v. Dover*," *Montana Law Review*, 68 (Spring, 2007): 7–57, accessed April 30, 2012, http://www.discovery.org/f/1372; David K. DeWolf, John G. West, Casey Luskin, and Jonathan Witt, *Traipsing Into Evolution: Intelligent Design and the Kitzmiller v. Dover Decision* (Seattle, WA: Discovery Institute Press, 2006).

29. DeWolf, West, Luskin, and Witt, *Traipsing Into Evolution*, 8.

30. For example, Discovery Institute's opposition to ID and calls to repeal Dover's ID policy are documented in DeWolf, West, and Luskin, "Intelligent Design Will Survive *Kitzmiller v. Dover*"; David K. DeWolf, John G. West, and Casey Luskin, "Rebuttal to Irons," *Montana Law Review*, 68 (Spring, 2007): 90–94, accessed April 30, 2012, http://www.discovery.org/f/1373.

31. *Kitzmiller v, Dover*, 400 F.Supp.2d 707, 746 (M.D. Pa. 2005).

32. Ibid., 742.

33. For example, after the ruling, former NCSE staff member Nicholas Matzke stated that ID "is probably toast." Emma Marris, "Intelligent Design Verdict Set to Sway Other Cases," *Nature*, 439 (January 5, 2006): 6–7. Likewise, Matzke and NCSE president Kevin Padian ebulliently declared "It's over for the Discovery Institute. Turn out the lights. The fat lady has sung." See Kevin Padian and Nick Matzke, "Discovery Institute tries to 'swift-boat' Judge Jones," NCSE.com, accessed April 30, 2012, http://ncse.com/creationism/general/discovery-institute-tries-to-swift-boat-judge-jones. For a response, see Casey Luskin, "Response to Matzke and Padian's Revisionist History and Gloat Parade," Evolution News & Views (January 20, 2006), accessed April 30, 2012, http://www.evolutionnews.org/2006/01/response_to_matzke_and_padians001832.html.

34. Jay D. Wexler, "*Kitzmiller* and the 'Is It Science?' Question," *First Amendment Law Review*, 5 (2006): 90, 93.

35. *Kitzmiller v. Dover*, 400 F.Supp.2d 707, 718 (M.D. Pa 2005).

36. For example, in 2009 University of Colorado Boulder philosopher Bradley Monton published a book subtitled "An Atheist Defends Intelligent Design." Sociologist Steve Fuller, philosopher and law professor Thomas Nagel, and mathematician David Berlinski are other noteworthy agnostic or atheist scholars who have defended ID as an argument that must be taken seriously. Philosopher Anthony Flew, a noteworthy atheist, also became a proponent of intelligent design a few years before his death in 2010. See Bradley Monton, *Seeking God in Science: An Atheist Defends Intelligent Design* (Peterborough, Canada: Broadview Press, 2009); Steve Fuller, *Dissent over Descent: Intelligent Design's Challenge to Darwinism* (Cambridge, UK: Icon Books, 2008); Thomas Nagel, *Mind and Cosmos: Why the Materialist Neo-Darwinian Conception of Nature Is Almost Certainly False* (Oxford, UK: Oxford University Press, 2012); David Berlinski, *The Deniable Darwin and Other Essays* (Seattle, WA: Discovery Institute Press, 2010).

37. While Newtonian mechanics still holds true under many circumstances, Einstein's theories of relativity showed that Newton's models are not always correct.

38. See Casey Luskin, "Richard Dawkins and Lawrence Krauss 'evangelize' for Evolution at Stanford," Evolution News & Views (March 10, 2008), accessed April 30, 2012, http://www.evolutionnews.org/2008/03/richard_dawkins_and_lawrence_k004976.html.

39. Richard Dawkins, "Is Science A Religion?," *Humanist* (January / February, 1997), accessed April 30, 2012, http://www.thehumanist.org/humanist/articles/dawkins.html.

40. "Free People From Superstition," *Freethought Today* (April, 2000), accessed April 30, 2012, http://www.ffrf.org/legacy/fttoday/2000/april2000/weinberg.html.

41. Geoff Brumfiel, "Who Has Designs on Your Students' Minds?," *Nature*, 434 (April 28, 2005): 1062–1065.

42. "Notable Signers," accessed April 30, 2012, http://www.americanhumanist.org/Who_We_Are/About_Humanism/Humanist_Manifesto_III/Notable_Signers.

43. "Humanist Manifesto III: Humanism and its Aspirations," American Humanist Association, accessed April 30, 2012, http://www.americanhumanist.org/Who_We_Are/About_Humanism/Humanist_Manifesto_III.

44. Edward J. Larson and Larry Witham, "Leading scientists still reject God," *Nature*, 394 (July 23, 1998): 313. See also Edward J. Larson and Larry Witham, "Scientists and Religion in America," *Scientific American* (September, 1999): 88–93.

45. Gregory W. Graffin and William B. Provine, "Evolution, Religion and Free Will," *American Scientist*, 95 (July / August, 2007): 294–297, accessed April 30, 2012, http://www.americanscientist.org/issues/pub/evolution-religion-and-free-will.

46. In recent years, the Darwin lobby has poured huge amounts of time, money, and other resources into trying to affect the outcome of evolution-education battles in states such as Kansas, Ohio, Texas, Florida, and Louisiana. Some noteworthy examples of lawsuits filed by the Darwin lobby include *Kitzmiller v. Dover, Selman v. Cobb County, Hurst v. Newman, Freiler v. Tangipahoa, and Edwards v. Aguillard.*

47. It is important to note that the *Kitzmiller v. Dover* case was not initiated by the ID movement. Discovery Institute strongly urged the Dover Area School District to drop its policy pushing ID into the classroom, and also discouraged them from going to court to defend the legally problematic policy. Additionally, the lawsuit was filed by Darwin lobbyists, not by ID proponents.

48. Only a couple lawsuits have been initiated by the ID movement. Examples include *American Freedom Alliance v. California Science Center* or *Coppedge v. Jet Propulsion Lab*, both of which sought to *defend* ID proponents against attacks on their free speech rights.

49. Jonathan Wells, *The Politically Incorrect Guide to Darwinism and Intelligent Design* (Washington D.C.: Regnery, 2006), 195.

50. This is discussed in Casey Luskin, "Redefining 'Science Literacy' As 'Acceptance of Evolution'," Evolution News & Views (August 11, 2011), accessed April 30, 2012, http://www.evolutionnews.org/2011/08/redefining_science_literacy_as049331.html.

51. William A. Dembski, "Dealing with the Backlash Against Intelligent Design" (April 14, 2004), accessed April 30, 2012, http://www.designinference.com/documents/2004.04.Backlash.htm.

Appendix A

1. See Douglas D. Axe, "Extreme Functional Sensitivity to Conservative Amino Acid Changes on Enzyme Exteriors," *Journal of Molecular Biology*, 301 (2000): 585–595; Douglas D. Axe, "Estimating the Prevalence of Protein Sequences Adopting Functional Enzyme Folds," *Journal of Molecular Biology*, 341 (2004): 1295–1315; Douglas D. Axe, "The Case Against a Darwinian Origin of Protein Folds," *BIO-Complexity*, 2010 (1): 1–12; Ann K. Gauger and Douglas D. Axe, "The Evolutionary Accessibility of New Enzyme Functions: A Case Study from the Biotin Pathway," *BIO-Complexity*, 2011 (1): 1–17. Ann K. Gauger, Stephanie Ebnet, Pamela F. Fahey, and Ralph Seelke, "Reductive Evolution Can Prevent Populations from Taking Simple Adaptive Paths to High Fitness," *BIO-Complexity*, 2010 (2): 1–9; Michael J. Behe, "Experimental Evolution, Loss-of-Function Mutations, and 'The First Rule of Adaptive Evolution'," *The Quarterly Review of Biology*, 85 (December, 2010): 419–445; Douglas D. Axe, "The Limits of Complex Adaptation: An Analysis Based on a Simple Model of Structured Bacterial Populations," *BIO-Complexity*, 2010(4): 1–10; Wolf-Ekkehard Lönnig, "Mutagenesis in *Physalis pubescens* L. ssp. *floridana*: Some further research on Dollo's Law and the Law of Recurrent Variation," *Floriculture and Biotechnology*, (2010): 1–21; Michael J. Behe and David W. Snoke, "Simulating Evolution by Gene Duplication of Protein Features That Require Multiple Amino Acid Residues," *Protein Science*, 13 (2004): 2651–2664; Michael J. Denton, Craig J. Marshall, and Michael Legge, "The Protein Folds as Platonic Forms: New Support for the pre-Darwinian Conception of Evolution by Natural Law," *Journal of Theoretical Biology*, 219 (2002): 325–342.

2. See Jonathan Wells, "Using Intelligent Design Theory to Guide Scientific Research," *Progress in Complexity, Information, and Design*, 3.1.2 (November, 2004); A. C. McIntosh, "Information and Entropy—Top-Down or Bottom-Up Development in Living Systems?," *International Journal of Design & Nature and Ecodynamics*, 4 (2009): 351–385; Josiah D. Seaman and John C. Sanford, "Skittle: A 2-Dimensional Genome Visualization Tool," *BMC Informatics*, 10 (2009): 451; Richard Sternberg and James A. Shapiro, "How repeated retroelements format genome function," *Cytogenetic and Genome Research*, 110 (2005): 108–116; James A. Shapiro, and Richard Sternberg, "Why repetitive DNA is essential to genome function," *Biological Reviews of the Cambridge Philosophical Society*, 80 (2005): 227–250; Richard Sternberg, "On the Roles of Repetitive DNA Elements in the Context of a Unified Genomic–Epigenetic System," *Annals of the NY Academy of Science*, 981 (2002): 154–188.

3. See Wolf-Ekkehard Lönnig, "Dynamic genomes, morphological stasis, and the origin of irreducible complexity," Dynamical Genetics, eds. Valerio Parisi, Valeria De Fonzo, and Filippo Aluffi-Pentini (Kerala, India, Research Signpost, 2004), 101–119; Paul Nelson and Jonathan Wells, "Homology in Biology," in *Darwinism, Design, and Public Education*, eds. John Angus Campbell and Stephen C. Meyer (East Lansing: Michigan State University Press, 2003), 303–322; John A. Davison, "A Prescribed Evolutionary Hypothesis," *Rivista di Biologia / Biology Forum*, 98 (2005): 155–166; Michael Sherman, "Universal Genome in the Origin of Metazoa: Thoughts About Evolution," *Cell Cycle*, 6 (August 1, 2007): 1873–1877; Albert D. G. de Roos, "Origins of introns based on the definition of exon modules and their conserved interfaces," *Bioinformatics*, 21 (2005): 2–9; Albert D. G. de Roos, "Conserved intron positions in ancient protein modules," *Biology Direct*, 2 (2007): 7; Albert D. G. de Roos, "The Origin of the Eukaryotic Cell Based on Conservation of Existing Interfaces," *Artificial Life*, 12 (2006): 513–523; Wolf-Ekkehard Lönnig, "Mutagenesis in *Physalis pubescens* L. ssp. *floridana*: Some further research on Dollo's Law and the Law of Recurrent Variation," *Floriculture and Biotechnology*, (2010): 1–21; Michael Sherman, "Universal Genome in the Origin of Metazoa: Thoughts About Evolution," *Cell Cycle*, 6 (August 1, 2007): 1873–1877; A. C. McIntosh, "Functional Information and Entropy in Living Systems," *Design and Nature III: Comparing Design in Nature with Science and Engineering*, 87 (WIT Transactions on Ecology and the Environment, Editor Brebbia, C. A., WIT Press, 2006).

4. See Jonathan Wells, "Do Centrioles Generate a Polar Ejection Force?," *Rivista di Biologia / Biology Forum*, 98 (2005): 71–96; Scott A. Minnich and Stephen C. Meyer. "Genetic Analysis of Coordinate Flagellar and Type III Regulatory Circuits in Pathogenic Bacteria," *Proceedings of the Second International Conference on Design & Nature, Rhodes Greece*, edited by M. W. Collins and C. A. Brebbia (WIT Press, 2004); A. C. McIntosh, "Information and Entropy—Top-Down or Bottom-Up Development in Living Systems?," *International Journal of Design & Nature and Ecodynamics*, 4 (2009): 351–385; D. Halsmer, J. Asper, N. Roman, and T. Todd, "The Coherence of an Engineered World," *International Journal of Design & Nature and Ecodynamics*, 4 (2009): 47–65; A. C. McIntosh, "Evidence of design in bird feathers and avian respiration," *International Journal of Design & Nature and Ecodynamics*, 4 (2009): 154–169; Wolf-Ekkehard Lönnig, Kurt Stüber, Heinz Saedler, Jeong Hee Kim, "Biodiversity and Dollo's Law: To What Extent can the Phenotypic Differences between *Misopates orontium* and *Antirrhinum majus* be Bridged by Mutagenesis," *Bioremediation, Biodiversity and Bioavailability*, 1 (2007): 1–30; Wolf-Ekkehard Lönnig, "Mutations: The Law of Recurrent Variation," *Floriculture, Ornamental and Plant Biotechnology*, 1 (2006): 601–607; Wolf-Ekkehard Lönnig and Heinz Saedler, "Chromosome Rearrangement and Transposable Elements," *Annual Review of Genetics*, 36 (2002): 389–410; Heinz-Albert Becker and Wolf-Ekkehard Lönnig, "Transposons: Eukaryotic," *Encyclopedia of Life Sciences* (John Wiley & Sons, 2005).

5. See Michael J. Behe and David W. Snoke, "Simulating Evolution by Gene Duplication of Protein Features That Require Multiple Amino Acid Residues," *Protein Science*, 13 (2004): 2651–2664; Ann K Gauger, Stephanie Ebnet, Pamela F Fahey, and Ralph Seelke, "Reductive Evolution Can Prevent Populations from Taking Simple Adaptive Paths to High Fitness," *BIO-Complexity*, 2010: 1–9; Ann K. Gauger and Douglas D. Axe, "The Evolutionary Accessibility of New Enzyme Functions: A Case Study from the Biotin Pathway," *BIO-Complexity*,

2011 (1): 1–17; Michael J. Behe, "Experimental Evolution, Loss-of-Function Mutations, and 'The First Rule of Adaptive Evolution'," *The Quarterly Review of Biology*, 85 (December, 2010): 419–445; Douglas D. Axe, "The Limits of Complex Adaptation: An Analysis Based on a Simple Model of Structured Bacterial Populations," *BIO-Complexity*, 2010 (4): 1–10; Wolf-Ekkehard Lönnig, "Mutagenesis in *Physalis pubescens* L. ssp. *floridana*: Some further research on Dollo's Law and the Law of Recurrent Variation," *Floriculture and Biotechnology*, (2010): 1–21; Douglas D. Axe, "The Case Against a Darwinian Origin of Protein Folds," *BIO-Complexity*, 2010 (1): 1–12; Douglas D. Axe, "Extreme Functional Sensitivity to Conservative Amino Acid Changes on Enzyme Exteriors," *Journal of Molecular Biology*, 301 (2000): 585–595; Douglas D. Axe, "Estimating the Prevalence of Protein Sequences Adopting Functional Enzyme Folds," *Journal of Molecular Biology*, 341 (2004): 1295–1315.

6. See A. C. McIntosh, "Information and Entropy—Top-Down or Bottom-Up Development in Living Systems?," *International Journal of Design & Nature and Ecodynamics*, 4 (2009): 351–385; Stephen C. Meyer, "The origin of biological information and the higher taxonomic categories," *Proceedings of the Biological Society of Washington*, 117 (2004): 213–239; Stephen C. Meyer, Marcus Ross, Paul Nelson, and Paul Chien, "The Cambrian Explosion: Biology's Big Bang," in *Darwinism, Design, and Public Education*, eds. John A. Campbell and Stephen C. Meyer (East Lansing, MI: Michigan State University Press, 2003); D. Halsmer, J. Asper, N. Roman, and T. Todd, "The Coherence of an Engineered World," *International Journal of Design & Nature and Ecodynamics*, 4 (2009): 47–65; A. C. McIntosh, "Evidence of design in bird feathers and avian respiration," *International Journal of Design & Nature and Ecodynamics*, 4 (2009): 154–169; Solomon Victor and Vijaya M. Nayak, "Evolutionary anticipation of the human heart," *Annals of the Royal College of Surgeons of England*, 82 (2000): 297–302; Solomon Victor, Vljaya M. Nayek, and Raveen Rajasingh, "Evolution of the Ventricles," *Texas Heart Institute Journal*, 26 (1999): 168–175.

7. See Richard v. Sternberg, "DNA Codes and Information: Formal Structures and Relational Causes," *Acta Biotheoretica*, 56 (September, 2008): 205–232; Ø. A. Voie, "Biological function and the genetic code are interdependent," *Chaos, Solitons and Fractals*, 28 (2006): 1000–1004; David L. Abel and Jack T. Trevors, "Self-organization vs. self-ordering events in life-origin models," *Physics of Life Reviews*, 3 (2006): 211–228; David L. Abel, "Constraints vs Controls," *The Open Cybernetics and Systemics Journal*, 4 (January 20, 2010): 14–27; David L. Abel, "The GS (genetic selection) Principle," *Frontiers in Bioscience*, 14 (January 1, 2010): 2959–2969; David L. Abel, "The Capabilities of Chaos and Complexity," *International Journal of Molecular Sciences*, 10 (2009): 247–291; David L. Abel, "The 'Cybernetic Cut': Progressing from Description to Prescription in Systems Theory," *The Open Cybernetics and Systemics Journal*, 2 (2008): 252–262; David L. Abel, *"Complexity, self-organization, and emergence at the edge of chaos in life-origin models," Journal of the Washington Academy of Sciences*, 93 (2007): 1–20; David L Abel and Jack T. Trevors, "More than Metaphor: Genomes are objective sign systems," *Journal of BioSemiotics*, 1 (2006): 253–267; Kirk Durston and David K. Y. Chiu, "A Functional Entropy Model for Biological Sequences," *Dynamics of Continuous, Discrete & Impulsive Systems: Series B Supplement* (2005); David L. Abel and Jack T. Trevors, "Three subsets of sequence complexity and their relevance to biopolymeric information," *Theoretical Biology and Medical Modeling*, 2 (August 11, 2005): 1–15; David L. Abel, "Is Life reducible to complexity?," *Fundamentals of Life*, Chapter 1.2 (2002); D. K. Y. Chiu and T. H. Lui, "Integrated Use of Multiple Interdependent Patterns for Biomolecular Sequence Analysis," *International Journal of Fuzzy Systems*, 4 (September, 2002): 766–775.

8. See Stephen C. Meyer, "The origin of biological information and the higher taxonomic categories," *Proceedings of the Biological Society of Washington*, 117 (2004): 213–239; William A. Dembski, *The Design Inference: Eliminating Chance through Small Probabilities* (Cambridge University Press, 1998); A. C. McIntosh, "Information and Entropy—Top-Down or Bottom-Up Development in Living Systems?," *International Journal of Design & Nature and Ecodynamics*, 4 (2009): 351–385; William A. Dembski and Robert J. Marks II, "Conservation of Information in Search: Measuring the Cost of Success," *IEEE Transactions on Systems, Man, and Cybernetics-Part A: Systems and Humans*, 39 (September, 2009): 1051–1061; Winston Ewert, William A. Dembski, and Robert J. Marks II, "Evolutionary Synthesis of Nand Logic: Dissecting a Digital Organism," *Proceedings of the 2009 IEEE International Conference on Systems, Man, and Cybernetics* (October, 2009); William A. Dembski and Robert J. Marks II, "Bernoulli's Principle of Insufficient Reason and Conservation of Information in Computer Search," *Proceedings of the 2009 IEEE International Conference on Systems, Man, and Cybernetics* (October, 2009); Winston Ewert, George Montanez, William Dembski and Robert J. Marks II, "Efficient Per Query Information Extraction from a Hamming Oracle," *42nd South Eastern Symposium on System Theory*, 290–297 (March, 2010): 290–297; Douglas D. Axe, Brendan W. Dixon, Philip Lu, "Stylus: A System for Evolutionary Experimentation Based on a Protein/Proteome Model with Non-Arbitrary Functional Constraints," *PLoS One*, 3 (June, 2008): e2246; Stephen C. Meyer, "The origin of biological information and the higher taxonomic categories," *Proceedings of the Biological Society of Washington*, 117 (2004): 213–239; Kirk K. Durston, David K. Y. Chiu, David L. Abel, Jack T. Trevors, "Measuring the functional sequence complexity of proteins," *Theoretical Biology and Medical Modeling*, 4 (2007): 47; David K. Y. Chiu,, and Thomas W. H. Lui, *"Integrated Use of Multiple Interdependent Patterns for Biomolecular Sequence Analysis," International Journal of Fuzzy Systems*, 4 (September, 2002): 766–775; Michael J. Behe, "Experimental Evolution, Loss-of-Function Mutations, and 'The First Rule of Adaptive Evolution'," *The Quarterly Review of Biology*, 85 (December, 2010): 419–445; Douglas D. Axe, "The Limits of Complex Adaptation: An Analysis Based on a Simple Model of Structured Bacterial Populations," *BIO-Complexity*, 2010: 1–10; George Montañez, Winston Ewert, William A. Dembski, and Robert J. Marks II, "A Vivisection of the ev Computer Organism: Identifying Sources of Active Information," *BIO-Complexity*, 2010 (3): 1–6; William A. Dembski and Robert J. Marks II, "The Search for a Search: Measuring the Information Cost of Higher Level Search," *Journal of Advanced Computational Intelligence and Intelligent Informatics*, 14 (2010): 475–486; Douglas D. Axe, *"The Case Against a Darwinian Origin of Protein Folds," BIO-Complexity*, 2010; David L. Abel, "Constraints vs Controls," *The Open Cybernetics and Systemics Journal*, 4 (January 20, 2010): 14–27; David L. Abel, "The GS (genetic selection) Principle," *Frontiers in Bioscience*, 14 (January 1, 2010): 2959–2969; David L. Abel, "The Universal Plausibility Metric (UPM) & Principle (UPP)," *Theoretical Biology and Medical Modelling*, 6 (2009); David L. Abel, "The Capabilities of Chaos and Complexity," *International Journal of Molecular Sciences*, 10 (2009): 247–291; David L. Abel, "The biosemiosis of prescriptive information," *Semiotica*, 174 (2009): 1–19; David L. Abel, "The 'Cybernetic Cut': Progressing from Description to Prescription in Systems Theory," *The Open Cybernetics and Systemics Journal*, 2 (2008): 252–262; David L. Abel, "Complexity, self-organization, and emergence at the edge of chaos in life-origin models," *Journal of the Washington Academy of Sciences*, 93 (2007): 1–20; Felipe Houat de Brito, Artur Noura Teixeira, Otávio Noura Teixeira, Roberto C. L. Oliveira, "A Fuzzy Intelligent Controller for Genetic Algorithm Parameters," in *Advances in Natural Computation*, eds. Licheng Jiao, Lipo Wang, Xinbo Gao, Jing Liu, Feng Wu (Springer-Verlag, 2006); Felipe Houat de Brito, Artur Noura Teixeira, Otávio Noura Teixeira, Roberto C. L. Oliveira, "A Fuzzy Approach to Control Genetic Algorithm Parameters," *SADIO Electronic Journal of Informatics and Operations Research*, 7 (2007): 12–23; David L. Abel and Jack T. Trevors, "Self-organization vs. self-ordering events in life-origin models," *Physics of Life Reviews*, 3 (2006): 211–228; Granville Sewell, "A Mathematician's View of Evolution," *The Mathematical Intelligencer*, 22 (2000); David L Abel and Jack T. Trevors, "More than Metaphor: Genomes are objective sign systems," *Journal of BioSemiotics*, 1 (2006): 253–267; Øyvind Albert Voie, "Biological function and the genetic code are interdependent," *Chaos, Solitons and Fractals*, 28 (2006): 1000–1004; David L. Abel and Jack T. Trevors, "Three subsets of sequence complexity and their relevance to biopolymeric information," *Theoretical Biology and Medical Modeling*, 2 (August 11, 2005): 1–15; Wolf-Ekkehard Lönnig, "Mutation Breeding, Evolution, and the Law of Recurrent Variation," *Recent Research Developments in Genetics & Breeding*, 2 (2005): 45–70; Douglas D. Axe, "Extreme Functional Sensitivity to Conservative Amino Acid Changes on Enzyme Exteriors," *Journal of Molecular Biology*, 301 (2000): 585–595; Douglas D. Axe, "Estimating the Prevalence of Protein Sequences Adopting Functional Enzyme Folds," *Journal of Molecular Biology*, 341 (2004): 1295–1315; Michael J. Behe and David W. Snoke, "Simulating Evolution by Gene Duplication of Protein Features That Require Multiple Amino Acid Residues," *Protein Science*, 13 (2004): 2651–2664; David L. Abel, "Is Life reducible to complexity?," *Fundamentals of Life*, Chapter 1.2 (2002); D.

K. Y. Chiu and T. H. Lui, "Integrated Use of Multiple Interdependent Patterns for Biomolecular Sequence Analysis," *International Journal of Fuzzy Systems*, 4 (September, 2002): 766–775; Granville Sewell, Postscript, in *Analysis of a Finite Element Method: PDE/PROTRAN* (Springer Verlag, 1985); A. C. McIntosh, "Functional Information and Entropy in Living Systems," *Design and Nature III: Comparing Design in Nature with Science and Engineering*, 87 (WIT Transactions on Ecology and the Environment, WIT Press, 2006).

9. See Wolf-Ekkehard Lönnig, "Dynamic genomes, morphological stasis, and the origin of irreducible complexity," Dynamical Genetics, eds. Valerio Parisi, Valeria De Fonzo, and Filippo Aluffi-Pentini (Kerala, India, Research Signpost, 2004), 101–119; Stephen C. Meyer, "The origin of biological information and the higher taxonomic categories," *Proceedings of the Biological Society of Washington*, 117 (2004): 213–239; Stephen C. Meyer, Marcus Ross, Paul Nelson, and Paul Chien, "The Cambrian Explosion: Biology's Big Bang," in *Darwinism, Design, and Public Education*, eds. John A. Campbell and Stephen C. Meyer (East Lansing, MI: Michigan State University Press, 2003).

10. See Guillermo Gonzalez *et al.*, "Refuges for Life in a Hostile Universe," *Scientific American* (October, 2001); D. Halsmer, J. Asper, N. Roman, T. Todd, "The Coherence of an Engineered World," *International Journal of Design & Nature and Ecodynamics*, 4 (2009): 47–65; Frank J. Tipler, "Intelligent Life in Cosmology," *International Journal of Astrobiology*, 2 (2003): 141–148,; Stanley L. Jaki, "Teaching of Transcendence in Physics," *American Journal of Physics*, 55 (October, 1987): 884–888; William G. Pollard, "Rumors of transcendence in physics," *American Journal of Physics*, 52 (October, 1984).

Appendix D

1. See Donald Voet, Judith G. Voet, and Charlotte W. Pratt, *Fundamentals of Biochemistry: Life at the Molecular Level*, 2nd ed. (Hoboken, NJ: John Wiley & Sons, 2006), 974.

2. See David Goodsell, "Aminoacyl-tRNA Synthetases," Molecule of the Month at Protein Data Bank (April, 2001), accessed May 2, 2012, http://www.rcsb.org/pdb/education_discussion/molecule_of_the_month/download/Aminoacyl-tRNASynthetases.pdf.

3. Some have cited the work of Michael Yarus as reportedly showing how the genetic code could have arisen out of the RNA world. But Yarus's experiments and arguments are beset by a myriad of problems and deficiencies, which are too technical for our discussion here. For a more complete discussion of Yarus's work and its problems, see Stephen Meyer and Paul Nelson, "Can the Origin of the Genetic Code Be Explained by Direct RNA Templating?," *BIO-Complexity*, 2011 (2): 1–10; Casey Luskin, "Premature Falsification: Critics of Signature in the Cell Rush to Judgment and Miss Meyer's Argument," *Salvo Magazine*, 14 (Fall 2010): 50–53.

4. J. T. Trevors and D. L. Abel, "Chance and necessity do not explain the origin of life," *Cell Biology International*, 28 (2004): 729–739.

5. Ibid.

Appendix E

1. James W. Valentine, *On the Origin of Phyla* (Chicago, IL: University of Chicago Press, 2004), 35 (internal citations removed).

2. Charles R. Marshall, "Explaining the Cambrian 'Explosion' of Animals," *Annual Review of Earth and Planetary Sciences*, 34 (2006): 355–384.

3. Kevin J. Peterson, Michael R. Dietrich and Mark A. McPeek, "MicroRNAs and metazoan macroevolution: insights into canalization, complexity, and the Cambrian explosion," *BioEssays*, 31 (2009): 736–747 (emphases in original, internal citation numbers removed).

4. J. William Schopf, "Solution to Darwin's dilemma: Discovery of the missing Precambrian record of life," *Proceedings of the National Academy of Sciences U.S.A.*, 97 (June 20, 2000): 6947–6953.

5. Peter Douglas Ward, *On Methuselah's Trail: Living Fossils and the Great Extinctions* (New York, NY: W. H. Freeman, 1992), 36.

6. Richard Fortey, "The Cambrian Explosion Exploded?," *Science*, 293 (July 20, 2001): 438–439.

7. Andrew H. Knoll and Sean B. Carroll, "Early animal Evolution: Emerging Views from Comparative Biology and Geology," *Science*, 284 (June 25, 1999): 2129–213.

8. Vicki Pearse, John Pearse, Mildred Buchsbaum, and Ralph Buchsbaum, *Living Invertebrates* (Palo Alto, CA: Blackwell Scientific Publications, 1987), 764.

9. James W. Valentine, David Jablonski, Douglas H. Erwin, "Fossils, molecules and embryos: new perspectives on the Cambrian explosion," *Development*, 126 (1999): 851–859.

10. Schopf, "Solution to Darwin's dilemma: Discovery of the missing Precambrian record of life," 6947–6953.

11. Ibid.

GLOSSARY

aaRS (AMINOACYL tRNA SYNTHETASE) ENZYME: A protein that attaches tRNA molecules to the proper amino acid according to the genetic code. Most cells use 20 different aaRS enzymes, one for each amino acid used in life. These aaRS molecules themselves are encoded by the genes in the DNA. This forms the essence of a "chicken-egg problem": aaRS enzymes themselves are necessary to perform the very task that constructs them.

ACADEMY: The community of scientists, writers, and scholars who have the goal of advancing science and literature.

ACQUIRED CHARACTERISTICS: See Inheritance of Acquired Characteristics.

ACTIN: Protein filaments that function as a "track" for the myosin molecular machine. Together, actin and myosin enable muscle contraction.

***AD HOMINEM* ATTACK:** A fallacious debating tactic intended to discredit a person rather than an argument.

ADAPTATION: A feature that enables an organism to survive and reproduce in its environment.

ADAPTIVE RADIATION: The supposed rapid diversification of a species after entering an empty habitat, or niche.

ARGUMENT FROM POPULARITY: Urging people to believe an idea because it is popular.

ALIMENTARY CANAL: See Digestive Tract.

ALTRUISM: Unselfish regard for the welfare of others.

AMETABOLISM: The simplest type of insect development, where the insect undergoes no metamorphosis, and the young insect (nymph) has essentially the same body plan as the adult. After emerging from the egg, the insect undergoes only changes in size, but not shape, as it matures.

AMINO ACIDS: Simple organic molecules that form the building blocks of proteins. While there are many amino acids, life generally uses only 20 specific ones to build proteins.

***ARCHAEOPTERYX*:** An extinct bird known from the late Jurassic period that has teeth, wing claws, and a bony tail.

***ARCHAEORAPTOR*:** A fossil that was originally claimed to be a link between birds and dinosaurs. It was later shown to be a forgery—a skillful fusion of a dinosaur fossil and a bird fossil.

***ARDIPITHECUS RAMIDUS* (a.k.a. "Ardi"):** A hominin fossil that was initially labeled the most recent common ancestor of apes and humans when it was first publicized in 2009. Later studies have found that Ardi was more similar to apes than to humans, and was not a human ancestor.

ARTIFACT HYPOTHESIS: The argument that the lack of transitional forms in the fossil record is strictly the result of an imperfect fossil record.

ARTIFICIAL SELECTION: The selection of certain animals by a breeder in order to increase desirable traits in a population or eliminate undesirable ones.

ASEXUAL REPRODUCTION: A method of reproduction where organisms reproduce by cloning themselves, with no mate required to produce offspring. It is considered to be the simplest and earliest form of reproduction.

ASTRONOMY: The study of celestial bodies and phenomena.

ATP (adenosine triphosphate): The primary energy-carrying molecule in all cells.

ATP SYNTHASE: A protein-based molecular machine that generates ATP. Powered by protons, it is composed of two spinning motors connected by an axle.

***AUSTRALOPITHECUS*:** An ape-like genus of small-brained hominins thought to have existed from about 4.2 million years ago until about 2 million years ago. In Latin, the name means "southern ape," since the first australopithecine fossils were found in southern Africa. There are two major groups of australopithecines, the larger "robust" forms and the smaller "gracile" forms. Most evolutionary paleoanthropologists claim humans are descended from the gracile forms, such as *Australopithecus afarensis*, which includes the famous fossil "Lucy."

AXON: The hair-like extension of a nerve cell which carries electrical impulses.

BACTERIAL FLAGELLUM: A propulsion system used by certain bacteria. It is composed of a rotary engine and a propeller, similar to the design of an outboard motor.

BASE PAIRS: In the stable, double-helix shape of DNA, adenine (A) only pairs with thymine (T) and cytosine (C) only pairs with guanine (G) in order to store the information. These combinations are called base pairs.

BERRA'S BLUNDER: The error of seizing on evidence for common design and mistakenly calling it evidence for common descent.

BIG BANG THEORY: A model of the universe's origin, which holds it is finite in size and age. According to this theory, the universe—including all space and time—originated with a single, powerful expansion event, and is still expanding.

BIOMIMETICS: A process of engineering where humans turn to nature for inspiration or guidance when creating technology.

BLACK BOX: A term scientists use to describe a system that they find interesting but whose inner workings they are unable to fully understand.

BODY PLAN: An organism's arrangement of organs and other parts that coordinate to perform functions and allow it to survive in its environment.

BULLYING TACTIC: Using outlandish rhetoric to play on emotions and force someone into agreement.

CAMBRIAN EXPLOSION: A geologically brief period about 530 mya, during which nearly all of the major animal phyla, including many diverse body plans, suddenly appeared.

CELL: The basic structural and functional unit of all living organisms, enclosed by a semipermeable plasma membrane.

CELL MEMBRANE: A semi-porous barrier surrounding every living cell, which allows water and nutrients in, lets waste products out, and keeps harmful elements from entering. It also keeps the cell's components together to allow cellular processes to take place.

CELL WALL: A rigid wall providing the boundary for plant cells. It limits cellular growth and restricts movement, but also provides protection and structural stability.

CHEMICAL EVOLUTION: The theory that chemicals in nature assembled through blind, unguided, chance chemical reactions to create life.

CHLOROPHYLL: A green pigment used by plants and other photosynthetic organisms to capture the energy from sunlight for photosynthesis.

CHLOROPLAST: A molecular machine found in green plants and algae which conducts photosynthesis using sunlight energy captured by chlorophyll. During photosynthesis, chloroplasts combine this energy with water and carbon dioxide to create free oxygen and sugars.

CHRYSALIS: The hardened outer layer of a butterfly pupa.

CIRCULAR REASONING: A fallacy of reasoning where the rules or starting assumptions permit only the desired results.

CIRCUMSTELLAR HABITABLE ZONE: A narrow band around a star that permits temperatures just right for liquid water.

COCOON: The silken encasement a pupa weaves around itself. Many homometabolous species spin a cocoon, including moths.

CODON: A three-string series of nucleotides in DNA forming a basic unit of the genetic code. The biochemical language of the genetic code uses codons to symbolize commands—start commands, stop commands, as well as codons that signify each of the 20 amino acids used in life.

COMMON ANCESTRY: See Common Descent.

COMMON DESCENT: The idea that all life forms are related and descended from a single common ancestor.

COMMON DESIGN: The idea that since designers regularly reuse parts that work in different designs, shared functional similarities between different organisms might result from intelligent design using a common blueprint.

COMPLEX AND SPECIFIED INFORMATION ("CSI"): Information that is both complex (meaning it is unlikely) and specified (meaning it matches an independent pattern).

CONNECTIVE TISSUE: One of the four tissue types that compose animal body plans. This tissue connects parts of the body, providing a framework in which other tissue types are embedded.

CONVERGENT EVOLUTION: Two or more species independently acquiring the same trait, supposedly through Darwinian evolution.

CO-OPTION: To take and use for another purpose. In evolutionary biology, it is a highly speculative mechanism where blind and unguided processes cause biological parts to be borrowed and used for another purpose.

COPERNICAN PRINCIPLE: The belief that Earth occupies no special place in our solar system, galaxy, or universe.

COSMOLOGICAL CONSTANT: A measure of the effect of dark energy on the expansion of the universe.

COSMOLOGY: The study of the universe as a whole—its structure, origin, and development.

CREATIONISM: The belief that the universe and all life were created by God as described in the book of Genesis in the Bible. Some forms of creationism accept the mainstream scientific view that the Earth and universe are billions of years old. In contrast, "young Earth" creationism holds that the Earth and universe are only a few thousand years old. What all forms of creationism share in common is that they start with religious texts and end with religious conclusions.

CSI: See Complex and Specified Information.

DARK ENERGY: The effect in the universe that acts to push galaxies away from one another, accelerating the expansion of the universe. The cosmological constant is a measurement of this effect.

DARWIN LOBBY: A coalition of scientific, educational, legal, and political activist groups that work to censor non-evolutionary viewpoints.

DARWINISM (original theory): The theory that all life shares common ancestry and evolved through descent with modification, driven by an unguided process of natural selection acting upon random variations. See also Neo-Darwinism.

DARWIN'S TEST OF EVOLUTION: In *The Origin of Species*, Darwin wrote: "If it could be demonstrated that any complex organ exists which could not possibly have been formed by numerous, successive, slight modifications, my theory would absolutely break down."

DENDRITES: Branches of a nerve cell that receive signals sent from other neurons.

DERMAL TISSUE: A plant tissue that covers and guards the plant. It generates a waxy substance that aids in water conservation and prevents disease by keeping out harmful chemicals or microorganisms. Small, adjustable openings within the derma (called stomata) allow the exchange of gases, such as oxygen and carbon dioxide, in and out of the plant.

DEVELOPMENTAL HOURGLASS: A model of vertebrate development which acknowledges that vertebrate embryos start differently, but claims that they pass through a highly similar stage midway through development.

DIGESTIVE TRACT: The pathway, lined with mucous membrane along its entire length, which food travels along from the start to the end of digestion.

DNA (deoxyribonucleic acid): A molecule found in every organism which consists of two long, intertwined chains of nucleotides that carry the information necessary to produce proteins in that organism.

DOPPLER EFFECT: A physical effect where sound waves are heard with a higher frequency when the source of the sound is moving toward the hearer, but with a lower frequency when moving away. Although light waves behave differently from sound waves, a similar effect takes place.

DYSTELEOLOGY: The view that nature was not intelligently designed, often based on the claim that some natural structures are flawed or functionless.

ECHOLOCATION: A sonar system used by some animals, such as bats and dolphins. Sounds are emitted and their echoes interpreted to locate objects.

ECTOTHERMIC: Cold-blooded organisms that regulate their temperatures by exchanging heat with the environment.

ELECTROMAGNETIC FORCE: One of the four fundamental forces in nature, this force is associated with electric and magnetic fields which govern the interaction of electrically charged particles.

EMBRYO: An animal in its early stages of development.

EMPIRICAL: Based on observation, experience, or experiment.

ENDOTHERMIC: The ability of warm-blooded animals to control their temperature by internal means.

ENTROPY: A measure of the amount of disorder in a system.

EPITHELIAL TISSUE: One of the four tissue types that make up animal bodies. This tissue protects and conducts materials within the body by covering or lining surfaces, and also absorbs and secretes substances. Skin is made of epithelial tissue.

EQUIVOCATION: Changing the meaning of words in the middle of an argument.

EVOLUTION: This term has three common meanings:

1. **Microevolution:** Small-scale changes in a population of organisms.

Macroevolution can be divided into two separate parts:

2. **Universal Common Descent:** The view that all organisms are related and are descended from a single common ancestor. It does not explain how those relationships developed.

3. Natural Selection: The view that an unguided process of natural selection acting upon random mutation has been the primary mechanism driving the evolution of life.

EVOLUTIONARY DEVELOPMENTAL BIOLOGY ("Evo-Devo"): A field that combines ideas from evolutionary biology and developmental biology to claim that changes to the master genes that control the development of an organism, such as *Hox* or *homeobox* genes, can cause large, abrupt changes in body plans.

EVOLUTIONIST: A person who believes in macroevolution, typically a supporter of neo-Darwinism.

FALLACY OF REASONING: An argument based on erroneous logic.

FALSE CHOICE: A logical fallacy that portrays two options as mutually exclusive when they really are not.

FALSIFIABLE: A common requirement of the scientific method that there must be a means to test whether a theory is false. For example, the multiverse theory is not scientific because there is no way to determine if it is false.

FINE-TUNING (OF THE UNIVERSE): The widely accepted idea that there are dozens of universal parameters—such as the masses of protons, neutrons, electrons, and the speed of light—that would render life impossible if they had slightly different values.

FITNESS: The likelihood that a gene, or other trait, will be passed on to the next generation.

FLAGELLUM: See Bacterial Flagellum.

FREE OXYGEN: Oxygen that is not combined with other elements (unlike the combination of oxygen with hydrogen in H_2O).

GALACTIC HABITABLE ZONE: The optimal location for life within our galaxy, midway between the center and the edge, which escapes the large zones of deadly radiation at the core, yet is far enough from the edge to contain the elements necessary for complex life. This region is where our solar system exists.

GALÁPAGOS ISLANDS: A group of volcanic islands off the Pacific coast of South America that became famous after Darwin studied the diversity of organisms that live on the islands.

GAPS-BASED REASONING: The tendency to assume the answer to unsolved scientific questions. "God of the gaps" reasoning assumes the answer lies in supernatural causes. "Materialism of the gaps" reasoning assumes the answer lies in strictly material causes. ID rejects gaps-based reasoning of all kinds, and suggests that scientists should follow the evidence where it leads—free from philosophical presuppositions.

GENE: A basic unit of heredity, typically understood as a section of DNA that contains assembly instructions for a particular protein.

GENERAL RELATIVITY: Albert Einstein's theory, proposed in 1915, that describes gravity as it relates to space and time.

GENETIC CODE: The set of rules used by cells to convert the genetic information in DNA or RNA into proteins. The genetic code is essentially the language in which the genetic information is written.

GENETIC FALLACY: A logical fallacy that attacks the origin of an argument rather than the argument itself.

GENETIC KNOCKOUT EXPERIMENT: An experiment where genes or gene sequences in a system are mutated so that they no longer function. This experimental method can be used for determining the gene's function, and also for testing irreducible complexity.

GENETIC MUTATION: A change in the nucleotide sequence of DNA. When a genetic mutation is in sex cells, it is heritable.

GENOME: The full complement of genetic information carried by an organism in its DNA.

GRAVITATIONAL CONSTANT: A universal constant that determines the force of gravitational attraction between bodies.

GROUND TISSUE: A type of plant tissue that makes up the bulk of the plant and also helps form the structural foundation of other plant tissues. It is the site of photosynthesis and food storage.

GUILT BY ASSOCIATION: A logical fallacy that claims, in effect, "A has some similarity to or association with B, therefore A is the same as B."

HAECKEL'S EMBRYO DRAWINGS: A set of inaccurate—some would say fraudulent—drawings (circa 1874) by the German biologist Ernst Haeckel, which overstate the degree of similarity between the embryos of certain vertebrates.

HEMIMETABOLISM (partial metamorphosis): A type of insect development involving gradual, progressive changes in form, where change occurs through a process of instars (periods of growth and change) and molts (sheddings of skin).

HEMOGLOBIN: The oxygen-transporting protein that gives red blood cells their color and carries oxygen to tissues.

HOATZIN: A bird living in the Amazonian marshland similar in size and some other characteristics to *Archaeopteryx*.

HOLOMETABOLISM (complete metamorphosis): The most common and complicated form of insect development, where ultimately the larva turns into a pupa, and undergoes complete transformation of the body plan.

HOMINID: The most popular view holds that hominids include humans, great apes (chimpanzees, gorillas, and orangutans), and all extinct species leading back to their alleged common ancestor.

HOMININ: Generally defined as humans and chimpanzees, and any extinct organisms leading back to their presumed most recent common ancestor.

***HOMO*:** Latin for "man" in Latin, *Homo* is a genus that includes *Homo erectus*, Neanderthals, our own species *Homo sapiens*, and several other extinct species.

***HOMO ERECTUS*:** An extinct member of the genus *Homo* that is generally believed to have lived from about 1.9 million years ago until 100,000 years ago, highly similar to modern humans below the neck. The name means "upright walking man." Its brain size is within the range of modern human variation.

***HOMO HABILIS*:** An extinct hominin species that some leading paleoanthropologists consider to have been erroneously placed in the *Homo* genus. Dated at about 1.9 million years ago, it was placed within *Homo* to imply an ancestral link. It is now thought by many to actually belong with *Australopithecus* since its anatomical characteristics better fit in that genus.

***HOMO SAPIENS*:** Latin for "wise man," this is the human species, the only living member of the genus *Homo*.

HOMOLOGY: Similarity of structure and position, but not necessarily function. Within neo-Darwinian theory, it means "similarity due to common ancestry."

HORIZONTAL GENE TRANSFER: A process by which microorganisms obtain genes through mechanisms other than inheritance from a parent, specifically by sharing and swapping genes with their neighbors.

HORMONES: Molecules secreted by a cell, gland, or organ to stimulate some function in another part of the organism.

***HOX* GENES:** Master genes that control the development of an organism.

HUMAN EXCEPTIONALISM: A view holding that the human race has unique and unparalleled moral, intellectual, and creative abilities.

HYPOTHESIS: (1) an educated guess; (2) a tentative explanation; (3) a claim that can be tested by comparing it to the evidence.

IDA: A fossil once heavily promoted to the public as a human ancestor, now thought to be a lower primate not pertinent to human origins.

INFLATION THEORY: A cosmological model proposing that during a tiny fraction of a second after the Big Bang the universe expanded at many times the velocity of light, after which the expansion slowed.

INFORMATION SEQUENCE PROBLEM: Unguided and chance processes could not properly order the information sequence in an information-carrying molecule (such as RNA or DNA) necessary for the first life. This would be a major obstacle for any materialistic hypothesis of the origin of life.

INFRASOUND: A low-frequency sound below the human ability to hear. Elephants use infrasound to communicate.

***INHERIT THE WIND* STEREOTYPE:** A stereotype commonly promoted by the media which holds that opponents of Darwinian evolution are ignorant bigots who want to shut down scientific thought, whereas evolutionists are intelligent, articulate, and open-minded people who want to advance it.

INHERITANCE OF ACQUIRED CHARACTERISTICS: The now-refuted theory that organisms can pass on traits to their offspring that they acquire during their lifetime, such as missing limbs or tails.

INORGANIC: Compounds or materials that are of inanimate or non-biological origin.

INSTAR: A period of growth and change during the larval stage of insects that undergo metamorphosis.

INTELLIGENT AGENT: A being with the ability to plan ahead and think with a goal in mind.

INTELLIGENT DESIGN: A scientific theory that holds that many features of the universe and life are best explained by an intelligent cause.

IRREDUCIBLE COMPLEXITY: A characteristic of a system composed of multiple interacting parts wherein the removal of any part causes the system to effectively cease functioning.

JUNK DNA: Non-coding DNA which allegedly has no function for the organism that carries it. Much of it is now known to have function.

***KALAM* ARGUMENT:** An argument for a first cause; the argument has three parts:

- Anything that begins to exist has a cause.
- The universe began to exist.
- Therefore the universe has a cause.

KIN SELECTION: A claim of evolutionary psychology which holds that individuals may neglect their own reproductive success to help close blood relatives raise their own offspring, allowing some of their own genes to be passed on.

LARVA: The juvenile form in insects before undergoing metamorphosis into adults.

LARYNGEAL NERVE: See Recurrent Laryngeal Nerve (RLN) or Superior Laryngeal Nerve (SLN).

LIGHT-YEAR: The distance light travels in one year (5.88 trillion miles).

LUCY: A fossil of the species *Australopithecus afarensis*, "Lucy" is perhaps the most famous australopithecine fossil because 40% of her bones were uncovered, making her one of the most complete known fossils among early hominins. Many have claimed that Lucy's species was ancestral to the genus *Homo*; however, the fossil record doesn't support the claim.

MACROEVOLUTION: Large-scale changes in populations of organisms, including the evolution of fundamentally new biological features. Typically this term also means that all life-forms descended from a single common ancestor through unguided natural processes.

MATERIALISM: The philosophical belief that the material world is the only reality that exists.

MEIOSIS: A process of cell division used to make sex cells in sexually reproducing organisms, where the daughter cells have half the chromosomes of the original parent cell.

MESSENGER RNA (mRNA): A strand of RNA that carries information from the DNA to the ribosome so it can be used in the construction of proteins. See also Transcription and Translation.

METAMORPHOSIS: A process of pre-programmed development where an organism changes its body plan.

METHODOLOGICAL NATURALISM: The belief that, whether or not the supernatural exists, we must pretend that it doesn't when practicing science.

MICROEVOLUTION: Small-scale changes in a population of organisms.

MILLER-UREY EXPERIMENTS: Experiments performed in the 1950s by Stanley Miller and his mentor, Harold Urey, which sparked electricity through gases, intending to simulate the atmosphere on early Earth. While some amino acids were created, it is now thought that the gases used in the experiments probably did not represent the atmosphere of the early Earth.

MITOSIS: A process of cell division where a cell clones itself, and the daughter cells have the same chromosomes as the parent cell.

MOLECULE-BASED TREE: A phylogenetic tree based on comparing DNA, RNA, or protein sequences in different organisms.

MOLT: The process by which an organism sheds its skin. Insects that undergo metamorphosis molt after each instar.

MORPHOLOGY: The form, structure, and body plan of an organism.

MORPHOLOGY-BASED TREE: A phylogenetic tree based on comparing physical characteristics, such as anatomical and structural similarities of different organisms.

MULTI-MUTATION FEATURE: Features that require multiple mutations before providing any benefit to the organism.

MULTIVERSE THEORY: A theory that there are an infinite number of universes, each with different values for its physical laws and constants, and that we occupy the one that happens to permit complex life. The theory is philosophical, not scientific, because there is no way to test it empirically.

MUSCLE TISSUE: One of the four types of tissue that make up animal bodies. It is composed of large cells called muscle fibers. There are three types of muscle tissues: smooth, skeletal, and cardiac.

MUTATION: See Genetic Mutation.

MYOSIN: A molecular machine that pulls itself along a track of actin filaments to transport cellular cargo, and also forms the basis for muscle contraction.

NATURAL SELECTION: An unguided natural process whereby organisms that are better suited to survive and reproduce tend to leave more offspring, passing on their traits to the next generation.

NATURALISM: See Philosophical Naturalism.

NEANDERTHAL: A member of the genus *Homo* that looked nearly identical to modern humans, except for their slightly larger and thicker skulls. Whether they went extinct is debated—some paleoanthropologists believe we interbred, are members of the same species, and that they were little different from modern humans.

NEO-DARWINISM: The theory that all life shares common ancestry, and evolved through descent with modification, driven by unguided natural selection acting upon random genetic mutations in DNA. This is basically a revision of the original theory of Darwinism, but it attempts to incorporate our modern knowledge of genetics, which was not understood in Darwin's time.

NERVOUS TISSUE: One of the four types of tissue that make up animal bodies. It is responsible for transmitting electrical signals within the organism.

NEURON: A single nerve cell, composed of a cell body and an axon.

NUCLEOTIDE: A molecule consisting of the combination of a nucleotide base, a sugar molecule (ribose), and a phosphate group. Nucleotides are the basic structural units of RNA and DNA.

NUCLEOTIDE BASES: Molecules that store the information in DNA and make it a "digital code." In DNA, they are adenine (A), cytosine (C), thymine (T), and guanine (G), whereas RNA uses uracil (U) instead of thymine.

NYMPH: An immature form in ametabolous or hemimetabolous insects, where the juvenile's body plan resembles the adult form. During this stage, the young insect primarily feeds and grows.

OCKHAM'S RAZOR: A logical principle, often used by scientists, that holds that the simplest explanation tends to be the correct one.

OSCILLATING UNIVERSE THEORY: A cosmological model that proposes that the universe perpetually expands, then collapses, and then expands again.

OXIDATION: The combination of a substance with oxygen.

PALEOANTHROPOLOGY: The field of science that studies the origin of humans.

PANDA'S THUMB: An elongated bone in the hand of the giant panda which it uses like an opposable thumb to strip leaves off bamboo stalks so it can eat the shoots. The bone is not a true digit. ID critics cite the "panda's thumb" as a supposed example of poor design even though it is quite efficient at serving its purpose.

PEER REVIEW: An imperfect process by which scientists double-check the work of their peers to make sure it has been done correctly.

PEPPERED MOTH (*Biston betularia*): A species of moth found in many parts of the world, known for having darker and lighter-colored forms.

PHARYNGULA STAGE: A mid-way stage of embryonic development where vertebrate embryos are said to have a similar appearance. Also called the phylotypic stage.

PHILOSOPHICAL NATURALISM: The philosophical belief that the material world is the only reality that exists. See also Materialism.

PHLOEM: Channels that conduct sugars and other materials down from the leaves to the rest of a plant.

PHOTOSYNTHESIS: A multi-step chemical process by which plants derive energy from the sun. During photosynthesis, chloroplasts combine this energy with water and carbon dioxide to create oxygen and sugars.

PHYLOGENY: A hypothesis describing the ancestral relationships among organisms.

PHYLUM: The highest level of taxonomic classification below kingdom. Among animals, each phylum has a unique body plan.

PILTDOWN MAN: A fossil skull discovered in England in 1912, once thought to be a human ancestor. It was later determined to be a forgery, pieced together from the jaw of an orangutan and the skull of a human.

PLASMA MEMBRANE: See Cell Membrane.

PLATE TECTONICS: A theory holding that the Earth's outermost layer (called the lithosphere) is divided up into plates which can move.

PREBIOTIC: Before the existence of biological life.

PREBIOTIC SOUP: See Primordial Soup.

PRIMORDIAL SOUP: A hypothetical sea of simple organic molecules on the early Earth where life formed by chance.

PROBIOTICS: Beneficial bacteria found in the intestines, which aid digestion.

PROFESSIONAL INTIMIDATION: Citing one's educational training or professional accomplishments to silence another viewpoint.

PROTEINS: Worker molecules that carry out many key biological functions. Proteins are composed of long chains of amino acids.

PROTOPLASM: The complex contents of a cell encapsulated by the cell membrane.

PSEUDOGENES: Supposedly broken, useless copies of once-functional genes.

PUNCTUATED EQUILIBRIUM ("PUNC EQ"): A theory of evolution where most change takes place in small populations over relatively short geological time periods. These hypothetical periods of rapid change are interspersed between long time-spans without change, called periods of stasis.

PUPA: The stage between larva and adult in holometabolous insects, in which the larva undergoes complete metamorphosis.

QUANTUM THEORY: A field of physics that describes matter and energy in terms of subatomic units called "quanta." In a quantum vacuum, electromagnetic waves and particles appear to pop in and out of existence.

QUESTIONING MOTIVES: A logical fallacy in which one attacks another person's reason for making an argument rather than the argument itself.

RECAPITULATION THEORY: An inaccurate evolutionary model of development which holds that the development of an organism (ontogeny) replays (recapitulates) its evolutionary history (phylogeny).

RECURRENT LARYNGEAL NERVE ("RLN"): A nerve that carries impulses from the brain to the larynx in many vertebrates, including humans. It takes a route that some evolutionists claim is poorly designed. See also Superior Laryngeal Nerve (SLN).

REDSHIFT: Light waves coming from a receding object have their frequency shifted down toward the red end of the visible light spectrum. Light waves from an approaching object will shift up toward the blue end. Edwin Hubble's research discovered a disproportionally high level of red light coming from virtually every galaxy and thus confirmed that galaxies are receding from one another.

RIBOSOME: A large, multi-part molecular machine responsible for translating genetic instructions in mRNA to assemble proteins.

RNA (ribonucleic acid): A molecule that carries genetic information much as does DNA, but replaces the nucleotide base thymine with uracil. Unlike DNA, it usually exists only as a single strand, and serves as a mobile information-transportation molecule within cells.

RNA WORLD: An origin-of-life hypothesis holding that life first appeared when a self-replicating RNA molecule arose by chance.

SCIENTIFIC METHOD: A process used to test ideas and reach valid conclusions about nature. It frequently involves making observations, forming a hypothesis, experimenting, and drawing tentative conclusions.

SEARCH FOR EXTRATERRESTRIAL INTELLIGENCE ("SETI"): A science research program that has spent tens of millions of dollars in an unsuccessful attempt to find extraterrestrial life. The project uses intelligent design reasoning in that it seeks to detect radio emissions sent by an extraterrestrial civilization.

SECOND LAW OF THERMODYNAMICS: A basic law of nature holding that entropy tends to increase in a closed system. This implies that the amount of usable energy in the universe will decrease over time, falsifying the oscillating universe theory.

SINGULARITY: A point where matter is so compressed by gravity that it has infinite density and almost no volume.

SOLAR SYSTEM: The sun and the planets and other celestial bodies that orbit it.

SPECIES: One of the basic units of taxonomic classification; it is the rank below genus. There are various definitions of the term, but a common one is a reproductively isolated population of organisms capable of interbreeding to produce fertile offspring.

SPECIFIED COMPLEXITY: A property exhibited by an event, object, or sequence that is unlikely (complex) and matches an independent pattern (specified).

SPONTANEOUS GENERATION: The sudden development of organisms from non-living matter. In essence, the idea is that life can come spontaneously from non-life.

STATIC UNIVERSE: A cosmological model which claims that the universe is eternal in age and constant in size, and thus its origin requires no explanation.

STEADY STATE THEORY (a.k.a. Static Universe): A cosmological model that proposed there was no actual beginning because the universe has been expanding eternally. The theory was refuted by the discovery of cosmic microwave background radiation.

STOMATA: Small adjustable openings, or pores, in the dermis of a plant which open and close to allow the passage of gases and water vapor.

STRAW MAN: A fallacious debating tactic in which the opponent's argument is misrepresented, and then the false version is attacked.

STRONG NUCLEAR FORCE: One of the four fundamental forces in nature, this force binds together protons and neutrons in the nucleus of an atom.

SUPERIOR LARYNGEAL NERVE ("SLN"): A nerve that provides impulses from the brain to the larynx, taking a direct route. See also Recurrent Laryngeal Nerve (RLN).

SURVIVAL OF THE FITTEST: A phrase commonly used to describe natural selection, where those members of a population that are better suited to survive and reproduce (being the most "fit") tend to leave more offspring, passing on their traits to the next generation.

SYSTEMATICS: The science of classification. In biology, it is the field that studies the relationships between groups of organisms.

TAXONOMIC CLASSIFICATION: The grouping of organisms into categories based on features that scientists consider similar. Modern evolutionary biologists attempt to classify organisms according to their presumed evolutionary relationships—often resulting in illogical groupings.

TELOMERE: The region of repetitive DNA sequences found at the end of a chromosome.

TENET: A basic principle or belief.

TETRAPOD: In general, a vertebrate animal with four limbs. There are a few vertebrates classified as tetrapods—such as snakes and whales—that do not have four limbs.

THEROPOD: A type of carnivorous, bipedal dinosaur with small forelimbs.

***TIKTAALIK ROSEAE*:** A fossil discovered in 2004 in the Canadian Arctic, alleged to be a transitional form between fish and tetrapods. A closer inspection shows that it had fishlike fins, a very different skeletal structure from tetrapod limbs, and no wrist bones.

TISSUE: A combination of cells in plants and animals that coordinates to perform specific roles, such as the absorption of nutrients, support of other structures, or production of particular chemicals.

TOP-DOWN DESIGN: A method of engineering where the final goal is determined first, and then lower levels of detail are identified that would be necessary to achieve that goal.

TRANSCRIPTION: The first main step in making proteins. Cellular machinery copies the information in a gene-coding section of DNA to a strand of messenger RNA (mRNA).

TRANSFER RNA ("tRNA"): A small RNA molecule that ferries needed amino acids to the ribosome so a protein can be assembled.

TRANSITIONAL FORM: An organism having characteristics that are intermediate between an ancestor and a descendant in an evolutionary sequence.

TRANSLATION: After transcription, the mRNA molecule is sent to the ribosome, the molecular machine that assembles the amino acid chain that forms a protein. To do this, the ribosome reads, follows, and "translates" the instructions carried on the mRNA strand to make a protein.

TREE OF LIFE: A diagram intended to show that all life is related, with a single trunk indicating the common ancestor and the branches representing different types of organism.

UNIVERSAL COMMON DESCENT: See Common Descent.

UNIVERSAL CONSTANTS: Physical quantities believed to be valid throughout the entire universe, e.g., the speed of light.

UNWARRANTED CONCLUSION: Over-extrapolating from the evidence to make an unrelated or unjustified argument.

VASCULAR TISSUE: A plant tissue composed of xylem and phloem, which carries fluids throughout the plant.

VERTEBRATE: An animal with a backbone or spinal column.

VESTIGIAL ORGAN: A biological structure that once had a function but lost part or all of that function through evolution.

VILLI: Tiny, carpet-like fingers in the small intestine that absorb nutrients.

WEAK NUCLEAR FORCE: One of the four fundamental forces in nature, this force is responsible for particle decay processes.

XYLEM: Channels in plants that conduct water and minerals from the roots upward into the plant.

YOUNG EARTH CREATIONISM: The belief that the Earth and the universe are only a few thousand years old.

ILLUSTRATION CREDITS

Cover

Front Cover: *Photo by Daniel Ripplinger. Used with permission.*

Chapter 1

Night Sky: *Copyright © isoga Fotolia.com. Used with permission.*

Railroad Crossing: *Illustration created by Valerie Gower. Copyright © Valerie Gower, 2012. Used with permission.*

Mountain (El Capitan): *iStock. Used with permission.*

Mount Rushmore: *iStock. Used with permission.*

Chapter 2

Charles Darwin Portrait (Figure 2-1): *Public Domain. Found at Wikipedia. Accessed September 27, 2012, http://en.wikipedia.org/wiki/File:Charles_Darwin_ seated_crop.jpg.*

Lamarckian Evolution (Figure 2-2): *Illustration created by Valerie Gower. Copyright © Valerie Gower, 2012.*

Taxonomic Classification (Figure 2-3): *Illustration created by Casey Luskin. Copyright © Casey Luskin, 2012. All Rights Reserved.*

Alfred Russel Wallace Portrait (Figure 2-4): *Public Domain. Found at Wikipedia. Accessed September 27, 2012, http://en.wikipedia.org/wiki/File:PSM_V11_ D140_Alfred_Russel_Wallace.jpg.*

The Circular Logic of Materialism (Figure 2-5): *Illustration created by Jens Jorgenson, Casey Luskin, Hallie Kemper, and Gary Kemper. Copyright © Casey Luskin, Hallie Kemper, and Gary Kemper, 2012. All Rights Reserved.*

Chapter 3:

Plato (Figure 3-1): *Copyright © Marie-Lan Nguyen (user: Jastrow) / Wikimedia Commons / CC BY 2.5. Used with permission under Creative Commons Attribution 2.5 Generic License. Usage not intended to imply endorsement by the author / licensor of the work. Accessed September 27, 2012, http://en.wikipedia.org/wiki/File:Plato_ Silanion_ Musei_ Capitolini_MC1377.jpg.*

Albert Einstein Portrait (Figure 3-2): *Public Domain. Found at Wikipedia. Accessed September 27, 2012, http://en.wikipedia.org/wiki/File:Einstein_1921_ portrait2.jpg.*

Albert Einstein and Edwin Hubble Verifying the Redshift (Figure 3-3): *Illustration created by Valerie Gower. Copyright © Valerie Gower, 2012. Used with permission.*

Chapter 4:

Solar System (Figure 4-2): *Illustration created by Valerie Gower. Copyright © Valerie Gower, 2012. Used with permission.*

Universe Creating Machine (Figure 4-3): *Illustration created by Valerie Gower. Copyright © Valerie Gower, 2012. Used with permission.*

Galaxy and Dart (Figure 4-4): *Illustration created by Valerie Gower. Copyright © Valerie Gower, 2012. Used with permission.*

Chapter 5:

The Lottery: *Illustration created by Valerie Gower. Copyright © Valerie Gower, 2012. Used with permission.*

Chapter 6:

Planet Earth (Figure 6-1): *Copyright © FrameAngel, Fotolia.com. Used with permission.*

Galactic Habitable Zone (Figure 6-2): *Illustration created by Valerie Gower. Copyright © Valerie Gower, 2012. Used with permission.*

Solar System and Circumstellar Habitable Zone (Figure 6-3): *Illustration created by Valerie Gower. Copyright © Valerie Gower, 2012. Used with permission.*

Chapter 7:

Protein are Composed of Amino Acids (Figure 7-1): *Illustration created by Valerie Gower. Copyright © Valerie Gower, 2012. Used with permission.*

The Cell as a Miniature Factory (Figure 7-3): *Illustration created by Valerie Gower after David S. Goodsell. The Machinery of Life, Figure 4.2 p. 55 (Second ed., Springer, 2009). Also available at http://mgl.scripps.edu/people/goodsell/illustration/public. Used with permission. Usage not intended to imply endorsement by the author. Illustration by David S. Goodsell, the Scripps Research Institute.*

Recipe for Mice: *iStock. Used with permission.*

The Cell as Darwin as Seen by Darwin (Figure 7-4): *Illustration created by Casey Luskin. Copyright © Casey Luskin, 2012. All Rights Reserved. Used with permission.*

The Miller-Urey Apparatus (Figure 7-5): *Copyright Jody F. Sjogren 2000. Used with permission. From Figure 2-1, Jonathan Wells, Icons of Evolution: Science or Myth? (Washington, D.C.: Regnery, 2000).*

Test Tube: *Illustration created by Valerie Gower. Copyright*

© Valerie Gower, 2012. Used with permission.

Chapter 8:

Nucleotides Bases and Codons (Figure 8-3): *Illustration created by Valerie Gower. Copyright © Valerie Gower, 2012. Used with permission.*

DNA and its Nucleotide Language (Figure 8-4): *Illustration created by Valerie Gower. Copyright © Valerie Gower, 2012. Used with permission.*

The Chicken and Egg Problem (Figure 8-5): *Copyright © Casey Luskin. 2012. All rights reserved. Used with permission.*

Chapter 9:

Levels of Structure in Proteins (Figure 9-2): *Illustration created by Valerie Gower. Copyright © Valerie Gower, 2012. Used with permission.*

Hemoglobin Protein (Figure 9-3): *Illustration created by Valerie Gower. Copyright © Valerie Gower, 2012. Used with permission.*

The 'Hand-in-Glove' Fit of Proteins (Figure 9-4): *Illustration created by Hallie Kemper and Jens Jorgensen. Copyright © Hallie Kemper, 2012. Used with permission.*

Bacterial Flagellum (Figure 9-5): *Illustration created by Valerie Gower. Copyright © Valerie Gower, 2012. Used with permission.*

Chapter 10:

Bacteria (Figure 10-1): *iStock. Used with permission.*
Dog (Figure 10-1): *iStock. Used with permission.*

Levels of Organization in a Complex Animal (Figure 10-2): *Illustration created by Casey Luskin, Hallie Kemper, and Gary Kemper. Copyright © Casey Luskin, Hallie Kemper, and Gary Kemper, 2012. All Rights Reserved. Used with permission.*

Leaves: *Photograph by Hallie Kemper. Copyright © Hallie Kemper, 2012. All Rights Reserved. Used with permission.*

Plant Cell (Figure 10-3): *Illustration created by Valerie Gower. Copyright © Valerie Gower, 2012. Used with permission.*

Herbaceous Plant (Figure 10-4): *Illustration created by Valerie Gower. Copyright © Valerie Gower, 2012. Used with permission.*

Bat: *iStock. Used with permission.*

Nerve Cell (Figure 10-5): *Illustration created by Valerie Gower. Copyright © Valerie Gower, 2012. Used with permission.*

Hummingbird: *Photo by Daniel Ripplinger. Used with permission.*

Fish: *iStock. Used with permission.*

Elephant: *iStock. Used with permission.*

Caterpillar Metamorphosis into Butterfly: *iStock. Used with permission.*

Chapter 11:

Human Eye (Figure 11-1): *Illustration created by Valerie Gower. Copyright © Valerie Gower, 2012. Used with permission.*

Human Female Reproductive System (Figure 11-2): *Illustration created by Valerie Gower. Copyright © Valerie Gower, 2012. Used with permission.*

Fertilization (Figure 11-3): *Illustration created by Valerie Gower. Copyright © Valerie Gower, 2012. Used with permission.*

Human Digestive System (Figure 11-4): *Illustration created by Valerie Gower. Copyright © Valerie Gower, 2012. Used with permission.*

Chapter 12:

Human Hand and Panda's Paw (Figure 12-1): *Illustration created by Valerie Gower. Copyright © Valerie Gower, 2012. Used with permission.*

Recurrent and Superior Laryngeal Nerves (Figure 12-2): *Illustration created by Valerie Gower. Copyright © Valerie Gower, 2012. Used with permission.*

Human Appendix (Figure 12-3): *Illustration created by Valerie Gower. Copyright © Valerie Gower, 2012. Used with permission.*

Chapter 13:

Part of Darwin's "Tree of Life" Illustration (Figure 13-1): *Public Domain. Found at Wikipedia. Accessed September 28, 2012, http://en.wikipedia.org/wiki/ File:Darwins_tree_of_life_1859.gif.*

Homology in Vertebrate Limbs (Figure 13-2): *Copyright Jody F. Sjogren, 2000. Used with permission. From Figure 4-1, Jonathan Wells, Icons of Evolution: Science or Myth? (Washington, D.C.: Regnery, 2000).*

Tangled Bush of Life (Figure 13-3): *From Figure 3, W. Ford Doolittle, "Phylogenetic Classification and the Universal Tree," Science, 284 (June 25, 1999): 2124-2128. Reprinted with permission from AAAS.*

Horizontal Gene Transfer (Figure 13-4): *Illustration created by Casey Luskin. Copyright © Casey Luskin, 2012. All Rights Reserved. Used with permission.*

"Convergent Evolution" in Saber-Toothed Cats (Figure 13-5): *Illustration created by Valerie Gower. Copyright © Valerie Gower, 2012. Used with permission.*

Berra's Blunder and Corvette "Evolution" (Figure 13-6): *Copyright Jody F. Sjogren, 2000. Used with permission. From Figure 4-3, Jonathan Wells, Icons of Evolution: Science or*

Myth? (Washington, D.C.: Regnery, 2000).

Chapter 14:

Haeckel's Embryo Drawings (Figure 14-1): *This version of Haeckel's drawings is from George Romane's 1892 book, Darwinism Illustrated, as used in Figure 5-1, Jonathan Wells, Icons of Evolution: Science or Myth? (Washington, D.C.: Regnery, 2000).*

Haeckel's Drawings vs. Actual Vertebrate Embryos (Figure 14-2): *Copyright Jody F. Sjogren, 2000. Used with permission. From Figure 5-2, Jonathan Wells, Icons of Evolution: Science or Myth? (Washington, D.C.: Regnery, 2000).*

Embryonic Hourglass (Figure 14-3): *Copyright Jody F. Sjogren, 2000. Used with permission. From Figure 5-4, Jonathan Wells, Icons of Evolution: Science or Myth? (Washington, D.C.: Regnery, 2000).*

Peppered Moth (Figure 14-4): *Illustration created by Valerie Gower. Copyright © Valerie Gower, 2012. Used with permission.*

Darwin's Finches (Figure 14-5): *Copyright Jody F. Sjogren, 2000. Used with permission. From Figure 8-1, Jonathan Wells, Icons of Evolution: Science or Myth? (Washington, D.C.: Regnery, 2000).*

Chapter 15:

Darwinian Gradual Evolution (Figure 15-1): *Figure by Jens Jorgensen and Casey Luskin. Copyright © Casey Luskin and Jens Jorgensen, 2012. All rights reserved. Used with permission.*

The Cambrian Explosion (Figure 15-3): *Figure by Jens Jorgensen and Casey Luskin. Copyright © Casey Luskin and Jens Jorgensen, 2012. All rights reserved. Used with permission.*

Punctuated Equilibrium (Figure 15-4): *Figure by Jens Jorgensen and Casey Luskin. Copyright © Casey Luskin and Jens Jorgensen, 2012. All rights reserved. Used with permission.*

Chapter 16:

***Archaeopteryx* (Figure 16-2):** *Copyright © H. Raab (User: Vesta) / Wikimedia Commons / CC BY-SA 3.0. Used with permission under Creative Attribution-Share Alike 3.0 Unported License. Usage not intended to imply endorsement by the author / licensor of the work. Accessed October 2, 2012, http://en.wikipedia.org/wiki/File:Archaeopteryx_ lithographica_(Berlin_specimen).jpg.*

Irreducibly Complex Structure of a Flight Feather (Figure 16-4): *Illustration created by Valerie Gower. Copyright © Valerie Gower, 2012. Used with permission.*

Three-Chambered Heart vs. Four-Chambered Heart (Figure 16-5): *Illustration created by Valerie Gower after heart illustration on page 131 of Stephen C. Meyer, Scott Minnich, Jonathan Moneymaker, Paul A. Nelson, and Ralph Seelke, Explore Evolution: The Arguments For and Against Neo- Darwinism (Melborne, Australia: Hill House, 2007). Copyright © Valerie Gower, 2012. Used with permission.*

Chapter 17:

Ape-to-Human Icon (Figure 17-1): *Copyright Jody F. Sjogren, 2000. Used with permission. From Figure 11-1, Jonathan Wells, Icons of Evolution: Science or Myth? (Washinton, D.C.: Regnery, 2000).*

Skeletons of Homo vs. Australopithecus (Figure 17-2): *Illustration From Figure 1, John Hawks et. al., "Population Bottlenecks and Pleistocene Human Evolution," Molecular Biology and Evolution, Copyright 2000, 17 (1): 2–22, by permission of the Society for Molecular Biology and Evolution.*

How Important are Skulls of "Intermediate" Size? (Figure 17-5): *Created by Jonathan Jones. Copyright © Discovery Institute, 2012. Used with permission.*

Chapter 18:

Tree-House: *iStock. Used with permission.*

Jelly beans: *iStock. Used with permission.*

Chapter 19:

How the Darwinian Lobby... (Figure 19-1): *Illustration by Jens Jorgensen and Casey Luskin. Copyright © Casey Luskin and Jens Jorgensen, 2012. All rights reserved. Used with permission.*

John Scopes (Figure 19-2): *Public Domain. Found at Wikipedia. Accessed October 4, 2012, http://en.wikipedia.org/wiki/File:John_t_scopes.jpg.*

Chapter 20:

Bacterial Flagellum: *Illustration created by Valerie Gower. Copyright © Valerie Gower, 2012. Used with permission.*

***Archaeopteryx*:** *Copyright © H. Raab (User: Vesta) / Wikimedia Commons / CC BY-SA 3.0. Used with permission under Creative Attribution-Share Alike 3.0 Unported License. Usage not intended to imply endorsement by the author / licensor of the work. Accessed October 2, 2012, http://en.wikipedia.org/wiki/File:Archaeopteryx_ lithographica_(Berlin_specimen).jpg.*

Darwin's "Tree of Life": *Public Domain. Found at Wikipedia. Accessed September 28, 2012, http://en.wikipedia.org/wiki/File:Darwins_tree_of_life_1859.gif.*

DNA Curved: *Illustration created by Valerie Gower. Copyright © Valerie Gower, 2012. Used with permission.*

INDEX

Q

R

S

ABOUT THE AUTHORS

Gary Kemper:

While Gary was still an undergraduate, testing revealed that he had an unusually high aptitude for scientific analysis. At the age of 19 he began a career as an aerospace systems analyst. He eventually moved on to the entertainment industry and learned the craft of writing. A former skeptic of intelligent design (ID), Gary became a strong supporter of ID after becoming aware of the enormous amount of academic and media misinformation on the subject.

Hallie Kemper:

Hallie is a longtime homeschool educator in California, who has taught classes in ecology, botany, and intelligent design with multiple homeschool education groups in the greater Los Angeles Area. Her unsuccessful search for an ID curriculum led her and her husband to write the first draft of this book. The addition of Casey Luskin to their writing team resulted in this final version.

Casey Luskin:

Casey was trained as both a scientist and an attorney, having earned his bachelor's and master's degrees in Earth sciences at the University of California at San Diego and a law degree from the University of San Diego. He formerly conducted scientific research at Scripps Institution for Oceanography and studied evolution extensively at both the undergraduate and graduate levels. In 2001, Casey cofounded the Intelligent Design and Evolution Awareness (IDEA) Center, helping high school and college students to learn about ID by forming IDEA Clubs. He now works as Research Coordinator at the Discovery Institute, the leading think tank promoting ID, assisting educators nationwide to teach evolution more accurately. He has lectured and published widely on ID in journals, books, and popular venues.